NEW DEVELOPMENTS IN ARRAY TECHNOLOGY AND APPLICATIONS

INTERNATIONAL ASTRONOMICAL UNION

UNION ASTRONOMIQUE INTERNATIONALE

NEW DEVELOPMENTS IN ARRAY TECHNOLOGY AND APPLICATIONS

PROCEEDINGS OF THE 167TH SYMPOSIUM OF THE
INTERNATIONAL ASTRONOMICAL UNION,
HELD IN THE HAGUE, THE NETHERLANDS, AUGUST 23–27, 1994

EDITED BY

A. G. DAVIS PHILIP

Union College and the Institute for Space Observations

KENNETH A. JANES

Boston University

and

ARTHUR R. UPGREN

Van Vleck Observatory, Wesleyan University

KLUWER ACADEMIC PUBLISHERS

DORDRECHT / BOSTON / LONDON

A C.I.P. Catalogue record for this book is available from the Library of Congress.

ISBN 0-7923-3639-9

*Published on behalf of
the International Astronomical Union
by
Kluwer Academic Publishers, P.O. Box 17, 3300 AA Dordrecht, The Netherlands.*

*Kluwer Academic Publishers incorporates
the publishing programmes of
D. Reidel, Martinus Nijhoff, Dr W. Junk and MTP Press.*

*Sold and distributed in the U.S.A. and Canada
by Kluwer Academic Publishers,
101 Philip Drive, Norwell, MA 02061, U.S.A.*

*In all other countries, sold and distributed
by Kluwer Academic Publishers Group,
P.O. Box 322, 3300 AH Dordrecht, The Netherlands.*

Printed on acid-free paper

All Rights Reserved
©1995 International Astronomical Union

No part of the material protected by this copyright notice may be reproduced or utilized in any form or by any means, electronic or mechanical including photocopying, recording or by any information storage and retrieval system, without written permission from the publisher.

Printed in the Netherlands

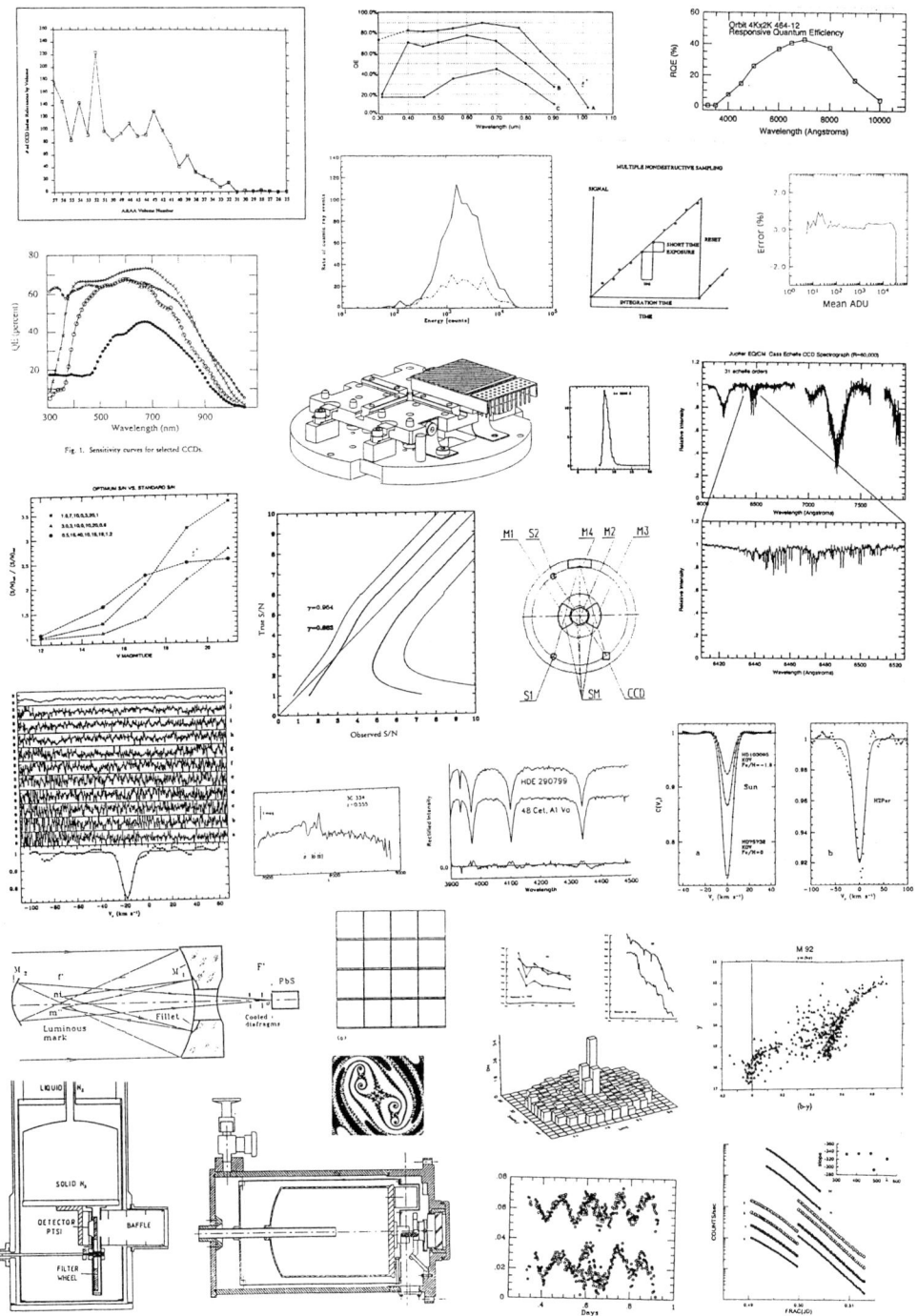

v

Table of Contents

Preface   xv

Section I - Review Papers

INTRODUCTION

    A. G. Davis Philip   3

ARRAY DETECTORS AND INSTRUMENTS FOR THE ESO VLT

    Sandro D'Odorico   9

DESIGN AND FABRICATION OF LARGE CCDs FOR THE KECK OBSERVATORY DEIMOS SPECTROGRAPH

    R. J. Stover, W. E. Brown, D. K. Gilmore and M. Wei   19

SCIENTIFIC CCD PROSPECTS FOR 1994 AND BEYOND

    Paul R. Jorden and A. P. Oates   27

NEW DEVELOPMENTS IN CCD TECHNOLOGY FOR THE UV-EUV SPECTRAL RANGE

    Giovanni Bonanno   39

CCD CONTROLLERS

    Robert W. Leach   49

CCDs IN ACTIVE ACQUISITION SYSTEMS

    A. Blecha   57

CCD DEVELOPMENT ACTIVITIES AT ESO

    O. Iwert   67

INFRARED ARRAY DETECTORS: PERFORMANCE AND PROSPECTS

    Ian S. McLean   69

THE LICK OBSERVATORY TWO MICRON CAMERA

  K. Gilmore, D. Rank and P. Temi   79

INFRARED ARRAYS AT THE EUROPEAN SOUTHERN OBSERVATORY

  G. Finger, G. Nicolini, P. Biereichel, M. Meyer and
  A. F. M. Moorwood   81

THE IMPACT OF INFRARED ARRAY TECHNOLOGY ON ASTRONOMY

  Giovanni G. Fazio   93

MONOLITHIC Si BOLOMETER ARRAYS: DETECTORS FOR FAR INFRARED AND SUBMILLIMETER DETECTION

  Harvey Moseley   95

OBSERVATIONAL CONCERNS AND TECHNIQUES FOR HIGH-BACKGROUND MID-INFRARED (5 - 20 micron) ARRAY IMAGING

  Daniel Y. Gezari   97

AN INFRARED CAMERA BASED ON A LARGE PtSi ARRAY

  I. S. Glass, K. Sekiguchi and Y. Nakada   109

PtSi IR ARRAY IN MOSAIC CONFIGURATION

  Munetaka Ueno, Fumiaki Tsumuraya and
  Yoshihiro Chikada   117

CCD PHOTOMETRY - PRESENT AND FUTURE

  Alistair R. Walker   123

CCD PHOTOMETRY: SOME BASIC CONCERNS

  C. Sterken   131

CHOOSING FILTERS TO MAKE CCD PHOTOMETRY TRANSFORMABLE TO OTHER DETECTORS

  Andrew T. Young   145

AUTOMATED CCD SCANNING FOR NEAR EARTH ASTEROIDS

        Robert Jedicke   157

CCD TIME SERIES PHOTOMETRY OF ASTRONOMICAL SOURCES

        Steve B. Howell   167

HIGH PRECISION STELLAR PHOTOMETRY WITH CCDs. I.

        A. J. Penny   173

HIGH PRECISION STELLAR PHOTOMETRY WITH CCDs. II.

        Michael S. Bessell   175

PRECISION DIFFERENTIAL CCD PHOTOMETRY

        L. A. Balona   187

ARRAY POLARIMETRY AND OPTICAL-DIFFERENCING PHOTOMETRY

        J. Tinbergen   197

COSMIC RAY EVENTS AND NATURAL RADIOACTIVITY IN CCD CRYOSTATS

        Ralph Florentin-Nielsen, Michael J. Andersen
        and Sven P. Nielsen   207

MOCAM: A 4k X 4k CCD MOSAIC FOR THE CANADA-FRANCE-HAWAII TELESCOPE PRIME FOCUS

        J. C. Cuillandre, Y. Melliers, R. Murowinski,
        D. Crampton, G. Luppino and R. Arsenault   213

ECHELLE SPECTROSCOPY WITH A CCD AT LOW SIGNAL-TO-NOISE RATIO

        Didier Queloz   221

NICMOS3 DETECTOR FOR SPECTROSCOPY

        L. Vanzi, A. Marconi and S. Gennari   231

SPECTRAL CLASSIFICATION WITH ARRAY DETECTORS

        C. J. Corbally   241

SPECTROSCOPIC OBSERVATIONS OF SOLAR SYSTEM OBJECTS: PUSHING THE LIMITS

        Anita L. Cochran   251

"VA-ET-VENT" ("BACK-AND-FORTH") CCD SPECTROSCOPY: A NEW WAY TO INCREASE THE LIMITING MAGNITUDE OF VERY LARGE TELESCOPES

        G. Soucail, J. C. Cuillandre, J. P. Picat and B. Fort   263

SOME PROBLEMS OF WIDE-FIELD ASTROMETRY WITH A SHORT-FOCUS CCD ASTROGRAPH

        I. S. Guseva   275

A DUAL CCD MOSAIC CAMERA SYSTEM SEARCHING FOR MASSIVE COMPACT HALO OBJECTS (MACHOs)

        Kem H. Cook   285

MULTI-FIBER SPECTROSCOPY WITH WIDE-FIELD TELESCOPES

        F. G. Watson   287

PERFORMANCE OF A 2048 x 2048 PIXEL THREE-SIDE-BUTTABLE CCD DESIGNED FOR LARGE FOCAL PLANES IN ASTRONOMY

        J. A. Cortiula   295

CCD MOSAIC DEVELOPMENT FOR LARGE OPTICAL TELESCOPES

        G. A. Luppino, M. R. Mezger and S. Miyazaki   297

HIGHLIGHTS OF IAU SYMPOSIUM No. 167

        Sandro D'Odorico   309

Section II - Poster Papers

THE TRANSPUTER BASED CCD CONTROLLER AT ESO

    Roland Reiss   317

DETECTOR CONTROLLERS FOR THE GALILEO TELESCOPE: A PROGRESS REPORT

    G. Bonanno, P. Bruno, R. Consentino, F. Bortoletto, M. D'Alessandro, D. Fantinel, A. Balesstra and P. Marcucci   319

THE RUTHERFORD-SAAO CCD CONTROLLERS AND THEIR APPLICATIONS

    I. S. Glass, D. B. Carter, G. F. Woodhouse, N. A. Waltham and G. M. Newton   321

CCD IMAGERS WITH ENHANCED UV SENSITIVITY FOR INDUSTRIAL AND SCIENTIFIC APPLICATIONS

    G. I. Vishnevsky, M. G. Vydrevich, L. Yu. Lazovsky, V. G. Kossov and S. S. Tataushchikov   323

THE LARGE-FIELD BRIGHT-STAR HIGH-PRECISION CCD PHOTOMETER OF BAO

    Shiyang Jiang   325

EFFECTS OF SHUTTER TIMING ON CCD PHOTOMETRY

    D. Galadi-Enriquez, C. Jordi and E. Trullols   327

ASTRONOMICAL APPLICATIONS OF CCDs IN HUNGARY: THE FIRST STEPS AND FUTURE PLANS

    G. Szécsényi-Nagy   329

A SIMPLE CCD-SYSTEM FOR SECONDARY ALIGNMENT OF THE SPECTRUM-UV SPACE TELESCOPE

    V. Didkovsky, N. V. Steshenko, P. I. Borzyak and A. I. Dolgushin   331

THE MODERNIZATION OF THE PULKOVO PHOTOGRAPHIC
(PHOTOELECTRIC) VERTICAL CIRCLE BY A CCD ARRAY

    G. A. Goncharov, B. K. Bagildinsky, E. V. Kornilov,
    D. D. Polojentsev, K. V. Rumyantsev and V. D. Shkutov 333

A NEW OCULAR MICROMETER FOR THE MAHIS

    T. R. Kirian, V. S. Korepanov and V. M. Grozdilov 335

PbS AND CCD ARRAY AUTOCOLLIMATION MICROMETERS
FOR THE INFRARED MERIDIAN CIRCLE

    V. N. Yershov 337

THE NOAO CCD CONTROLLER - ARCON

    Alistair R. Walker 339

INTENSIFIED ELECTRON-BOMBARDED CCD IMAGES FOR
INDUSTRIAL AND SCIENTIFIC APPLICATIONS

    I. Dalinenko, G. Vishnevsky, V. Kossov, L. Lasovsky,
    G. Kuzmin and A. Malyarov 341

THE CCDs AT ESO: A SYSTEMATIC TESTING PROGRAM

    T. M. C. Abbott 343

DEVELOPMENT OF A 7000 x 4000 PIXEL MOSAIC CCD CAMERA

    Nobunari Kashikawa, Masafumi Yagi, Naoki Yasuda,
    Sadanori Okamura, Kazuhiro Shimasaku, Mamoru
    Doi and Maki Sekiguchi 345

THE FEASIBILITY OF A CCD FOR AN ASTROMETRIC REFRACTOR

    A. R. Upgren, Alice Morales, Jose Herrero, J. W.
    Griese, III, J. M. Vincent and John T. Lee 347

STELLAR POSITIONS FROM CCD IMAGES

    A. R. Upgren, C. Abad and J. Stock 349

ON THE ACCURACY OF CCD AND PHOTOGRAPHIC
OBSERVATIONS OF ASTEROIDS AND THEIR CURRENT
ORBIT DETERMINATIONS

    O. P. Bykov   351

ON THE PROBLEM OF STANDARD FIELDS FOR CCD ASTROMETRY

    I. S. Guseva   353

COMBINED VISUAL AND NEAR-IR DIGITAL PHOTOMETRY: THE
VERY YOUNG CLUSTER WESTERLUND 2

    M. D. Guarnieri, M. Gai, G. Massone,
    M. G. Lattanzi, U. Munari and A. Moneti   355

A DUAL CCD MOSAIC CAMERA SYSTEM SEARCHING FOR
MASSIVE-COMPACT HALO OBJECTS (MACHOs)

    Kem H. Cook   357

JHK PHOTOMETRY OF EXTRAGALACTIC SOURCES USING AN
INFRARED ARRAY CAMERA

    I. S. McLean, T. Liu and H. Tepliz   359

PRECISION CCD PHOTOMETRY OF THE HORIZONTAL BRANCH

    A. G. Davis Philip   361

CCD PHOTOMETRY OF THE M 67 CLUSTER IN THE VILNIUS
PHOTOMETRIC SYSTEM

    R. Boyle, V. Straižys, F. Vrba,
    F. Smriglio and A. Dasgupta   363

ESTIMATION OF THE ERRORS INVOLVED IN THE INTENSITY
MEASUREMENT OF LOW S/N RATIO EMISSION LINES

    Claudia Rola and Didier Pelat   365

EEV AND ELECTRON CORP. VIRTUAL PHASE CCDs IN THE
NEAR IR REGION, He λ 10830 Å

    A. G. Shcherbakov, Z. A. Shcherbakova, I. Ilyin and
    I. Tuuominen   367

STELLAR PHOTOMETRY WITH A PERFECT CCD

        Bjarne Thomsen and Frank Grundahl Jensen   369

APPLICATIONS OF A REALISTIC MODEL FOR CCD IMAGING

        Steve B. Howell and William J. Merline   371

3D: THE NEW NEAR-INFRARED FIELD IMAGING SPECTROMETER

        L. E. Tacconi-Garmen, L. Weitzel, M. Cameron, S. Drapatz, R. Genzel, A. Krabbe, H. Kroker and N. Thatte   373

TECHNICAL CCD SYSTEMS FOR THE ESO VERY LARGE TELESCOPE

        J. Erich   375

POSTER DISCUSSIONS   377

INDEX   381

NAME INDEX   383

OBJECT INDEX   388

SUBJECT INDEX   390

PREFACE

At the XIX General Assembly in Bueonos Aires it was announced that the format of future General Assemblies would be changed and that Symposia would be held concurrently with the Assemblies and not as satellite meetings as had been the custom previously. On the flight back to New York I thought about what topic might be good for a meeting at the XXII General Assemby. The most important topic to me was the new technology that has developed as a result of the invention of CCDs and Arrays. CCD and Array detectors have become the detectors of choice at optical observatories over the world. Direct imaging, photometry and spectroscopy are all vastly improved as a result. Observations can be made of faint and crowded objects that one could not even consider measuring before the advent of the CCD. The accuracy of photometric and spectroscopic measures has increased dramatically. Small telescopes can now do research projects that used to be possible only on the larger telescopes. I thought that the General Assembly in The Hague would be an excellent forum for the meeting and would have a great potential for interactions between the members of many different commissions, astronomers working with the devices and people who design and manufacture the devices.

Commission 45 (Stellar Classification) was the official commission sponsor of the meeting. Co-sponsors were Commissions 8, 9, 15, 16, 20, 24, 25, 28, 29, 30, 37 and 42. The institutions that sponsored the meeting were:

The Institute for Space Observations	Schenectady, New York
The European Southern Observatory	Garching, Germany
Union College	Schenectady, New York
Van Vleck Observatory	Middletown, Connecticut.

The Scientific Organizing Committee was composed of the following:

A. G. Davis Philip	Chair
Sandro D'Odorico	Co-Chair
Michael Bessell	Australia
Andre Blecha	Switzerland
Roberto Bonanno	Italy
Edward Bowell	USA
Martin Cullum	Germany
Lief Helmer	Denmark
Hugo Levato	Argentina
David Monet	USA
Lloyd Robinson	USA
Marcello Rodono	Italy
Richard West	Germany
Andy Young	USA

The Local Organizing Committee was composed of the following:

Jaap Tinbergen	Chair
Christiaan Sterken	Belgium

No List of Participants is available for printing in the proceedings volume. The IAU presented a list of 300 astronomers who had indicated their interest in attending the meeting but we have no way of determining which people actually were present.

Special thanks are due Mary Bongiovanni who collected the discussion sheets, typed them up and then passed them back to speakers for corrections. The discussions were in almost final form by the end of the meeting. The IAU Executive Committee assigned some Dutch astronomers to help us in running the daily affairs of the meeting. They were a great help and I want to thank Marco Kouwenhoven, Frank Molster, Nick Schutgens, Robert van Eysbergen and Pieter Veen for their fine work. Sandro D'Odorico was a great help in organizing the meeting and was in charge of the European side of things. The European Southern Observatory supported the meeting financially and also printed abstract booklets, containing abstracts of oral and poster papers, for distribution to participants. Immo Appenzeller, who opened the meeting with a short speech of welcome, gave very useful advice in the early stages of planning for the meeting. My two co-editors found many things to correct, from grammer, spelling to problems with references. Kristina Philip also helped as a proofreader.

The final discussion was prepared at the Moletai Observatory in Lithuania and mailed from there to authors. Papers were submitted on floppy disks or sent via email to Schenectady. WordPerfect 6.0b was used to edit the ASCII versions of the submitted papers. Proofs were were sent back to each author and then authors emailed their corrections back to Schenectady. This cooperation between the editors and authors is a necessary part of preparing an accurate transcript of the meeting and all who participated in the effort are thanked for their help.

The cover photo shows a CAD view of part of the new ESO detector head (taken from D'Odorico's paper, page 15).

Schenectady
April, 1995

A. G. Davis Philip

LIST OF CONTRIBUTORS

T.M.C. Abbott
ESO
Casilla 19001
Santiago 19
Chile
TABBOTT@ESO.ORG

L.A. Balona
South African Astronomical Observatory
P.O. Box 9
Observatory
7935 Capetown
South Africa
LAB@MV.SAAO.AC.ZA

Michael S. Bessell
Mount Stromlo and Siding Spring Observatories
Weston Creek PO, ACT 2611
Australia
BESSELL@MIDGET.ANU.EDU.AU

A. Blecha
Observatoire de Genève
51 Ch. des Mailettes
1290 Sauverny
Switzerland
BLECHA@OBS.UNIGE.CH

Giovanni Bonanno
Osservatorio Astrofisico di Catania
Viale A. Doria 6
92125 Catania
Italy
GBONANNO@CT.ASTRO.IT

R. Boyle
Vatican Observatory Group
Steward Observatory
University of Arizona
Tucson, Arizona 85721
U.S.A.
RBOYLE@AS.ARIZONA.EDU

O.P. Bykov
Pulkovo Observatory
196140 St Petersburg
Russia
OBYK@GOARAN.SPB.SU

Anita L. Cochran
University of Texas
McDonald Observatory
Austin, Texas 78712
U.S.A.
ANITA@ZINFANDEL.AS.UTEXAS.EDU

Kem H. Cook
Lawrence Livermore National Laboratory
MS L-401 PO Box 808
Livermore, California 93405
U.S.A.
KCOOK@IMAGER.LLNL.GOV

C.J. Corbally
Vatican Observatory
Steward Observatory
University of Arizona
Tucson, Arizona 85721
U.S.A.
CCORBALLY@AS.ARIZONA.EDU

J.A. Cortiula
Thomson-CSF Semiconducteurs Specifiques
BP 123
F-38521 St Egreve
France

J.C. Cuillandre
Observatoire Midi-Pyrénées
URA-CNRS 285
14 Avenue E. Belin
F-31400 Toulouse
France
CUILLAND@SRVDEC.OBS-MIP.FR

I. Dalinenko
Electron Research Institute
St Petersburg
Russia

L.V. Didkovsky
Crimean Astrophysical Observatory
Nauchny, Crimea
Ukraine
DIDKOVSKY@CRAO.CRIMEA.UA

J. Ehrich
Jena-Optronik GmbH
Göschwitzer Strasse 33
D-07745 Jena-Göschwitz
Germany

Giovanni G. Fazio
Harvard-Smithsonian Center for Astrophysics
Cambridge, Massachusetts 02138
U.S.A.
FAZIO@CFA.HARVARD.EDU

G. Finger
European Southern Observatory
Garching bei München
Germany
GFINGER@ESO.ORG

Ralph Florentin-Nielsen
Niels Bohr Institute for Astronomy
Copenhagen University Observatory
Copenhagen
Denmark

D. Galadí-Enríquez
Dept. d'Astronomia i Meteorologia
Universitat de Barcelona
Avda. Diagonal 647
E-08028 Barcelona
Spain
CARMEN@FACJNO.UB.ES

K. Gilmore
Lick Observatory/UCO UCSC
Santa Cruz, California 95064
U.S.A.
GILMORE@HELIOS.UCSC.EDU

I.S. Glass
South African Astronomical Observatory
P.O. Box 9
Observatory
7935 Capetown
South Africa
ISG@SAAO.AC.ZA

Daniel Y. Gezari
NASA/Goddard Space Flight Center
Code 685
Greenbelt, Maryland 20771
U.S.A.
GEZARI@STARS.GSFC.NASA.GOV

G.A. Goncharov
Pulkovo Observatory
196140 St Petersburg
Russia
GOSHA@GAORAN.SPB.SU

M.D. Guarnieri
Osservatorio Astronomico di Torino
Torino
Italy
MDGUARNIERI@TO.ASTRO.IT

I.S. Guseva
Pulkovo Observatory
196140 St Petersburg
Russia
ISG@GAORAN.SPB.SU

Steve B. Howell
Planetary Science Institute
Tucson, Arizona 85705
U.S.A.
HOWELL@PSI.EDU

O. Iwert
European Southern Observatory
Garching bei München
Germany
OIWERT@ESO.ORG

Robert Jedicke
Lunar and Planetary Laboratory
University of Arizona
Tucson, Arizona 85721
U.S.A.
JEDICKE@PIRL.LPL.ARIZONA.EDU

Shiyang Jiang
Beijing Astronomical Observatory
Beijing 10080
People's Republic of China

Paul R. Jorden
Royal Greenwich Observatory
Madingly Road
Cambridge, CB3 0EZ
U.K.
PRJ@MAILAST.CAM.AC.UK

Nobunari Kashikawa
University of Tokyo
Bunkyo-ku
Tokyo 113
Japan
BKASHIK@C1.MTK.NAO.AC.JP

T.R. Kirian
Pulkovo Observatory
196140 St Petersbrug
Russia
KTR@GAORAN.SPB.SU

Robert W. Leach
Department of Astronomy
San Diego State University
San Diego, California 02182
U.S.A.
LEACH@MINTAKA.SDSU.EDU

G.A. Luppino
Institute for Astronomy
University of Hawaii
Honolulu, Hawaii 96822
U.S.A.
GER@GALILEO.IFA.HAWAII.EDU

Ian S. McLean
Departments of Astronomy and Physics
UCLA
Los Angeles, California 90024
U.S.A.
MCLEAN@BONNIE.ASTRO.UCLA.EDU

Harvey Moseley
Laboratory for Astronomy and Solar Physics
Goddard Space Flight Center
Greenbelt, Maryland 20771
U.S.A.
MOSELEY@STARS.GSFC.NASA.GOV

Sandro D'Odorico
European Southern Observatory
Garching bei München
Germany
SDODORIC@ESO.ORG

A.J. Penny
Rutherford Appleton Laboratory
Building R25/R68
Chilton, Didcot OX11 0QX
U.K.
AJP@STARLINK.RUTHERFORD.AC.UK

A.G. Davis Philip
1125 Oxford Place
Schenectady, New York 12308
U.S.A.
AGDP@GAR.UNION.EDU

Didier Queloz
Observatoire de Genève
51 Ch. des Maillettes
1290 Sauverny
Switzerland
QUELOZ@SCSUN.UNIGE.CH

Roland Reiss
European Southern Observatory
Garching bei München
Germany
RREISS@ESO.ORG

Claudia Rola
DAEC, Observatoire de Meudon
Meudon
France
ROLA@GIN.OBSPM.FR

A.G. Shcherbakov
Crimean Astrophysical Observatory
Nauchny, Crimea
Ukraine
SHERB@CRAO.CRIMEA.UA

G. Soucail
Observatoire Midi-Pyrénées
URA-CNRS 285
14 Avenue E. Belin
F-31400 Toulouse
France
SOUCAIL@SRVDEC.OBS-MIP.FR

Christiaan Sterken
IAAG, University of Brussels
Brussels
Belgium
CSTERKEN@ISI.VUB.AC.BE

Richard J. Stover
Lick Observatory
University of California
Santa Cruz, California 95064
U.S.A.
RICHARD@UCSCLOA.UCSC.EDU

Gabor Szécsényi-Nagy
Department of Astronomy
Eötvös Loránd University
Ludovika ter 2
H-1083 Budapest
Hungary
SZENA@LUDENS.ELTE.HU

L.E. Tacconi-Garman
Max-Planck-Institut für Extraterrestrische
  Physik
Garching bei München
Germany
LOWEL%MPE12@MPE.MPE-GARCHING.MPG.DE

Bjarne Thomsen
Institute of Physics and Astronomy
Aarhus University
Ny Munkegade
Dk-8000 Aarhus C.
Denmark
BT@OBS.AAU.DK

Jaap Tinbergen
Kapteyn Sterrewacht
Mensingweg 20
9301 KA Roden
The Netherlands
TINBERGEN@KSW.RUG.NL

Munetaka Ueno
Department of Earth Science and Astronomy
University of Tokyo
Tokyo
Japan
UENO@KYOHOU.C.U-TOKYO.AC.JP

A.R. Upgren
Van Vleck Observatory
Wesleyan University
Middletown, Connecticut 06457
U.S.A.
AUPGREN@EAGLE.WESLEYAN.EDU

Leonardo Vanzi
Universita degli Studi di Firenze
Firenze
Italy
VANZI@ARCETRI.ASTRO.IT

G.I. Vishnevsky
Electron Research Institute
St Petersburg
Russia
VISH@ISTA.SPB.SU

Alistair R. Walker
Cerro Tololo Inter-American Observatory
Casilla 603
La Serena
Chile
AWALKER@NOAO.EDU

F.G. Watson
Royal Greenwich Observatory
Cambridge
U.K.
FGW@MAIL.AST.CAM.AC.UK

Y.N. Yershov
Pulkovo Observatory
196140 St Petersbrug
Russia
YERSH@GAORAN.SPB.SU

Andrew T. Young
Astronomy Department
San Diego State University
San Diego, California 92182-0540
U.S.A.
ATY@MINTAKA.SDSU.EDU

Review Papers

Chairs of the Sessions

I	A. G. Davis Philip
II	Alistair Walker
III	Martin Cullum
IV	Andre Blecha
V	Ian McLean
VI	Giovanni Fazio
VII	Allen Penny
VIII	Christiaan Sterken
IX	Andy Young
X	A. G. Davis Philip
XI	Chris Corbally
XII	Michael Bessell
XIII	Anita Cochran
XIV	Richard Stover
XV	Irina Guseva

INTRODUCTION

A. G. Davis Philip

Union College and Institute for Space Observations

ABSTRACT: A short introduction to the subject of the meeting, IAU Symposium No. 167, New Developments in Array Technology and Applications is given. CCD and Array detectors have become the detectors of choice at optical observatories all over the world. Direct imaging, photometry and spectroscopy are all vastly improved as a result. Thirteen IAU Commissions joined in sponsoring this meeting which indicates the wide interest in this subject. In the five days of the symposium the following topics were discussed: New Developments in CCD Technology, New Developments in IR Detector Arrays, Direct Imaging with CCDs and Other Arrays, Spectroscopy with CCDs and Other Arrays and Large Field Imaging with Array Mosaics. A few papers concerning Astrometry with CCDs were given in the poster sessions. Scientific results were also presented in the poster sessions.

CCD and Array Detectors have become the detectors of choice at optical observatories over the world. Direct imaging, photometry and spectroscopy are all vastly improved as a result. Recently especially dramatic improvements have been made in the area of IR arrays. Now observations are routinely made of faint and crowded objects under a wide range of wavelengths that one could not even consider measuring before the advent of CCDs. The accuracy of photometric and spectroscopic measures has increased dramatically. However, some of the papers at this symposium point out that there are still problems in obtaining the highest precision from CCD observations.

The first CCDs were used in 1975 and in the 1980s they spread throughout the astronomical community. Now almost all the direct imaging done at optical telescopes is done with some form of array detector. In view of these developments it was felt that the XXII General Assembly would be a good place for a discussion of New Developments in Array Technology and Applications. Thirteen IAU Commissions have joined in sponsoring the meeting and this number of supporting commissions is evidence of the wide interest in this subject by the astronomical community. In the five days of the symposium we will discuss the following topics: New Developments in CCD Technology, New Developments in IR Detector Arrays, Direct Imaging with CCDs and Other Arrays, Spectroscopy with CCDs and Other Arrays and Large Field Imaging with Array Mosaics. A few papers on Astrometry with CCDs will be given in the poster sessions, as will papers on hardware, software, techniques and scientific results.

I made a tabulation in A&AA of all the references listed under CCD (----) in the subject index for volumes 25 to 57, covering the years 1979 to 1993. The number of references plotted in Fig. 1 is a lower limit to the number of articles on CCD techniques and results. Not all the CCD articles are indexed under a CCD heading. But the plot gives a good indication of the

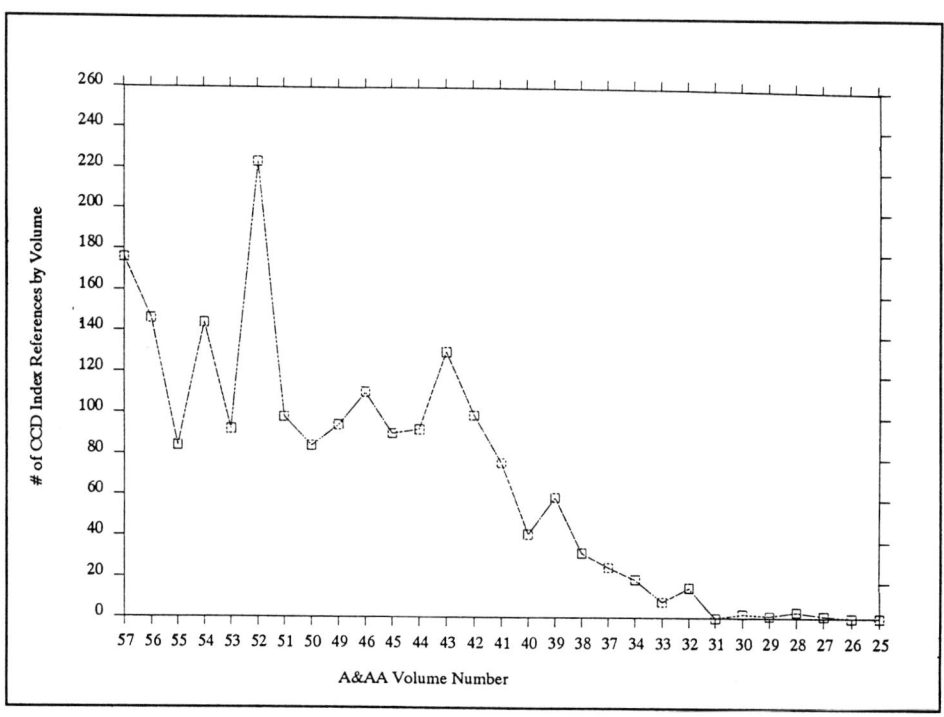

Fig. 1. The number of Index CCD References in A&AA Volumes (1979 - 1993)

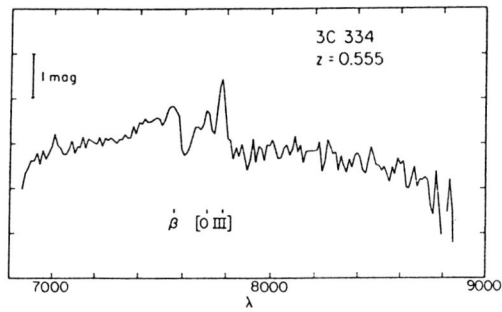

Fig. 2. A CCD spectrum of the quasar 3C 334 (from Oke 1978)

TABLE 1

A List of Some CCD Meetings (1979 - 1994)

---

21 - 23 Nov., 1979, ESO Workshop on Two Dimensional Photometry, at Noordwijkerhout, The Netherlands, P. Crane and K. Kjar, eds.

10 - 11 June, 1981, Solid State Imagers for Astronomy, at Harvard Smithsonian Center for Astrophysics, Proceedings SPIE 290, John c. Geary and David W. Latham, eds.

8 - 10 Sept., 1981, Instrumentation for Astronomy with Large Optical Telescopes, IAU Colloquium No. 67, at Zelenchukskaya, USSR, Colin M. Humphries, ed., D. Reidel, Dordrecht

8 - 10 Mar., 1982, Instrumentation in Astronomy. IV., at Tucson, Arizona, David L. Crawford, ed., Proceedings SPIE 331

7 - 9 Sept., 1993, Instrumentation in Astronomy. V., at London, England, Alec Boksenberg and David L. Crawford, eds., Proceedings SPIE 445

22 - 23 Aug., 1985, Solid State Imaging Arrays, at San Diego, Cal., Keith N. Pettyjohns and Eustace L. Dereniak, eds., Proceedings SPIE 570

4 - Mar., 1986, Instrumentation in Astronomy. V., at Tucson, Arizona, David L. Crawford, ed., Proceedings SPIE 627

17 - 19 June, 1986, The Optimization of the Use of CCD Detectors in Astronomy, at Haute Provence, France, J. -P. Baluteau and S. D'Odorico, eds., ESO Conf. and Workshop Proc. No. 25

21 - 24 Mar., 1988, ESO Conference on Very Large Telescopes and Their Instrumentation, at Garching, Germany, M. -H. Ulrich, ed.

6 - 8 Sept., 1989, CCDs in Astronomy, at Tucson, Arizona, George H. Jacoby, ed., ASP Conf. Series, Vol. 8

15 - 17 Mar., 1990, CCDs in Astronomy. II., at Charleston, South Carolina, A. G. Davis Philip and Saul J. Adelman, eds., L. Davis Press, Schenectady

25 - 27 Feb. 1991, Charge-Coupled Devices and Solid State Optical Sensors. II. at San Jose, Cal., Morely M. Blouke, ed., Proceedings SPIE 1447

7 - 9 June, 1991, IEEE Charge-Coupled Devices Workshop, University of Waterloo, Waterloo, Canada, Savvas G. Chamberlin, organizer

27 - 30 Apr., 1992, Progress in Telescope and Instrumentation Technologies, at ESO, Garching, Germany, M. -H. Ulrich, ed.

15 - 17 Nov., 1993, Calibrating the Hubble Space Telescope, at STScI, Baltimore, Maryland, J. Chris Blades and Samantha J. Osmer, eds.

13 - 14 Mar., 1994, Instrumentation in Astronomy. VIII. at Kona, Hawaii, David L. Crawford and Eric R. Craine, eds., Proceedings SPIE 2198

rapid increase of CCD literature in the last decade and a half.

A search was made to find meetings that were mainly concerned with CCD Technology from 1979 to 1994. Table 1 (on page 5) lists 16 meetings for which proceedings have been published. Other workshops have been held with no proceedings published. The list in Table 1 will serve as a guide for persons interested in following developments in this field.

The earliest CCD result that I came across during my search was a paper by J. B. Oke (1978) in the Journal of the Royal Astronomical Society of Canada which showed CCD spectra taken of the quasars 3C 380 and 3C 334 in the wavelength region 7000 Å to 9000 Å. A copy of the spectrum for 3C 334 is shown in Fig. 2.

The first commercially available chips were made by Fairchild which made chips up to 488 x 380 pixels with noise levels of 30 electrons. Texas Instruments made a 400 x 500 thin chip with noise levels of 10 -15 electrons. RCA came out with a 512 x 320 pixel, back illuminated chip. By 1980 TI was making 800 x 800 pixel devices. By 1985 the noise level was down to 4 - 15 electrons. In each of these devices the silicon circuit stores electrons produced by the photons incident on the chip as charge. At the end of the exposure the charge is moved to the edge of the chip, row by row, where it is read off.

In the mid 1980's IR arrays appeared (32 x 32 pixels). In 1987 buttable CCDs were being designed as a way of increasing the size of the detector. In 1989 Jacoby, at Kitt Peak, was able to report that CCD detectors were being used in 97% of the optical observations, up from 2% in 1979 (Jacoby 19990). Now the standard chip at many observatories is a 2048 x 2048 pixel chip. Chips of 4096 x 4096 pixels have been made but the success rate is not nearly as high as a run of the 2048 x 2048 pixel chips. Mosaics of chips have been made as a way of increasing the size of the area of sky which can be investigated. Chips are being treated to increase the UV sensitivity. Thinning, backside charging, antireflection coatings are all techniques for increasing the sensitivity of CCD chips.

The largest telescopes can now make measures of faint objects with accuracies impossible to obtain with the earlier technology. And telescopes of modest aperture can do work that used to be done only by the largest telescopes. Small observatories and amateur astronomers can do research work of high quality - so the improvement in observing techniques has been an improvement across the board, for all investigators involved in astronomical observations. As many people have pointed out, the CCD is the "People's Detector".

At this meeting we will learn of CCD activities at the large telescopes, new developments in infrared arrays, advances and concerns with CCD photometry and spectroscopy, developments in the creation of large mosaics and related topics. We will begin with the opening paper of the scientific sessions, by Sandro D'Odorico of ESO with his talk on Array Detectors for the ESO Very Large Telescope. Before he starts I would like to express my thanks for his efforts as co-chairman of the SOC and the great help he has given me in organizing this conference. Thanks are also due to ESO which provided the abstract booklets containing the abstracts of the invited review papers and of the poster papers and support to bring my secretary, Mary Bongiovanni to The Hague to take care of the transcription of the discussion following each paper.

REFERENCES

Jacoby, G. H. ed. 1990 in CCDs in Astronomy, ASP Conf. Ser. 8, p. ix
Oke, J. B. 1978 Journal of the Royal Astronomical Soc. of Canada 72, 121

ARRAY DETECTORS AND INSTRUMENTS FOR THE ESO VLT

Sandro D'Odorico

European Southern Observatory

ABSTRACT: Two-D array detectors are one of the key components of the instruments being built for the four ESO 8-m telescopes, the VLT project. Three optical and two infrared instruments are under construction. One faint object imager and multi-slit spectrograph to operate in the spectral region 350 - 1000 nm (acronym FORS) is based on a 2048^2 high-efficiency CCD. A two-channel echelle spectrograph to work in the 300 - 1000 nm region (UVRS) is built around a mosaic of 2048^2, 15 μm CCDs. A multifiber, new spectrograph is based on a 4096 x 2048, 16 μm CCD. The two infrared high-resolution instruments, one imager/long-slit spectrometer (ISAAC) and one high-resolution camera are designed to work with adaptive optics and in the speckle mode (CONICA). Both are designed to incorporate IR arrays up to a format of 1024^2 and will work in the spectral region of 1 - 5 μm. The main properties of the instruments and of detectors are presented and their close interdependence is illustrated.

1. THE VLT PROJECT OF ESO AND ITS INSTRUMENTATION

The VLT project of the European Southern Observatory foresees the construction of four 8-m telescopes to work between the atmospheric cutoff in the UV and 20 μm, complemented by an interferometric array of three 2-m class telescopes. The project was funded at the end of 1987 and it is now being executed. The new observatory is being built on the Paranal Peak, a new astronomical site in the Atacama Desert in the north of Chile. First light at the first 8-m telescope is foreseen in 1998.

The VLT alt-azimuth 8-m unit telescope is described by Enard (1994) and its optics by Diericks (1994). In the first implementation every unit telescope will be equipped with two Nasmyth foci and a Cassegrain focus. At the time of this writing (November, 1994) there are six instruments under construction for the VLT. They are listed in Table 1 together with their main characteristics. More data can be found in the ESO publication "Instruments for the ESO VLT" edited by A. Moorwood (1994). There are another nine instruments under study for a possible implementation at the telescope (previous report and D'Odorico 1994). For the six approved first VLT instruments we plan to usee CCD detectors up to a format of 2048^2, 4096 x 2048 and 4096^2 for FORS, FUEGOS and UVES respectively and infrared arrays of 256 (current baseline) or 1024^2 (possible upgrade) for ISAAC and CONICA.

2. PROPERTIES OF MODERN CCD DETECTORS

The paper by Finger et al. (1995) in these Proceedings presents an updated view of infrared arrays and I refer to it for a summary of their properties. In the following, I discuss the key properties of current CCD detectors with reference to the ESO experience in this area.

TABLE 1
Approved VLT Instruments[+]

Name	P.I.	Focus	Modes	Spectral Coverage (μm)	Spectral Resolution ($R\theta$)	Max. Field (Imaging)	Pixel Scale (")	Array Size (pixels)	Notes
ISAAC	A. Moorwood (ESO)	Nasmyth UT1	Imaging Grating Spectroscopy	1-5	500-3000	128"x128"	0.06-0.5	2×256^2	2 channels (1-2.5, 2.5-5μm). Upgrade to 2×1024^2 arrays possible
FORS*	I. Appenzeller (Heidelberg Obs.)	Cass. UT1	Imaging Grism Spectroscopy Polarimetry	0.35-1	300-3000	6'.8x6'.8	0.2-0.1	2048^2	19 slits of 22" length positionable in the field
CONICA	R. Lenzen (MPI Heidelberg)	Nasmyth UT1	Imaging at diffraction limit. 2D spectroscopic imaging	1-5	2000	26"x26"	.1-.013	256^2	Combined with adaptive optics unit. Spectroscopic imaging with Fabry-Perot. Upgrade to 1024^2 array under consideration
UVES	H. Dekker (ESO)	Nasmyth UT2	Echelle Spectroscopy	0.3-1	40000	—	0.215 blue .155 red	4096x2048 2048^2	Red and blue channels. Upgrade of red channel to 4096^2 CCD possible
FUEGOS	P. Felenbok (Meudon)	Nasmyth UT3	Multi-fibre Spectroscopy, Area Spectroscopy	0.35-1	30000-1500	—	0.2-0.7	4096x2048	Field of view of multifibre 28'; of area spectroscopy 5" and 18"

[+] Status as of November 94.
* A copy is under construction for UT3. In FORS2 polarimetry is substituted by an echelle mode.
ISAAC: Infrared Spectrometer and Array Camera; FORS: Focal Reducer/Low Dispersion Spectrograph; CONICA: High resolution Near Infrared Camera; UVES: UV-Visual Echelle Spectrograph; FUEGOS: Multi-Fiber Area Spectrograph

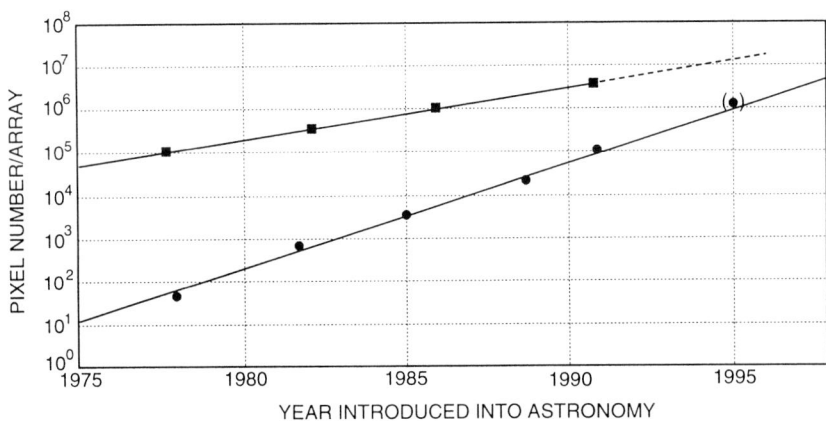

Fig. 1: Maximum number of pixels in array detectors at optical and infrared wavelengths (square and circles respectively) as a function of time. At optical wavelengths, no thinned device is routinely available at format larger than 2048^2 although 4096 x 2048 format have been produced. At infrared wavelength, two manufacturers have announced 1024^2 arrays for delivery in 1995.

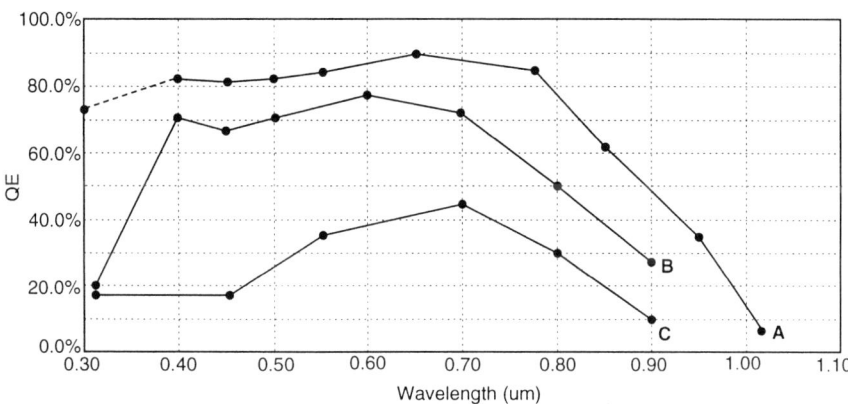

Fig. 2: Quantum efficiency curves of three ESO CCD devices. Curve A is a LORAL 2688 x 512, 15μm pixel CCD thinned for ESO by M. Lesser at the University of Arizona. The measurements have been taken after UV flooding. Curve B is a thinned 1024^2, 24μm SITe CCD. Curve C is a lumogen coated LORAL 2048^2, 15μm CCD. The last curve is indicative of the devices in use on the WF/PC 2 of the Hubble Space Telescope.

2.1 Pixel Size and Global Size

Fig. 1 (on the previous page) shows how the maximum number of pixels/array has been steadily increasing for CCDs and for infrared arrays. For optical detectors the 2048^2 format has become a standard product, but larger formats such as 4096^2 and 4096 x 2048 are still in prototyping stage. The pixel size provided in the latest products is 15 μm, with the exception of the SITe 2048^2 and 1024^2 devices which have 24 μm pixels. The choice of pixel size and CCD formats is dictated by the silicon technology which for most manufacturers is presently based on the use of four-inch wafers. Four 2048^2, 15 μm CCD devices can be fitted in a single silicon wafer thus ensuring a relatively high yield to the process. When the scientific application requires a detector size larger than 30 x 30 mm, it is planned to achieve that by mosaicing of 2048 or 4096 x 2048 buttable CCDs.

For infrared arrays the change has been even more spectacular, with an increase in the number of pixels by four orders of magnitude in the last two decades.

2.2 Quantum Efficiency

As the telescopes are getting larger and the instruments more efficient it has become imperative that detectors operate at QEs close to 100%. Fig. 2 (on the previous page) shows three typical QE curves for three CCDs used on ESO instruments. The high values neded for modern ground-based instruments can be obtained with thinned CCDs with a surface treatment and an antireflection coating only. Curve A, although obtained after flooding the cold CCD with UV light, sets a mark which should be the goal for future devices. If the CCD is warmed up, the high QE is reduced and the process has to be repeated. Future efforts should concentrate in obtaining a stable process for enhancing the sensitivity (Lesser 1994)

2.3 Read-Out-Noise and Dark Current Noise

While size and QE are possibly the two key properties of CCD detectors, read-out-noise and dark current play an important role in certain regimes of observations. At large telescopes this is the case for those observing modes where the detector noise dominates photon noise or sky backgound photon statistics, such as high resolution spectroscopy of faint objects. In this case the gain by going to a larger telescope is particularly relevant. For a given exposure time and limiting magnitude, the S/N ratio to be obtained is proportional to D^2, while it is proportional to D in the cases where the noise is dominated by photon statistics of the object or the sky.

With the improvement of on-chip electronics, the read-out-noise has considerably improved in the last two decades. Values of 30 e^- rms were the standard in 1980 while values $\leq$ 5 e^- are regularly achieved nowadays. This performance is measured at sampling rates of 20 - 50 μs/pixel. More recently, those low values have been measured at faster rates as well (see Luppino et al. 1995 in these Proceedings).

The Poisson noise of thermally generated charge (dark current) is also significant in the observing modes which require very long integrations of faint sources. Again at large telescopes this is the case in high resolution spectroscopy. Dark current values can be reduced by operating the CCD at very low temperature but this often leads to a reduction of the QE. Dark current has been considerably reduced in modern CCDs when operated in Multi-Pinned-Phase mode (Janesick et al. 1989). At ESO we regularly achieve dark current values of less than 10 2re$^-$/hour at temperature of -120° C with devices operated in MPP mode.

2.4 Linearity, Full Well Capacity, Radiation Hits and Flatness

In astronomical applications other CCD properties have to be taken into account carefully. Linearity over a broad dynamic range is usually quoted as one of the main advantages of CCD detectors with respect to both photographic plates and photon-counting detectors. Actually the manufacturers do not carry out any highly accurate tests of the linearity when a new device is produced. Linearity is also related to the CCD associated control electronics and it has to be verified eventually by the final user. Very few measurements on CCD linearity can be found in the literature. As these measurements require a stable laboratory setup and tedious testing, the ultimate deviations from a linear behavior in CCDs are poorly known. At ESO the linearity of the CCDs is measured at the time of characterization and later on monitored during the operation at the telescope, as described by Abbott (1995) in these proceedings. Fig. 3 shows the measured deviations from linearity for a SITe 1024^2 thinned CCD. Apart from the measurements uncertainties at low level, the linearity is confirmed to better than 0.3% over four decades of recorded intensity values.

Radiation hits are charges produced by radioactive processes in the vicinity of the CCD, including the silicon itself or byproducts, typically muons, of interactions of cosmic particles with the Earth's atmosphere. The resulting charge can vary from a few electrons to a few hundreds and is normally concentrated on one to two pixels. There is no significant change in the event count rate when the CCD is moved from sea level to the observatory and the rates for a given CCD type are relatively constant. Out of 11 1024^2 and 512^2 thinned SITe CCDs received at ESO between 1990 and 1994, we obtain a radiation hit rate of 3.1 ±0.2 events/cm^2/min. At these rates the effect does degrade the signal to noise ratio of many types of observations and effectively limits the exposure time. On Loral unthinned CCDs we measure typically 1.3 ±0.1 events in the same units, in agreement with the lower QE of these devices. A more dedicated effort should be made to pin down the possible local source of the contaminating radiation, which might well originate to a large fraction in the chip package or in the dewar materials.

A final property of CCDs which is very relevant to many astronomical applications is their flatness. Deviation from flatness may originate in the process of mounting the thinned CCDs on a substrate or in the poor quality of the substrate itself. When operated with fast optics these deviations can affect the final optical quality of the instrument. The specifications of future thinned CCDs call for deviations from flatness smaller than ±15 μm. The current SITe 2048^2, 24 μm thinned CCDs on the contrary show a difference of about 0.2 mm between the center and the corners of the device. This curved detector plane must be taken into account in the optical design of the instruments which use this type of detector.

In view of the large number of VLT instruments which will rely on the use of CCDs, ESO has initiated a number of developments to secure high performance, standard detector systems to be operational on the VLT at first light. The main actions are reported in the next sections.

3.1 Development of CCDs

There are four VLT instruments under construction which require CCD detectors. For the two FORS instruments, ESO has already procured two SITe 2048^2 thinned CCDs. The 24 μm pixels match well to the scale of the instrument giving a 0.20 arcsec/pixel and a total field of 6.8' x 6.8'. A change of collimator in the instrument provides a high resolution mode with a scale of 0.1 arcsec/pixel over a smaller field. It is interesting to note that for these Focal Reducer/Spectrographs which will be installed at the Cassegrain foci of UT1 and UT3, there is no need for a CCD with more pixels. The field covered by the 2048^2 SITe CCD is very close to the maximum field for an instrument at the Cassegrain focus. Some of the lenses in the focal reducer are also close to the maximum size which can be manufactured in the required FK54 glass material. The QE of the SITe CCDs (typically curve B in Fig. 2) is well matched to the transmission of this instrument which is high only above 400 nm. The expected read-out-noise of 5 e$^-$ typical for these chips is fully acceptable because the vast majority of observations with this instrument will be sky-background limited.

The other two VLT instruments UVES and FUEGOS are based on CCDs up to a format of 4096^2 and 15 μm pixels. High QE and the lowest possible read-out-noise are also required for the high resolution spectroscopy observations planned with these instruments. ESO is following two parallel routes for the procurement of the detectors for these instruments. Since 1992 we contracted the development of a 2048^2, 15 μm thinned CCD, which can be butted on three sides, to Thomson CSF. The specifications for these CCDs (Iwert 1994) call for high performance devices which could be used to build up mosaics of two x two or larger. This new product identification is 7397 M. The delivery of five prototypes is currently planned for the first quarter of 1995. In parallel, ESO has obtained through a foundry effort with LORAL SOUTH (the former Ford Aerospace) a number of front side 2048^2, 15 μm CCDs also buttable on three sides (original design by J. Geary and D. Bredthauer). ESO has contacted to M. Lesser of the Steward Observatory the thinning, packaging and antireflection treatment of the best CCDs of this lot. The first thinned devices are due for delivery at the end of 1994. The results obtained by Lesser on similar devices (see Fig. 2, curve A0), are encouraging. The CCDs will be mounted on the same package planned for the TH 7397 M CCDs.

For future VLT instruments ESO is considering participation in other development for the realization of high quality CCDs. There are two of them which appear well matched to our requirements for VLT instruments. Lincoln Lab at MIT is under discussion as a potential source of prototyping a 4096 x 2048, 15 μm thinned device with high speed, low noise amplifiers, as reported by Luppino et al. (1995) at this meeting. The British company EEV is also entering into the development of a 4096 x 2048, 13.5 μm thinned device for a consortium led by RGO.

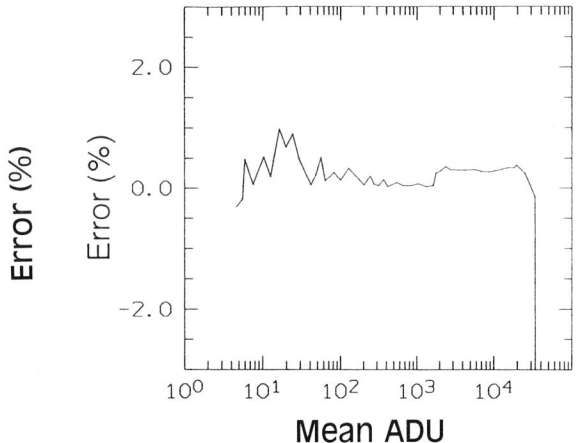

Fig. 3. Percentage deviations from the linearity for a SITe 1024^2 thinned CCD recently tested at ESO as a function of intensity (One ADU = six e$^-$). Noise below 200 ADU is due to the uncertainties at low signal levels. The step at about 1200 ADU is due to a calibration offset in the measuring device. Full well capacity of CCDs (defined as the charge value at which the deviation from linearity exceeds 1%) is related to the pixel size and to the mode of operation. With read-out-noise lower than five e$^-$ and 16 bit ADCs it is possible to work at full resolution with full well capacities up to 150,000 - 200,000 e$^-$. These values are normally achieved with 24 μm CCDs. Devices operated in MPP and with smaller pixel size are in our experience limited to less than 100,000 e$^-$.

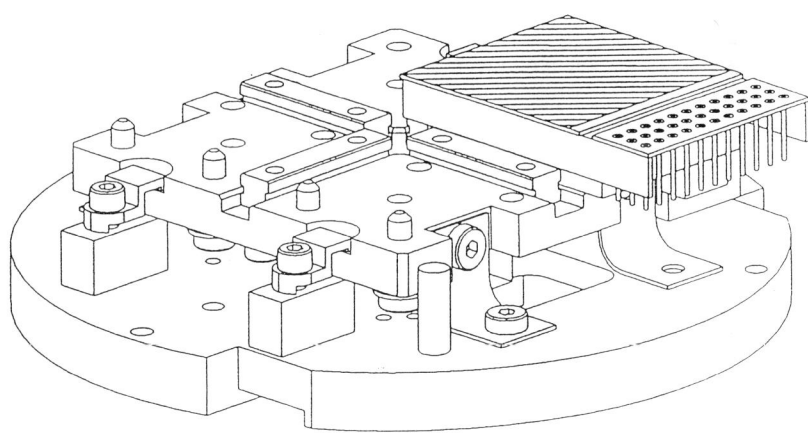

Fig. 4: A CAD view of part of the new ESO detector head in the configuration prepared for a mosaic of two x two large, buttable CCDs. One of the packaged devices is also shown in place. The design foresees the building up of a mosaic giving an overall flatness of better that 30 μm peak-to-peak with a minimum of adjustment required for the CCD alignment.

3.2 The New ESO Detector Controller ACE

A new CCD controller based on the use of transputer modules and DSPs has been developed at ESO to be used with single large size CCD or with mosaics of them as foreseen for the VLT instruments. The new controller, ACE, is described by Reiss (1994). The design has been driven by the need to handle with speed and low system noise modern large size detectors. The controller had also to be compact and to produce less than 25 watts. In parallel to controller development, ESO has designed a new CCD mounting head and associated cryogenics. The new standard head is designed to be simpler to integrate than the current one, with the possibility of fine adjustment of tilt and position of the detector surface while still keeping the very high stability which is required for long integration. It can accomodate mosaics of up to two x two 2048^2 CCDs for a total of 16 outputs. The detector front head in the mosaic case is shown in the CAD picture reproduced in Fig. 4.

REFERENCES

Abbott, T. 1995 in IAU Symposium No.167, New Developments in Array Technology and Applications, A. G. D. Philip, K. A. Janes and A. R. Upgren, eds., Kluwer Academic Pub., Dordrecht, p. 343

Diericks, P. 1994 ESO VLT Technical Report 526

D'Odorico, S. 1994 ESO Scientific Report No.15

Enard, D. 1994 Proc SPIE 2199, 394

Finger, G., Nicolini, G., Biereichal, P., Meyer, M. and Moorwood, A. F. M. 1995 in IAU Symposium No. 167, New Developments in Array Technology and Applications, A. G. D. Philip, K. A. Janes and A. R. Upgren, eds., Kluwer Academic Pub., Dordrecht, p. 81

Iwert, O. 1994 VLT-ESO 13600-0268

Janesick, J., Elliott, T., Fraschetti G., Collins, S., Blouke, M. and Corrie B. 1989 Proc. SPIE 1071, 15

Lesser, M. P. 1994 Proc. SPIE 2198, 782

Luppino, G. A., Metzger, M. R. and Miyazaki, S. 1995 in IAU Symposium No. 167, New Developments in Array Technology and Applications, A. G. D. Philip, K. A. Janes and A. R. Upgren, eds., Kluwer Academic Pub., Dordrecht, p. 297

Moorwood, A., ed. 1994 Instruments for the ESO VLT, ESO, Garching

Reiss, R. 1994 Proc. SPIE 2198, 895

DISCUSSION

PHILIP: The VLT will operate first as four separate telescopes but there are plans to operate it as one telescope. When will this be done?

D'ODORICO: The layout of the observatory includes light ducts which lead from the unit telescopes to a central laboratory. Both coherent light addition (interferometry) and incoherent combination are foreseen. At present this part of the program will be implemented starting in 2000. Preparatory studies are continuing.

COHEN: Are you building one of each instrument or four of them?

D'ODORICO: While the intial instrument plan foresaw a large amount of duplication, at present one instrument only, the Focal Reducer Spectrograph, is being built in two eventually identical copies which will go to telescopes one and three.

COHEN: Is ESO spending money on CCD development? If so, how much and with what companies? You have spoken only of commerically available CCDs.

D'ODORICO: In the past two years, ESO has spent on average half a million DM/year in CCD development with industry or resource laboratories. For the details, please refer to the paper by Iwert.

DESIGN AND FABRICATION OF LARGE CCDs FOR THE KECK OBSERVATORY DEIMOS SPECTROGRAPH

R. J. Stover, W. E. Brown, D. K. Gilmore and M. Wei

University of California Observatories/Lick Observatory

ABSTRACT: The Keck II Deep Imaging Multi-Object Spectrograph (DEIMOS) is a general purpose, faint object, multi-slit, double-beam spectrograph which offers wide spectral coverage, high spectral resolution, high throughput, and long slit length on the sky. This powerful instrument will be the principal optical spectrograph on the Keck II telescope. DEIMOS is optimized for faint-object spectroscopy of individual point sources, low-surface-brightness extended objects, or widely distributed samples of faint objects on the sky. To obtain high resolution (~ 1 Å) and wide spectral coverage (up to 5000 Å) the spectrograph uses wide angle cameras and large CCD detectors with many pixels.

This paper describes some of the work being carried out to obtain the CCD detectors required for the DEIMOS spectrograph. In addition, results are presented on the fabrication and characterization of a 4k x 2k three-side buttable CCD produced by Orbit Semiconductor, a silicon foundry in San Jose, California. This CCD was fabricated to test the ability of Orbit to produce high quality scientific CCDs with the characteristics required for detectors to be used in DEIMOS and other optical instruments of the Keck Observatory.

1. THE DETECTORS

Light entering the DEIMOS spectrograph through a slit mask strikes a single collimator, and is then divided into two beams by a V-shaped "tent" mirror. After striking a grating (or flat mirror in direct-imaging mode), each beam enters a wide-angle camera and is imaged onto a large detector mosaic. Each mosaic will be 8k x 8K pixels consisting of eight 4k x 2k detectors with 15 μm pixels. The details of the DEIMOS spectrograph design will be presented at other conferences. Here we focus only on the detectors.

A mosaic of CCDs is required for DEIMOS since an 8k x 8k array of 15 μm pixels can not be produced in a single, monolithic CCD with current CCD fabrication facilities anywhere. On a 100 mm silicon wafer a 4k x 4k CCD can be produced. However, for this application we chose a 4k x 2k format because of the higher fabrication yield likely with the smaller device. To form the large mosaic the devices must be three-side buttable, with a single serial register along a short edge. The spectrum runs in the long dimension of the CCDs and we want to minimize spectral loss in the gaps between CCDs. Therefore, our goal is to have no more than a 200 μm gap along the top edge of the CCDs. Along the other sides (the 4k-long edge) we want to have no more than a 1000 μm gap. The spectrograph design is making good use of the dead areas in the focal plane created by these wider gaps by placing there the mechanical members that define the shape of the slit mask. Table 1 lists some of the specifications for the CCDs.

TABLE 1

Specifications for CCDs

Geometry	2048 x 4096 15 μm pixels
	Three-edge abuttable with serial along short edge
	Dead region < 100 μm on top (2048) edge
	< 500 μm on 4096 edge
	Flatness: ±5 μm at -100° C
	Thinned to approx. 15 μm (assuming standard epitaxial wafers)
Mosaic	Each CCD to be individually packaged so they can be inserted and removed from a larger mosaic
	Gaps: < 200 μm on top and < 1000 μm on edges
Flatness	Entire mosaic maintains ±5 μm spec.
Amplifier	Readout noise < 5 e^- (2 e^- preferred) at 100 k pixels/sec
Linearity	< 5% deviation over full dynamic range of the CCD
RQE	> 60% at 350 nm (goal)
	> 90% peak RQE somewhere in the range 500 - 700 nm
	> 50% at 900 nm
	> 15% at 1000 nm (goal)
	< 5% variation over entire CCD
	< Stable RQE
Charge	CTE > 0.99999 for 1600 e^- signals
	Nor more than 100 pixels that exhibit low level CTE problems
	Full well > 150,000 e^-
	Dark current < 0.005 e^-/sec at -100° C
Fringing	Monochromatic interference fringing amplitude < 10% peak-to-peak

2. CCD FABRICATION

The CCDs described in Table 1 do not yet exist, yet are vital if DEIMOS is to achieve its full scientific potential. For this reason we can not just wait to see if they become available. We must do everything we can within our budget limitations to try to make sure they become available. Keck Observatory has chosen to work on several fronts. On one front, Keck is looking for a "full service" vendor who can supply us completed, working devices. A survey of potential suppliers identified MIT/Lincoln National Laboratory as the best candidate who could fabricate the CCDs at reasonable cost, and with a reasonable expectation of success within the time frame of the DEIMOS project. Lincoln has demonstrated capabilities in CCD mosaics, thinned CCDs, flat CCDs, and very low readout noise. To develop the Lincoln potential an international consortium has been established to fund a one year engineering effort at approximately $500 K total investment. This consortium is coordinated by Gerard Luppino of the University of Hawaii.

On a second front, Keck is actively funding the development of the 4k x 2k CCD through foundry sources. If we can acquire the CCDs we need through foundry sources, the total cost for the large number of devices we need could be significantly less than the cost from a "full service" vendor. With the eventual disruption of the Loral, Newport Beach facilities in mind, we began to look for an alternative foundry. At the same time, Orbit Semiconductor, a major silicon foundry, approached us and proposed a collaborative development effort, which would help them develop their scientific CCD capabilities. As an ambitious first test of their CCD fabrication abilities we decided to design a 100 mm wafer with one 4k x 2k CCD and two 2k x 2k CCDs. The remainder of this paper describes some of the results from these first wafer runs. As we will show, the devices produced by Orbit are very high quality scientific CCD imagers.

The design of the 4k x 2k CCD was done with the requirement of the DEIMOS CCDs in mind. The buttability requirement was met, with a dead space of about 200 μm between the saw blade line and the edge of the imaging area, in the long dimension of the device, and dead space of about 45 μm from the saw line to the imaging area along the top of the device. The design also includes split serial clocking, with identical amplifiers at each end of the horizontal shift register, which is located along the bottom of the device. MPP and notch implant technologies are also used to reduce dark current and improve charge transfer efficiency (CTE) respectively.

After fabrication the wafers were tested for DC shorts at Orbit and then delivered to Lick Observatory where we performed a detailed wafer probe examination of all working devices. After evaluating the devices on the wafer, we selected an engineering grade 4k x 2k device (device #464-12) for our initial lab tests. The wafer was diced, and then glued with Epo-Tek 301-2 epoxy to a piece of Kovar and subsequently electrically wire-bonded to a small printed circuit board.

3. PERFORMANCE CHARACTERISTICS

We performed all of the tests in a liquid nitrogen-cooled dewar. Our electronics includes a 20 x gain preamplifier located outside the dewar, and a correlated-double-sampling amplifier with 16 s sample times. The CCD was cooled to 120° C for all tests. Our measured performance characteristics for device 464-12 are summarized in Table 2, and discussed briefly in the following paragraphs.

3.1 Dark Current

Dark current, due to thermally generated electrons, is a limitation on read noise performance and is therefore an important parameter in astronomical applications where signal

TABLE 2

Measured Performance for an Orbit 4k x 2k CCD

4k x 2k CCD 464-12 Measured Performance	
Read Noise	< 4 electrons
Dark Current	<< 0.001 e$^-$/pixel/sec (MPP mode)
Horizontal CTE	> 0.999997 per phase (see Fig. 1)
Full Well (MPP)	70,000 e$^-$, (non MPP): 170,000 e$^-$
Low level localized traps	59 total (see Fig. 2)
RQE	Typical thick CCD response (see Fig. 3)

levels may be very low. In a 1000 second dark integration we found an average dark current of < 1 e$^-$ when the device was operated with MPP clocks. In fact, further tests with much longer integrations will be necessary before we can reliably determine just how low the dark current is. Spurious charge generated by running the vertical clocks can be reduced to a negligible level by setting the vertical clock levels to +two volts and -six volts.

3.2 Read Noise

Read noise places a lower limit on the smallest charge packets that can be detected and measured by the CCD. For low light level observations, the read noise value becomes a very important value in determining the CCD's limitations. To measure the read noise we clock the serial register so that any charge is transferred away from the output amplifier. All other signal processing is the same. With 16 s amplifier sample times the read noise per pixel was 3.5 e$^-$. (The intrinsic noise of our system with input grounded is less the 1 e$^-$.) We hope to maintain this noise figure, while increasing our readout speed, by incorporating the first stage of the preamplifier on the CCD package.

3.3 CTE

To measure charge transfer efficiency we illuminate the CCD with an Fe-55 x-ray source, each absorbed x-ray generating 1620 electrons. By comparing the measured charge packet with the expected charge packet the CTE can be measured accurately. The plot to the right (Fig. 1) shows the distribution of observed events, as a function of column number. (Multiply the vertical axis by 1.4 to convert the digital numbers shown to electrons.) If all of the charge produced from each absorbed X-ray were collected by a single pixel and if the CTE were perfect, then all of the events would fall on a single horizontal line in the figure. In fact, the charge from X-rays absorbed deep in the CCD may diffuse into two or more pixels so there are many more "split" events than single-pixel events. In the Figure the single pixel events fall on a (nearly) horizontal line at a level of about 1160. We fit a straight line to these events, and we measure the slope of the fitted line. The slope is a direct measure of the CTE, and for this

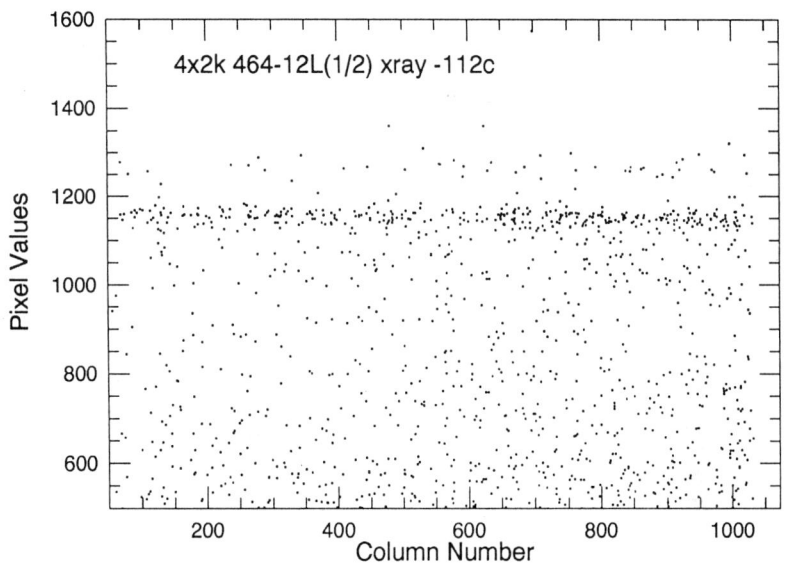

Fig. 1. CTE measured using Fe-55 X-rays

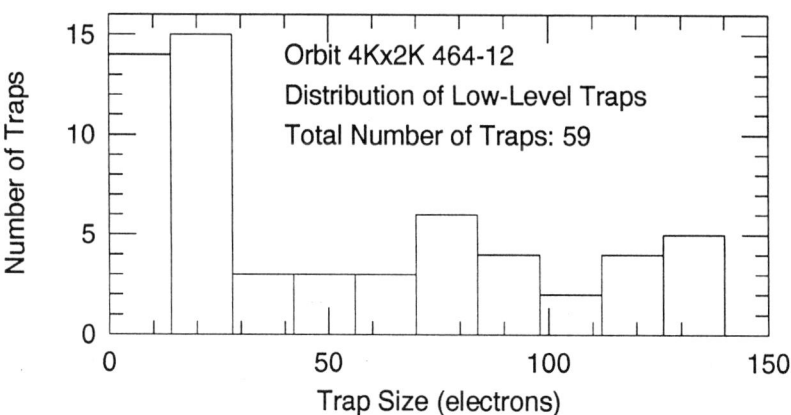

Fig. 2. Size Distribution of Localized Traps

device we found a CTE of 0.999997 per phase.

3.4 Full Well

We measured the full well capacity of the CCD in MPP mode, with all vertical clocks inverted during integration, and in non-MPP mode, with one vertical clock held high during integration. The measured full well was 70,000 e$^-$ in MPP mode and 170,000 e$^-$ in non-MPP mode. Since several different MPP doping levels were tried on different wafers in this run, these results can't be taken as an indication of the "typical" full well. As we have the opportunity to measure more devices we will learn how doping levels effect full well capacity.

3.5 Localized Traps

We measure low-level traps by taking a uniform illumination exposure that results in about 300 e$^-$ per pixel. Then, before reading out the image we shift the charge vertically on the CCD. First the charge is shifted in one direction by 24 rows and then is shifted back in the other direction by 24 rows. This is repeated 40 times before the image is read out. A low level trap shows up as a pixel with excess charge (where charge has been trapped on each shift) and one or more pixels with lower than expected charge (where the charge has been lost). The size of the trap is measured by dividing the size of the excess charge by the number of shifts. Fig. 2 shows the size distribution determined for the 4k x 2k CCD. There were a total of 59 traps located on the CCD by this method. This is an exceptionally small number of traps for such a large device. We found a consistent number of traps (about half) in the Orbit 2k x 2k CCDs we have measured.

3.6 RQE

We measured responsive quantum efficiency using a set of interference filters and a calibrated photodiode. Fig. 3 shows the measured RQE, which we find to be typical of thick CCDs.

4. CONCLUSIONS

The work being done with Lincoln Laboratory and Orbit Semiconductor represent an aggressive effort to produce the CCDs needed for the DEIMOS spectrograph. However, none of this work prevents Keck from considering other options, should they become available. For instance, it may be possible that SITe or some other full service vendor may develop an appropriate CCD, and it may be possible that one of Loral's other fabrication facilities may prove to be technically viable at reasonable cost. We will continue to monitor possibilities.

We began the project with Orbit Semiconductor as a test to see if they could produce the CCDs we need for the DEIMOS project. We still have some work to do, but our results show that Orbit can indeed produce very large, high quality scientific CCDs. The CCD mosaics for the DEIMOS spectrograph will have to be thinned, backside treated CCDs, and they will have to be maintained flat and co-planar to within about ten micrometers over the entire mosaic.

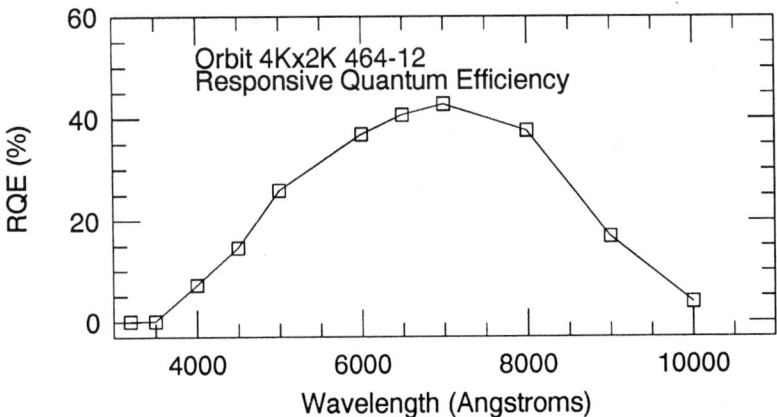

Fig. 3. Measured RQE for for thick 4k x 2k CCD

These requirements present formidable technical challenges which we are now beginning to address.

ACKNOWLEDGMENTS

We would like to thank the University of Hawaii, Institute for Astronomy for providing partial financial support for this project. We would also like to thank the Santa Cruz Institute for Particle Physics for allowing us to use their wire-bonder.

DISCUSSION

PHILIP: What is the readout time of a 2k x 4k chip?

STOVER: This depends how many amplifiers are used and the integration time per pixel. with our current system, and reading only one amplifier it takes about four minutes. Our goal is to read much faster and to use both amplifiers. Other groups have reported problems reading multiple amplifiers. If we cannot read multiple amplifiers then low-noise high-speed readout becomes even more important.

COHEN: We are currently working hard on fixing the negative cross talk in the multi-amplifier readout electronics. We should have that completed within the next two months.

STOVER: Other groups have reported similar problems too.

QUESTION: Do your 10% fringing specifications refer to front-side or back-side illumination, and to what wavelength does it refer?

STOVER: It refers to back-side illumination and is the highest amplitude at any wavelength.

WALKER: Please comment on the yield of Orbit 2k x 4k devices obtained so far.

STOVER: We had five working 4k x 2k CCDs with only one really good device. After Orbit did our wafer runs they did several more on their own to try to improve yield. The next three wafer runs will show whether or not Orbit can provide these CCDs with reasonably high yield.

QUESTION: Will you have the facility to cold test the CCDs before thinning?

STOVER: Yes, we can cool the wafers to about -10° C to test them.

JORDEN: What is the Keck budget for CCDs? And what devices are to be provided for this money?

STOVER: Keck is spending about $100,000 US on the Lincoln Labs effort, $100,000 on the foundry effort and $100,000 on the CCD thinning effort. Once all of these development efforts are concluded there should be one to two million dollars for actual CCD acquisition. We need about 24 science grade devices.

McBREEN: What is the current situation with Loral producing large CCDs?

STOVER: Loral, Newport Beach has stopped taking orders. Further work may be possible at Loral's Milpitas facility. Loral has also purchased a fabrication facility from IBM. This facility uses 5-inch wafers, so larger CCDs or more CCDs per wafer may be possible. However, it is still not clear what the costs at either of these facilities may be. We may do some more work with Loral after these issues are resolved.

McBREEN: Are you considering large frame transfer CCDs?

STOVER: Yes, the Lincoln Lab CCDs will have frame transfer. This is being done to satisfy interests of some of the consortium members. However, the DEIMOS spectrograph does not use this feature, and since it will increase the gaps between the CCDs in a mosaic, frame transfer is not included in the orbit semiconductor design.

SCIENTIFIC CCD PROSPECTS FOR 1994 AND BEYOND

Paul R. Jorden and A. P. Oates

Royal Greenwich Observatory

ABSTRACT. We review the present status of CCD use at major observatories, discuss the various developments that are leading to improved sensors. The choice of devices is quite extensive, but far from stable! A look at progress over the last few years helps to put present and future developments in perspective. Several areas of active CCD development are presented and discussed.

1. INTRODUCTION AND CURRENT STATUS OF OPTICAL CCDS.

There is no doubt that for the last decade CCDs have been the principal optical sensor at most observatories. The number of manufacturers, and of devices has increased over this period, as illustrated in Table 1. The table lists a snapshot of the main CCD suppliers and indicates the primary sensors that have been in widespread use at each time period. Prototype, or uncommon types are not shown here.

TABLE 1

Principal CCD manufacturers, and devices, 1980-1994

Year (no of device types)	Manufacturer	Device Formats	Pixel Size (microns)	Read Noise (e- rms)	Peak QE (%)
1980 (3)	RCA	320 x 512	30 x 30	80	70
	GEC	385 x 578	22 x 22	~20	40
	Fairchild	380 x 488	18 x 30	30	12
1985 (~10)	RCA	320 x 512	30 x 30	60	70
	EEV/GEC	385 x 578	22 x 22	10	50
	Thomson	385 x 576	23 x 23	10	40
	TI	800 x 800	15 x 15	5-10	70
	Fairchild	380 x 480	18 x 32	30	12
1990 (~20)	EEV (GEC)	385 x 578, 1242 x 1152	22 x 22	~4	50
	Thomson	385 x 576, 1024 x 1024	23, 19	~5	40
	Ford	516 x 516	15 x 15	~5	40
	Reticon	400 x 1200	27 x 27	~5	40
	Tektronix	512 x 512	27 x 27	~10	45
1994 (~30)	EEV	1242 x 1152 & oth	22 x 22	3/4	50/80
	Thomson	1024 x 1024 & oth	19 x 19	~4	40
	Loral (Ford)	2048 x 2048 & oth	15 x 15	4/5	40 (80)
	Reticon	400 x 1200 & oth	27 x 27	~5	40
	SITe (Tek.)	2048 x 2048 & oth	24 x 24	~4	70
	[Kodak]	2048 x 2048 & oth	9 x 9	~10?	~40

As a prelude to a discussion of future developments we shall review here the types of devices and performances that can be obtained today. The routine supply of good quality scientific CCDs is somewhat precarious, with a limited number of sources. Within the last few years the use of foundry manufacturers has developed - particularly with many devices having been made by Ford/Loral. Foundries offer the possibility of supplying a large quantity of devices at manageable costs; however, the yield and performance of such devices are not guaranteed.

The following tables present a broad picture of CCD usage by observatories around the world; quite a few different types of CCD are in use, obtained through several routes. Table 2 illustrates the wide range of devices known to be offered by major observatories. There may be a few omissions, but on the whole it should at least present a snapshot of current use. Most observatories are in the process of trying to upgrade to larger CCDs, so the picture could change quite a lot in the next year or so.

TABLE 2

Parameters of CCDs, used by major observatories (ESO, CTIO, KPNO, RGO, AAO).

CCD TYPE	Pixel size (Micron)	Format X	Format Y	Image size (mm) X	Image size (mm) Y	Status
GEC P8603/ EEV-02-06	22.0	385	578	8.5	12.7	Comm.
RCA SID501 (&640 x 1024)	30.0	320	512	9.6	15.4	Obs.
EEV CCD-05-30 (P88300)	22.5	1242	1152	27.9	25.9	Comm.
Tektronix TK512 (SITe)	27.0	512	512	13.8	13.8	Comm.
Tektronix TK1024 (SITe)	24.0	1024	1024	24.6	24.6	Comm.
Tektronix TK2048 (SITe)	24.0	2048	2048	49.2	49.2	Comm.
Reticon RA1200	27.0	400	1200	10.8	32.4	Comm.
Thomson TH7896 (TH1024)	19.0	1024	1024	19.5	19.5	Comm.
Loral (Ford) NOAO-3 k x 1 k	15.0	3072	1024	46.1	15.4	Found
Loral (Ford) SAO-Lick 2 k x 2 k	15.0	2048	2048	30.7	30.7	Found
TI-800	15.0	800	800	12.0	12.0	Obs.

As a further illustration of present-day status the Table 3 shows the principal parameters of one of the main camera types that we operate at the international Roque de los Muchachos Observatory on La Palma. This camera, utilizing the Tek1024 thinned CCD, exhibits performance that is typical of modern, high-grade sensors. We currently use CCDs for 94% of all observing on ING telescopes.

To conclude this section, Table 4 presents a list of most CCDs that are of interest for large-area (> 256 x 256 pixels) scientific imaging; the scene does change quite rapidly so it is unlikely to be an exhaustive list. The use of foundry sources for CCDs makes it particularly hard to be definitive about the exact number and type of devices that are around. Obsolete sensors (even if still in use) are not listed. Included in the table are many devices that are in prototype form, and not routinely available; some will become commercially available, some will not! Commercially available devices are shown with an *.

TABLE 3

An example of CCD parameters - The ING TEK1 CCD.

Detector type	TK1024, backside-thinned
Format	1024 x 1024
Pixel size	24 x 24 micron
Array size	25 x 25 mm
Readout noise	4-6 e⁻ rms
Operating temperature	180 K
Dark current	8 e⁻ pixel^{-1} hour^{-1}
Pixel sampling times	10-80 microsec (variable)
Quantum efficiency	75% (peak), 55% (400 nm)
Charge transfer efficiency	0.99999 (X & Y)
Cosmetics (grade-1 device)	excellent: no defective columns, ~10 single pixel traps
Cosmic ray rate	~2 events min^{-1} cm^{-2}

2. AREAS OF ACTIVE DEVELOPMENT, AND PROJECTIONS INTO THE NEAR FUTURE.

In this section various areas of current activity and interest will be introduced. Since several topics are due to be presented in more detail elsewhere in this Symposium, only a limited discussion will be given here. A few interesting features, that some users may not be aware of, are presented and some specific new developments are highlighted.

2.1 Mosaics - Multiple Chips, Multiple Outputs, Higher Speeds.

The requirements of large focal plane areas are such no single CCD is big enough. The largest sensors presently available have a 2k x 2k format; a 4k x 4k size appears unrealistic at the current state of technical development. A mosaic of modest sized CCDs is considered to be the realistic solution to this problem. There are several developments of 4k x 2k format devices at present; most current designs envisage using 2k x 4k three-edge buttable chips (a *VLT-CCD*) to create large focal plane arrays.

This VLT CCD is currently planned for at least the following telescopes - Keck, Gemini, ESO-VLT, Subaru, as well as many smaller ones (RGO, AAO, Galileo etc.). Within the next year or so we should hope to see the successful introduction of at least some devices of this format, although we should not underestimate the difficulties of achieving high performance (eg Quantum Efficiency) from such large devices.

With the use of large-format devices it becomes more and more important to achieve higher speeds of readout, in order to minimise total frame readout times. All large-format devices have multiple outputs (two or four) to facilitate multiple channel readout.

With improved baseline readout noise performance, it is possible to increase the speeds of readout and still achieve low readout noise. Whereas 10 e⁻ was about the lowest possible readout noise a decade ago, it is now possible to achieve ~2 e⁻ noise (at ~ 50 kHz pixel rates), and < 10 e⁻ is still achieved - but at much higher readout speeds than before. The paper by Luppino (1995, this Symposium) discusses such mosaics in more detail.

TABLE 4

A selection of scientific CCDs (1994).

CCD TYPE	Pixel size (microns)	Format X	Format Y	Image size X mm	Image size Y	Status
GEC P8603/ EEV-02-06*	22.0	385	578	8.5	12.7	Comm. Thick/thin
EEV CCD-05-20 (ex P88200)*	22.5	770	1152	17.3	25.9	Comm. Thick only
EEV CCD-05-30 (ex P88300)*	22.5	1242	1152	27.9	25.9	"
EEV CCD-05-50 (ex P88500)*	22.5	2186	1152	49.2	25.9	"
EEV CCD-15-11*	27.0	1024	256	27.6	6.9	Comm., new
EEV CCD-15-80 (small pixel)	13.5	2048	4096	27.6	55.3	Under development
Tektronix TK512*	27.0	512	512	13.8	13.8	Comm. Thick/thin
Tektronix TK1024*	24.0	1024	1024	24.6	24.6	Comm. Thick/thin
Tektronix TK2048*	24.0	2048	2048	49.2	49.2	Comm. Thick/thin
Kodak KAF0400*	9.0	768	512	6.9	4.6	Available, thick
Kodak KAF1400*	6.8	1340	1037	9.1	7.1	"
Kodak KAF1600*	9.0	1552	1032	14.0	9.3	"
Kodak KAF4200*	9.0	2048	2048	18.4	18.4	"
Reticon RA0512*	27.0	512	512	13.8	13.8	Comm? Thick/thin
Reticon RA1024*	13.5	1024	1024	13.8	13.8	"
Reticon RA1200*	27.0	400	1200	10.8	32.4	"
Reticon RA2000	13.5	2048	2048	27.6	27.6	"
Thomson TH7883*	23.0	384	578	8.8	13.3	Comm. Thick only
Thomson TH7895*	19.0	512	512	9.7	9.7	Comm. thin?
Thomson TH7896*	19.0	1024	1024	19.5	19.5	"
Thomson TH7897M	15.0	2048	2048	30.7	30.7	Under development
Ford-512 (Photometrics)	15.0	512	512	7.7	7.7	Made - foundry run.
Ford-1024 (Craf/Cass.-JPL)	12.0	1024	1024	12.3	12.3	"
Ford-1024	15.0	1024	1024	15.4	15.4	"
Ford-SAO-Lick 2048	15.0	2048	2048	30.7	30.7	"
Ford-NOAO-3 k * 1 k	15.0	3072	1024	46.1	15.4	"
Ford-SAO	15.0	2688	512	40.3	7.7	"
Ford-Loral	15.0	4096	2048	61.4	30.7	"
Dalsa CA-D9-5120	12.0	5120	5120	61.4	61.4	Available only as cameras
Dalsa CA-D9-2048	12.0	2048	2048	24.6	24.6	"
Dalsa CA-D9-1024	24.0	1024	1024	24.6	24.6	"
Dalsa CA-D2-1024	10.0	1024	1024	10.2	10.2	"
MIT Lincoln Lab. CCID-7	27.0	420	420	11.3	11.3	Not comm. avail.
LL CCID-10	24.0	1024	1024	24.6	24.6	"
LL CCID-14	18.0	1280	1280	32.4	32.4	"
LL CCID-16	24.0	1960	2560	47.0	61.4	Prototype

2.2 Higher Speeds, Lower Noise and Adaptive Optics.

A particular area of recent development has been that of low-noise at high speeds. Lincoln Labs have demonstrated that it is possible to make a CCD which yields a readout noise $\sim 6/7$ e^- at one MHz pixel rates (Burke et al. 1991); this noise was previously only achieved at 50 kHz pixel rates. By designing the output circuitry of the CCD to have a very high charge-to-voltage conversion factor it is possible to get a lower effective noise during readout - this is achieved at the expense of a more limited total charge-handling range. It is interesting to note that such a design can yield a one to two e^- readout noise at normal slow-scan rates (20 - 50 kHz).

SCIENTIFIC CCD PROSPECTS FOR 1994 AND BEYOND

At present (for example the TK1024 device) a readout noise of four to five e^- is routine, with somewhat lower figures being possible at very slow rates. It is now clear that noise figures of two to three e^- ought to be common in the next few years, as the higher-gain amplifier designs are introduced. The high QE of CCDs, coupled with such low noise, means that traditional (intensified) photon-counting systems have little advantage for most common astronomical applications.

The main interest in higher speed CCDs has developed from the requirements of adaptive optics systems. Here, it is crucial to have a high sensitivity at high frame rates ($\sim$ one kHz). High sensitivity is achieved through the use of a backside-thinned device, together with low readout noise at this high speed. Several CCD suppliers are developing higher speed sensors - these have a use in some industrial markets as well as for scientific imaging. For Adaptive Optics a thinned 64 x 64 format CCD, with $\sim$ five e^- noise at one MHz pixel rates is anticipated. The non-commercial Lincoln Laboratory has made such a device, and other manufacturers are expected (hoped?) to follow suit.

2.3 Different Chip Designs - Larger Areas, Higher QE, Lower Noise, Smaller Pixels.

Since the advent of foundry runs it has been possible for novel chips to be designed, built, and made available to the community (just). Examples of these include buttable-chips, multiple-output designs, and even the Geary polar-coordinate device (Geary and Luppino 1994). Ford-Aerospace (now Fairchild-Loral) made the first foundry CCDs for astronomy, and more recently the Orbit foundry has made similar devices (see Stover et al. 1995, this Symposium). This source of supply is now well established; however the disadvantages are that the yield is uncertain, and quality is not assured. Devices have to be thinned elsewhere in order to obtain a high quantum efficiency. With the uncertainty in supply of good large CCDs, several foundry sources are currently being investigated.

Higher quantum efficiency is always much in demand. The technique of manufacturing thinned, backside-illuminated devices is still proving difficult for all manufacturers. RCA supplied such devices in 1980, TI and Tektronix have supplied many such devices since then, and other manufacturers have developed thinning techniques. However, it is still proving difficult to make and supply large-area, high quality thinned devices. Mike Lesser has been particularly active in thinning chips from foundry sources. He has also demonstrated that very high QE (100%) can be achieved - by selective AR-coating of the silicon at certain wavelengths (Lesser 1994). It is also feasible to optimize the coating for red or blue regions, depending on the instrumental application required. As an aside, an optimized blue coating means that the chip will have a finite red-wavelength reflection loss - which results in a significant degree of fringing at the longer wavelengths.

CCD designers and manufacturers have been able to reduce readout noise, mainly as a result of experience and empirical design. The output transistors on CCDs can now yield very low noise, such that four to five e^- is often routinely attained from commercial chips. Noise figures of two to three e^- have been seen in some commercial chips and noise figures in the region of one to two e^- appear possible with certain limitations. A novel skipper amplifier has been produced which can potentially reduce noise by multiple sampling (Chandler et al. 1990), but it seems to have little benefit now that the basic CCD noise levels can be reduced so much.

2.4 Lower Dark Current-MPP Devices and Wobble-Clocks.

Many CCDs are now available in MPP (multi-phase-pinned) form; this allows a reduction of dark current by about a factor of 100. The devices require operation in a slightly different manner, and have a somewhat lower full-well capacity. The use of MPP devices allows the possibility of operating at higher temperatures than otherwise possible. In the case of cryogenic systems, a higher operating temperature can yield a better red-response. In other cases it can allow Peltier rather than cryogenic cooling, or even no cooling at all. It is also possible to minimise dark current by modulating CCD clocks during the exposure period (Thorne et al. 1990, Burke and Gajar 1991); this technique is particularly beneficial for non MPP devices

2.5 Improved Red Response; Deep-Depletion Devices.

Some devices have been made with a higher resistivity silicon (eg 1000 ohm-cm, c.f. the normal 20 - 50 ohm-cm). This gives a deeper depletion region within the device. If the device is manufactured with a thicker active region this allows it to have a greater absorption of near-infrared wavelengths. The quantum efficiency can be increased at wavelengths longer than about 700 nm; it also has benefits for x-ray charge detection (for example see Holland et al. 1992).

2.6 Antiblooming.

CCD imagers traditionally suffer from blooming or image spread when over-exposed. TV-camera sensors are often built with internal antiblooming structures, but this is not appropriate for scientific sensors. However, by modulating CCD clocks during the exposure period, it is possible to attenuate strongly or even eliminate the undesirable effects of pixel over-exposure; (see Neely and Janesick, 1993). The technique may need some care to implement, but could be quite powerful for selected astronomical applications. Of course, over-exposure of part of the CCD array can still lead to internal reflections and the risk of scattered light.

2.7 Improved Electronics - More Compact, Lower Power, More Dynamic Range.

The improvements in CCD design need to be matched by improvements in electronics systems. There are three main areas requiring development:

a) Multi-channel systems required for faster parallel readouts.
b) Higher speed readout and data-transfer systems required - for cryogenic and adaptive-optics systems.
c) 18-bit A-D converters and electronic systems are needed to take advantage of the wide inherent dynamic range of CCDs. (An 18-bit (262144) range allows us to have 260 K e^- signal range, with a two e^- readout noise, and one e^- quantisation unit). Most systems (currently 16-bit), with a one e^- digitization unit, only allow a full-range of 65536 e^-.

3. RGO CCD ACTIVITIES.

To conclude this review, we will take the opportunity to present a brief overview of current areas of RGO CCD development:

SCIENTIFIC CCD PROSPECTS FOR 1994 AND BEYOND

3.1 WYFFOS

This is a multi-fiber spectrograph for the William Herschel Telescope (4.2-m) - it is due for completion at the end of 1994. It has a robotic fiber-positioner (AUTOFIB) at the prime focus; this feeds a custom-designed optimised spectrograph which resides at the Nasmyth (GHRIL) platform. The detector is a thinned Tek1024 CCD embedded in a Schmidt cryogenic camera. (See Bingham et al. 1994).

3.2 INT PF Mosaic

This project, in collaboration with US and Netherlands groups aims to provide a large-area CCD focal plane for the Isaac Newton Telescope (2.5-m). Four thinned Loral CCD chips will provide a 4 k x 4 k mosaic format, covering a 25 x 25 arcmin FOV, with 0.37 arcsec pixels. A fifth co-planar CCD will be used to provide autoguiding for the telescope. An outline of the camera-head design is shown in Fig. 1; the whole instrument will be described fully in a future publication. The camera is to be used for surveys and supernovae searches, and is scheduled for completion at the end of 1995.

3.3 Large-Format CCD Upgrades

We are engaged in a program to enhance most of our CCD imaging facilities by providing larger, improved sensors. We have a batch of Loral 2 k x 2 k devices that has been manufactured, and is currently undergoing thinning (by Mike Lesser). We have also just placed a contract (with EEV) for 4 k x 2 k thinned CCDs. Our requirements are for larger areas, small pixels, lower noise, and good (thinned) quantum efficiency.

3.4 Intra-Pixel Response Measurements

We have recently analyzed the internal structure of CCDs in some detail (see Jorden et al. 1994). Several CCDs were tested in the laboratory to determine how response changes with intra-pixel sampling position. The internal architecture has an appreciable influence on the response of a CCD that have implications when the image in undersampled; this can be important in many areas of use - including photometry, spectral extraction, astrometry, centroiding (AO systems).

3.5 QE as a function of temperature.

We have analysed the effects of operating at different temperatures. Operating temperature is well known to influence the far-red response of a CCD, but we also found response changes at other wavelengths. This has been relevant recently in discussions about achieving good UV response - at ambient temperatures (where many manufacturers check the devices), and when cryogenically cooled (where we need to operate them). Fig. 2 shows spectral response curves (of ING telescope CCDs) that are typical of frontside and backside performance (at a cryogenic temperature). Fig. 3 shows measurements of A TK1024 thin CCD, and demonstrates the variation of QE with temperature from 173-293K.

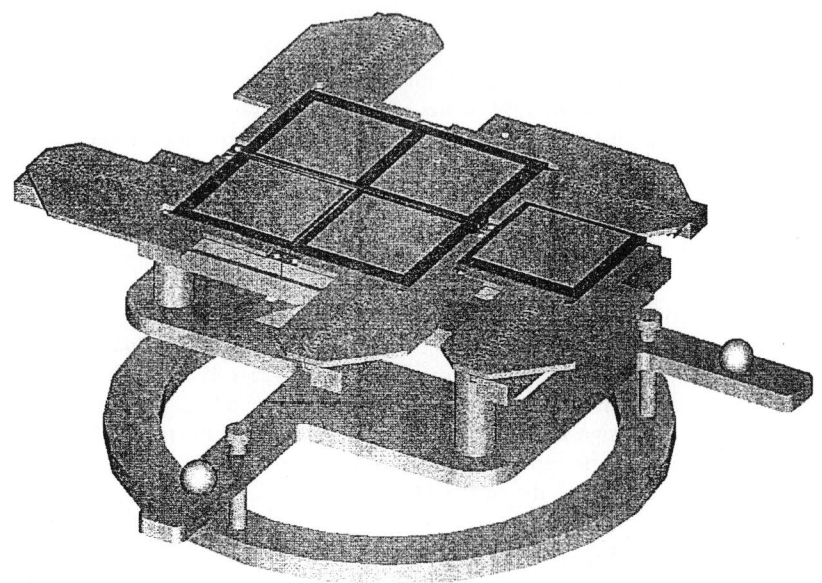

Fig. 1. The Isaac Newton Telescope Prime Focus Mosaic CCD camera.

3.6 Readout-Noise as a Function of Speed.

Prompted by an analysis of the requirements for a wavefront sensor, we have been operating CCDs at higher speeds than usual for cryogenic systems. Adaptive optics, and other wavefront sensors require much higher readout speeds than is normal in slow-scan systems. The readout noise does not always scale in a well-behaved way with frequency. The techniques required for higher speed operation have been examined.

3.7 Wavefront Sensors - for Gemini and UK AO

We have been involved in analyzing the requirements for two major new projects. The UK is embarking on a project to provide AO systems on its WHT and UKIRT telescopes; detailed atmospheric site monitoring will be followed by design and construction of suitable AO systems. The RGO is also responsible for a Gemini workpackage which will lead to the construction of an A&G unit for the 8-m telescopes. This unit will have a set of sophisticated wavefront sensors and acquisition facilities.

ACKNOWLEDGMENTS

We are pleased to acknowledge the support of many RGO colleagues, particularly Dave Banks, John Churchill, Derek Ives, Andrew Johnson, John Maclean, and Percy Terry.

SCIENTIFIC CCD PROSPECTS FOR 1994 AND BEYOND

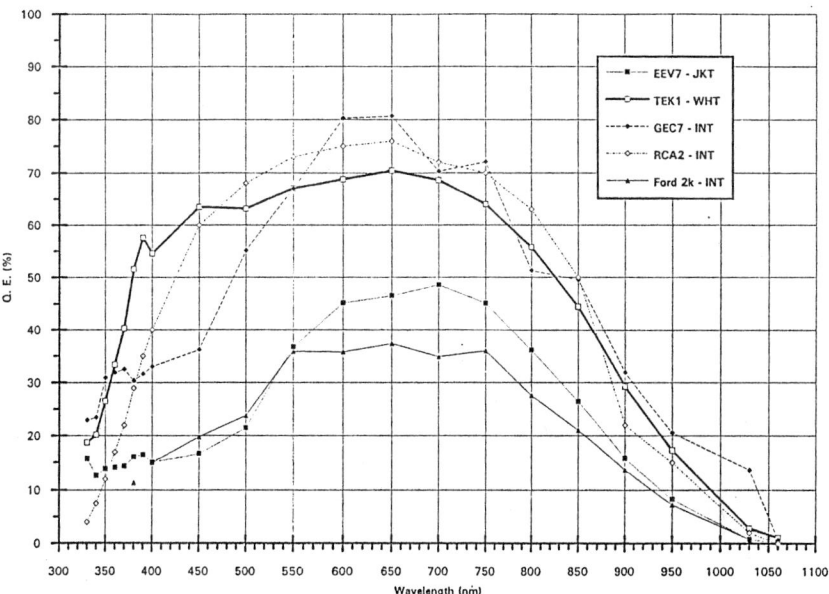

Fig. 2. Spectral response of ING CCD sensors, at cryogenic temperature.

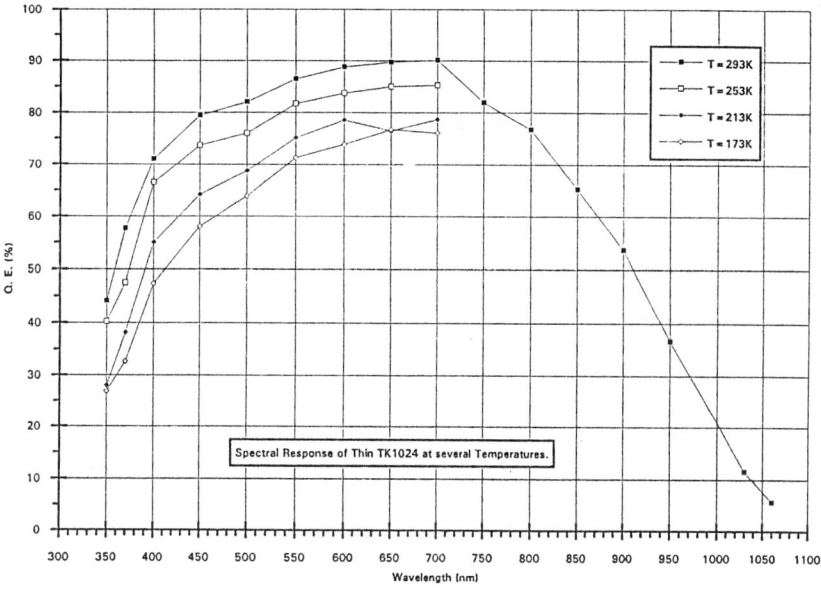

Fig. 3. Variation of spectral response (of thin TK1024 CCD) with temperature.

REFERENCES

Bingham, R. G., Gellatly, D. W., Jenkins, C. R., Worswick, S. P. 1994 proc. SPIE 2198, 56
Burke, B. E., Gajar, S. A. 1991 IEEE Trans. Electron Devices 38, 285
Burke, B. E., Mountain, R. W., Harrison, D. C., Bautz, M. W., Doty, J. P., Ricker, G. R. and Daniels, P. J. 1991 IEEE Trans. Electron Devices 38, 1069
Chandler, E.C ., Bredthauer, R. A., Janesick, J. R., Westphal, J. A. and Gunn, J. E. 1990 Proc. SPIE 1242.
Geary, J. and Luppino, G. A. 1994 Proc. SPIE 2201, 588
Holland, A., Turner, M. J. L., Wells, A. and Burt, D. J. 1992 Proc ESA Symposium on Photon Detectors for Space, ESA-SP-356.
Jorden, P. R., Deltorn, J-M. and Oates, A. P. 1994 Proc. SPIE 2198, 836
Lesser M.P., 1994 Proc SPIE 2198, 782.
Luppino, G. A. 1995 in IAU Symposium No. 167, New Developments in Array Technology and Applications, A. G. D. Philip, K. A. Janes and A. R. Upgren, eds., Kluwer Academic Pub. Dordrecht, p 297
Neely, A. W. and Janesick, J. R. 1993 PASP 105, 1330
Stover, R., Wei, M., Gilmore, K. and Brown, W. 1995 in IAU Symposium No. 167, New Developments in Array Technology and Applications, A. G. D. Philip, K. A. Janes and A. R. Upgren, eds., Kluwer Academic Pub. Dordrecht, p 19
Thorne, D. J., Waltham, N. R., Newton, G. M., van Breda, I. G., Fisher, M. and Rudd, P. J. 1990 Proc. SPIE 1235, 400

DISCUSSION

STOVER: Do you know of a commercial source for small, thinned CCDs suitable for autoguiders?

JORDEN: EEV has a 600 x 400 CCD which is used in RG0 autoguiders. We have not felt any necessity for smaller formats - in fact we use the full 400 x 600 formal to acquire quide stars.

STOVER: Have you found that using "wobble clocks" increased read noise due to extra generated electrons?

JORDEN: This technique is used in the autoguiders at RGO. In cryogenically cooled CCDs this should not be a problem because the clock switching needs to occur only once every few minutes.

STOVER: Do you have some ideas concerning the source of inter-pixel sensitivity variations in the thinned (TEK) CCD?

JORDEN: We were surprised to detect a response modulation (one%) from thinned (TEK) CCDs. We believe it is a result of the influence of the depletion region reaching to the backside of the chip and influencing charge collection (see reference: Jorden et al. 1994, Proc SPIE 2198, 836).

BLECHA: You mentioned 18 bit converters. In fact we do not need such precision since for high signal the shot noise dominates. Can you comment on the use of non-linear encoding scheme?

JORDEN: LOG-converters are very slow. An 18 bit linear converter would be the obvious system upgrade. They would provide a straightforward solution to extending the dynamic range. I agree that full precision is not needed in the presence of shot-noise.

WEAVER: Zoran Ninkov at RIT has reported non-linearities below 1500 e^- in Kodak CCDs. Have you heard of similar problems with other chips? Do the engineers routinely check for this before releasing them to astronomers?

JORDEN: No. That 1500 e^- level is very low. We always check our linearity over the whole operating range before installing CCDs on the telescope.

NEW DEVELOPMENTS IN CCD TECHNOLOGY FOR THE UV-EUV SPECTRAL RANGE

Giovanni Bonanno

Catania Astrophysical Observatory

ABSTRACT: Lately, Charge Coupled Device (CCD) detectors have had great advances both in the visible and in the X-ray spectral range. However, the technology applied to these devices in the ultraviolet (UV) spectral region has not developed as well, because of some problems connected with the interaction between UV radiation and the materials typically used in semiconductor technology. In our laboratory the ultraviolet response of some UV-enhanced CCDs has been investigated. In particular, the quantum efficiency of coronene and lumigen coated and of back-illuminated ion implanted CCDs have been measured in the 304 - 11000 Å spectral range. Very interesting results have been found, mainly for a one ion implanted CCD with quantum efficiency values of more than 60% at 304 Å. Some measurements of the response uniformity of this spectral region have also been made. The results obtained encourage the possible use of these detectors in ultraviolet astronomy with very good performance.

1. INTRODUCTION

Charge Coupled Devices (CCDs) are widely used as imaging detectors both in the optical and in the X-ray spectral range. In many observatories various CCD cameras are operated in the optical. There are very good results in laboratories such as JPL (Janesick et al. 1989, 1990) and Lockheed Palo Alto Research (Stern et al. 1986) in the USA, and Space Research at Utrecht (Verhoeve et al. 1991, Bailey et al. 1990) and Leicester University (McCarthy and Wells 1992) in Europe. The use of these devices in the ultraviolet (UV) spectral region is obviously very attractive especially for space instruments. But, CCD technology in this spectral region presents some problems connected with the interaction between UV radiation and silicon. As shown in Fig. 1 the absorption depth in the 1000 - 4000 Å region is very low and then the interaction occurs on the first layers of the material. For example, the relatively thick gate oxide layer (500 - 1200 Å) on the CCD front face strongly absorbs UV radiation, thus it is impossible to use front-illuminated CCDs for direct detection of UV radiation.

Many solutions have been tried: the most common technique is to sublimate a phosphor under vacuum conditions, such as "lumigen" on the sensitive surface (Bredthauer et al. 1988). This is done both for front-illuminated and thinned back-illuminated CCDs, down-converting the UV radiation and the absorption due to the thick polysilicon gate can be avoided. In back-illuminated CCDs, other problems have to be solved. After the device is thinned, a "native" back-oxide layer forms naturally which contains a lot of "dangling bonds" creating surface states. Since there are positive charges associated with them, a depletion region at the back of the CCD forms (backside well) and, as stated before, owing to the extremely short path length (< 100 Å) in silicon for UV photons, the charge is photogenerated very near to the CCD backside, so that the recombination is highly probable and a lot of signal charge can be lost.

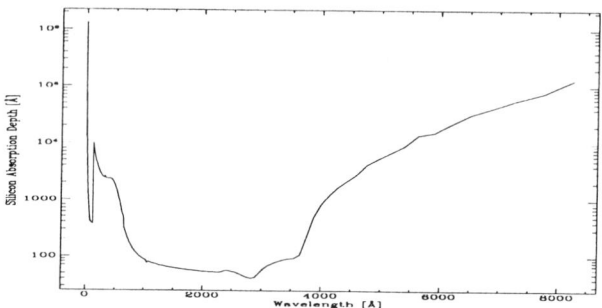

Fig. 1. Silicon Absorption Depth vs wavelength in the 1.2 - 8500 Å region.

Some methods have been developed to reduce surface state problems using different "back-accumulation" techniques. They are based on the hypothesis that the back surface potential can be *pinned* by heavily populating it with holes which create an electric field that directs the photogenerated charges away from the back surface (Janesick et al. 1989). The methods can be divided into three categories: backside charging or UV flooding (Janesick et al. 1985), flash gate techniques (Janesick et al. 1989) and ion-implantation and successive laser-annealing.

The last technique is based on the fact that the effect of a p$^+$ implantation on the backside lowers the backside well width (Janesick et al. 1985) to values of the order of 10 Å or less. Moreover, the implantation damages the crystal lattice sites, so that it is necessary to anneal and reconstruct the crystal structure. To do this a pulsed laser annealing of the CCD back surface is adopted (Bailey et al. 1990); if the process is optimal, at the end, no significant residual backside well should be present.

There is only one small problem that can be still present: if some ions penetrate deeply into the surface they cannot be reached by annealing; in such a way, the associated crystal damage acts like traps for the charge signal, reducing the QE performances. Thus, to minimize this problem, a low energy ion-implantation technique must be used.

Another new technique named *"anodic etching"* has been developed by English Electric Valve (EEV) and it consists in the removal of the oxide layer by selective *etching*. This technique allows the boron-implanted layer to be thinned in a controlled way (by means of an appropriate voltage during the anodization). Anodic etching could be applied either before or after laser annealing. Thus the deep penetrated boron atoms, which are left unannealed in the standard technique, could be reached now. Another technique that is under investigation at EEV is the *ultra low energy boron implantation* (E < 2 KeV) that leaves no unannealed boron and can be combined with *"anodic etching"*.

In the following sections, I will describe the results obtained from measurements of the QE and uniformity response performances in the 300 - 11000 Å spectral range of some UV-enhanced EEV CCDs. I selected one per type: one is front-illuminated and phosphor coated, the others are thinned back-illuminated with the surface treated as: phosphor coated, ion-implanted and ion-implanted with the "anodic etching".

2. CCD CHIPS

We have tested several EEV CCDs of the CCD02-06 (385 x 576 pixels, 22 μm pixel size) series. Table 1 summarizes the relevant characteristics of the four selected CCDs.

TABLE 1

Characteristics of Four Selected CCDs

CCD TYPE	ILL.	SURFACE TREATMENT	NOTES
Thick	Front	Lumigen vacuum sublimation	scientific grade
Thinned	Back	Coronene coating	scientific grade
Thinned	Back	Ion-implanted laser annealed	E = 10 KeV, C.= 10^{15} cm^2
Thinned	Back	Ion-implanted laser annealed	as above, anodic etched

The back-illuminated ion-implanted CCDs are of the same type as those furnished by EEV for the characterization tests of the detector for the Reflection Grating Spectrometer on the XMM mission (Verhoeve et al. 1992); the only difference is that this CCD has a lower resistivity (20 ohm x cm instead of 1500 ohm x cm). To achieve the optimal p$^+$ layer distribution, the energy of the implanted ions was only 10 KeV, the concentration was 10^{14} - 10^{15} cm^{-2} and the implanted angle of incidence was increased with respect to the typical angle of 7°. to avoid channeling of implanted boron to great depth: in such a way, the peak of the ion distribution should be, theoretically, at only about 0.05 μm inside the CCD. It is important to note that in the last CCD, EEV has applied the "*anodic etching*" after laser annealing. At a scan time of 100 μs per pixel all the CCDs show a readout noise smaller than 10 electrons rms.

3. EXPERIMENTAL SYSTEM

To evaluate the CCDs performances in the 300 - 11000 Å region, two different test facilities have been used. One performs QE measurements in the 300 - 2500 Å range and is located in Padua (Department of Electronics and Informatics) (Bonanno et al. 1992, Naletto et al. 1992) and the other is located in Catania (in our Observatory) and can characterize cooled CCDs in the 1150-11000 Å range. In these laboratories measurements of the response uniformity are also possible (REFERENCES 2) [11,12]XXXX

The vacuum facility used for the EUV spectral range measurements is shown schematically in Fig. 2: it consists essentially in a normal incidence monochromator and a calibration chamber.

The monochromator has a Johnson-Onaka configuration with an f/10 aperture and a toroidal reflection grating (600 l/mm, Pt coated). The entrance and exit slit are fixed and are 100 μm wide and three mm long. It is pumped by a turbomolecular pump which allows it to

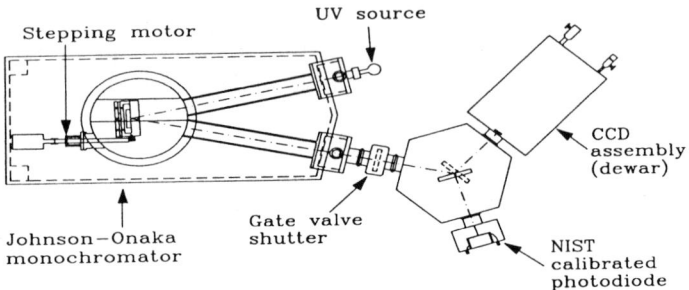

Fig. 2. Scheme of the calibration facility used for the QE evaluation in the EUV.

reach a pressure of about 2×10^{-6} mbar. As sources, both a 30 Watt Hamamatsu Deuterium lamp sealed with a MgF2 window emitting in the 1150 - 2500 Å spectral range and an open hollow cathode lamp to cover the range down to 300 Å have been used. The lines used are 304 Å (He II), 461 Å (Ne II), 584 Å (He I), 735 Å (Ne I) and 920 Å (Ar I). The calibration chamber accommodates a rotating Pt coated toroidal mirror mounted at its center which can focus the monochromatized beam on two symmetric positions, where both the CCD and the calibrated photodiode are mounted. To evaluate the QE, a pair of NIST calibrated photodiodes have been used: one sealed for the 1164-2537 Å range and the other windowless working in the 50-1216 Å region.

The system used for the 1150 - 11000 Å spectral range measurements is shown schematically in Fig. 3: it is based essentially in a precision 0.39-meter focal length vacuum monochromator. This last, thanks to a triple grating turret that can accommodate gratings blazed at appropriate wavelengths, allows good and automated measurements in the above mentioned operating range. Some off-axis paraboloidal mirrors are used to focus the beam on the monochromator. As sources, we use a Tungsten lamp to cover the 4000 - 11000 Å region and a 150 Watt Deuterium lamp sealed with a MgF2 window emitting in the 1150 - 3000 Å spectral range. To cover the 3000 - 4000 Å range, we actually use some spectral lamps, but in the future a Xenon lamp will be used. A movable plane mirror selects the source. To evaluate the QE of the CCDs in the spectral range of interest, three NIST calibrated photodiodes have been used: one sealed with a MgF2 window for the 1164 - 2537 Å range and the others, a quartz window working in the 2000 - 11000 Å region.

Actually, we use the system assembled on an optical bench without the vacuum chamber reducing the operating wavelength range to 2000 - 11000 Å. The system configuration allows simultaneous measurements to be made on both detector and calibrated photodiode, reducing the errors due to the source variability during the measurements.

The CCD controller used for these measurements has been made in our laboratory (Bonanno and Di Benedetto 1990) and in the future we will use the one designed for the Italian National Telescope "Galileo" (see Bonnano (1995) for the essential characteristics). The acquired images are analyzed using two packages, IDL and IRAF running in Unix-machine workstations.

NEW DEVELOPMENTS IN CCD TECHNOLOGY FOR THE UV-EUV

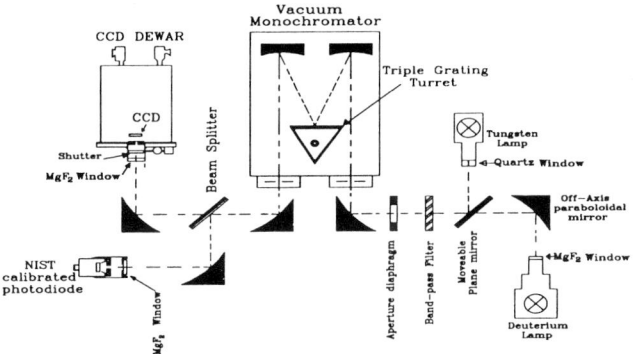

Fig. 3. Scheme of the calibration facility used for the QE evaluation in the 1150 - 11000 Å spectral range. The system is mounted on an optical bench and is presently used for the 2000 -11000 Å spectral range because the optical components are not housed in a vacuum chamber.

4. QUANTUM EFFICIENCY MEASUREMENTS

The QE measurement is conceptually easy: it is necessary to compare the CCD signal (electrons) with the signal measured by a calibrated photodiode at each wavelength. But the real measurement is not so simple, particularly at wavelengths shorter than 2000 Å. Since the sources and their characteristics are different for the regions above and below 1200 Å, different methods of evaluating the QE are adopted for the various cases in the two laboratories (Bonnano et al. 1992, Naletto et al. 1992). Particular care has been taken in the evaluation of the CCD conversion factor F, expressed in e$^-$/DN, because the accuracy of this parameter influences all data. It must be pointed out that in the UV spectral region the QE of CCDs is defined as the product of the effective CCD QE; that is, the fraction of incident photons producing one or more e$^-$h pairs that are collected in the CCD pixel wells, multiplied by the so called electron yield; that is, the number of e$^-$h pairs produced per interacting photon (usually it is assumed that the e$^-$h pairs are generated by an energy of 3.65 eV).

4.1 300 - 2500 Å Spectral Range

The results of the QE measurements in the 300 - 2500 Å region are shown in Fig. 4 and refer to data acquired about one hour after the cooling of the CCDs.

The measured QEs of both front-illuminated and back-illuminated coated CCDs, have a similar behavior. They show values greater than 10% at longer wavelengths and a minimum of the order of 1 - 1.5% in the 700 - 1200 Å range, and both show an increase at the shortest wavelength. A different behavior can be noted on the two ion-implanted devices: first of all, they show a QE greater than 10% in a larger spectral range (900 - 2500 Å), at wavelengths shorter than 700 Å the QE increases quickly and in particular, the "etched" one has a QE greater than 60% at 304 Å. A difference of 25% can be noted between the two ion-implanted CCDs in the 400 - 2000 Å that drops to 10% at longer wavelengths. Moreover, it is very interesting to observe that the two QE curves converge in the UV region: this fact agrees fairly

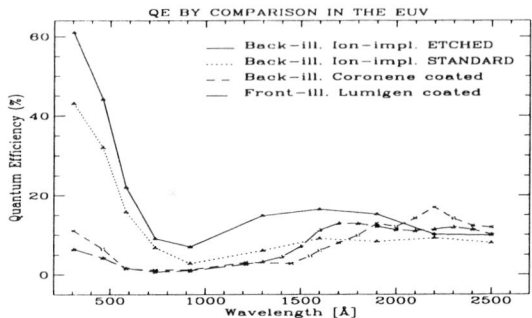

Fig. 4. Comparison between the quantum efficiency data for the various CCDs in the 300 - 2500 Å spectral range.

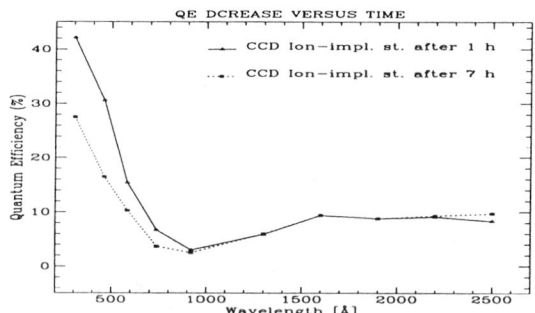

Fig. 5. Quantum efficiency decrease vs time for the thinned back-illuminated ion-implanted "standard" in the 300 - 2500 Å spectral range.

well with the QE measures obtained in the range 2500 - 11000 Å (see Fig. 6). The minimum values are showed in the range between 700 Å and 1200 Å and are 10% for the "etched" and 7% for the "standard".

During the QE measurements, a rather interesting phenomenon has been observed In practice, the QE decreases in time when the CCDs are cooled, to restart at high values after a cooling cycle. In Fig. 5 the QE data, taken at different times after the cooling of the CCD thinned back-illuminated ion-implanted "standard", is shown. After seven hours the QE at 584 Å from 15% drops to 10%.

Some tests were made to find whether this decrease was connected with the UV illumination or it if is due to other effects. There is only one obvious conclusion: the QE decrease in time is mainly related to the non perfect vacuum conditions. In fact the CCD, when cooled, acts as a cryo-pump and some contaminants present in the vacuum chamber are frozen over the CCD.

NEW DEVELOPMENTS IN CCD TECHNOLOGY FOR THE UV-EUV

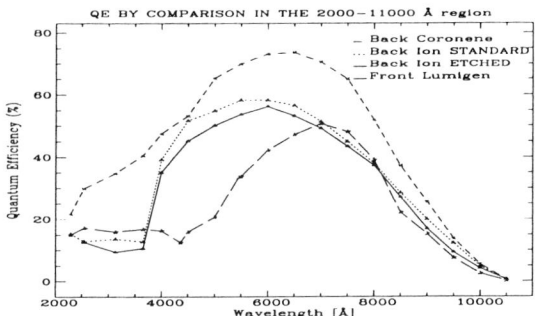

Fig. 6. Comparison between the quantum efficiency data for the various CCDs in the 2000 - 11000 Å spectral range.

4.2 2000-11000 Å Spectral Range

The results of the QE measurements in the 2000 - 11000 Å region are compared in Fig. 6. For the sake of clarity, we decided not to show the error bars, which are at most not greater than 10% in the UV region, and much lower at the other wavelengths.

In the 4000 - 10000 Å range all the CCDs, either front- or back-illuminated, show a "typical" QE. In contrast, in the 2000 - 4000 Å range we note a different behavior for the various CCDs. The QE of the *front lumigen* CCD is higher than 15% and is almost constant in that spectral range. The *back coronene coated* CCD shows a higher QE. But, I have to point out that, unfortunately, the QE decreases with the number of times the CCD is operated. The two curves of the *back ion* CCDs are slightly shifted within a 10% difference, and, being both within their error bars, we can conclude that they are comparable. Thus, from these results we can say that the *"anodic etching"* technique improves the QE mainly in the EUV spectral region, leaving it practically unchanged at longer wavelengths.

5. UNIFORMITY MEASUREMENTS

To study the uniformity of the CCDs response and its color dependence we used a system constituted essentially by an integrating sphere that is illuminated through interferential filters by different sources in the 2000 - 11000 Å range. Instead, in the 300 - 2500 Å range, a ground glass inclined at 45° to collect the diffuse radiation of the beam coming out from the monochromator is inserted in the calibration chamber. To be sure of the validity of this method, we have been checked that the fluorescent contribution to diffuse light due to the ground glass is negligible with respect to that of the UV.

Here we report the analysis at only two UV wavelengths: $\lambda = 1600$ Å and $\lambda = 2200$ Å for the front-illuminated CCD and the back-illuminated ion-implanted "standard" CCD. The pixel to pixel variations have been estimated by dividing the CCD areas into 10 x 10 pixel sub-areas (to minimize the large-scale variations due to the source), and evaluating the mean signal

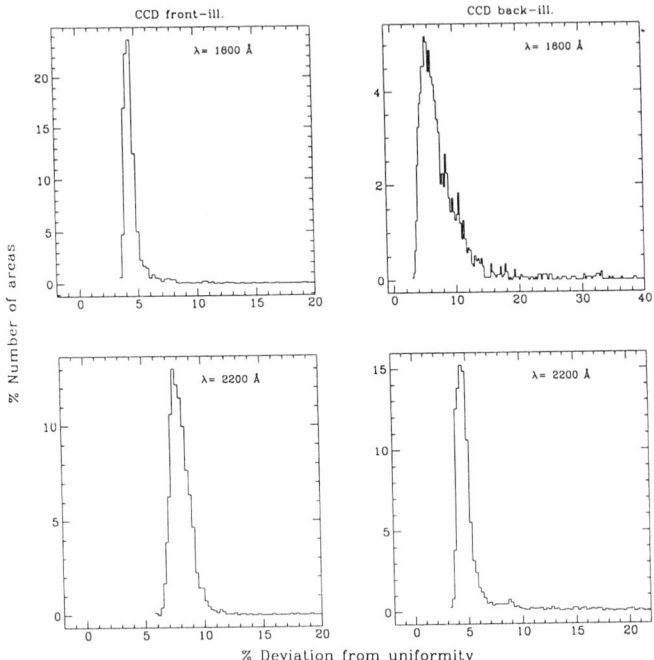

Fig. 7. Deviation of uniformity at λ = 1600 Å and λ = 2200 Å for the front-illuminated CCD and the back-illuminated ion-implanted "standard" CCD.

m and the standard deviation σ for each sub-area. The deviation from uniformity in any sub-area is given by the ratio $2\sigma/m$.

Fig. 7 shows the histograms relating the number of the sub-areas to their deviation from uniformity. As can be noted, the front-illuminated CCD shows an opposite behavior with respect to the back-illuminated ion-implanted "standard" CCD. While at λ = 1600 Å almost the totality of the considered subareas of the front-illuminated CCD have a narrow distribution around 5% of non-uniformity, at λ = 2200 Å the distribution is enlarged and around 8% of uniformity. Exactly the contrary is observed for the back-illuminated ion-implanted "standard" CCD. This is mainly due to the absorption depth of the materials at these wavelengths.

6. CONCLUSIONS

The QE measurements of the selected CCDs show that both the techniques of coating the CCD surface with some scintillators and of ion-implantation followed by laser annealing are effective in having a rather good QE in the UV spectral region.

The coated CCDs have shown a reasonable QE in the region in which the scintillator efficiently converts the UV radiation, while at shorter wavelengths they show a rather poor QE. Furthermore the QE depends on the number of times the CCD is operated.

The ion-implanted CCD has shown a very high QE, especially at the shortest wavelengths, and a uniformity of response which is good at longer wavelengths; this was an unexpected result, so other measurements have to be done to better understand it. The *etching* process is used to increase the EUV sensitivity of thinned back-illuminated and ion-implanted CCDs. In particular, since QE values as high as 70% have been measured at the shortest wavelengths, taking into account also the electron yield and the very low noise shown by currently available devices, it is possible to assert that in the spectral range below approximately 600 Å these detectors can work practically in a near photon-counting regime.

The problem of the QE decrease with time, due to residual contaminants inside the vacuum chamber, still remains, but it can be strongly reduced by improving the vacuum and by operating the devices at higher temperatures permitted by new devices that can operate in "inverted mode". In fact Multi Pinned Phase (MPP) devices do not need to be cooled to very low temperatures, thus avoiding the condensation of the contaminants over the sensitive surface. A more accurate evaluation of the QE of MPP CCDs in the EUV will be done in the future.

In conclusion, the results obtained show the real possibility of adopting in the near future these devices as optimal image sensors in the UV and EUV spectral range. In this respect, an effort is already in progress, both as regards the development of new CCD manufacturing technologies by EEV, and the improvement in the test facilities available in our laboratories.

REFERENCES

Bailey, P., Castelli, C., Cross, M., van Essen, P., Holland, A., Jansen, F., de Korte, P., Lumb, D., Pool. P. and Verhoeve, P. 1990 Proc. SPIE 1344, 356

Bonanno, G., Bruno, P., Consentino, R., Bortoletto, F., D'Alessandro, M., Fantinel, D., Balestra, A. and Marucci, P. 1995 in IAU Symposium No. 167, New Developments in Array Technology and Applications, A. G. D. Philip, K. A. Janes and A. R. Upgren, eds. Kluwer Academic Pub., Dordrecht, p. 319

Bonanno, G. and Di Benedetto, R. 1990 PASP 102, 835

Bonanno, G., Naletto, G., and Tondello, G. 1992 ESA Special Publication, 356, 233

Brinkman, A. C., Aarts, H. J. M., Burt, D. and Pool, P. 1992 ESA Special Pub. 356, 75

Janesick, J. R., Campbell, D., Elliot, T. and Daud, T. 1987 Opt. Eng. 26 (9), 852

Bredthauer, R. A., Chandler, C. E., Janesick, J. R., McCurnin, T. W. and Sims, G. R. 1988 in Instrumentation for Ground-Based Optical Astronomy, Present and Future, L. Robinson ed., Springer-Verlag, p. 486

Janesick, J. R., Elliot, T., Daud, T., McCarthy, J. and Blouke, M. 1985 in Solid States Imaging Arrays, K. N. Prettyjohns, and E. L. Dereniak, E. L. eds., Proc. SPIE, 570, 46

Janesick, J. R., Elliot, T., Dingizian, A., Bredthauer, R. A., Chandler, C. E., Westphal, J. A. and Gunn, J. E. 1990 Proc. SPIE 1242, 223

Janesick, J., Elliot, T., Fraschetti, G., Collins, S., Blouke, M. and Corrie, B. 1989 Proc. SPIE 1071, 153

McCarthy, K. J. and Wells, A. A. 1992 Proc. SPIE 1743, 211

Naletto, G., Tondello, G., Villoresi, P., Bonanno, G., Cali, A., Di Benedetto, R. and Scuderi, S. 1992 Proc. SPIE, 1743, 199

Stern, R. A., Catura, R. C., Blouke, M. N. and Winzenread, M. 1986 Proc. SPIE 627, 583

Verhoeve, P. W. A. M., den Boggende, A. J. F., Jansen, F. A., de Korte, P. A. J., Burt, D., Castelli , C. and Holland, A. D. 1991 in Astronomical Application, Conf. on Photoelectronic Image Devices 1991 - Section 1, p. 25

Verhoeve, P. W. A. M., Jansen, F. A., de Korte, P. A. J., den Boggende, A. J. F., Bonanno, G., Cali, A., Di Benedetto, R. and Scuderi, S. 1992 Proc. SPIE 1743, 223

DISCUSSION

STOVER: At Lick we are investigating two additional techniques to address the problems of backside treatments. First, we are having wafers fabricated which have the highly doped boron layer built into the silicon. This makes thinning the CCD easier. In addition, it should provide backside accumulation properties similar to ion implanting with annealing. Second, we are planning to experiment with n-type CCDs. CCDs made with this type of material will not have the backside accumulation problems of standard CCDs.

BONANNO: Very well. I agree completely to experiment with n-type CCDs. I hope we will see the results of these processes in the future.

COHEN: Since no one from JPL appears to be here, I should like to point out that Paula Grunthaner of the Microdevices Lab is working on this problem and is using a different technique. This has some advantages and some disadvantages - the process requires very high vacuum and cleanliness and is inconsistent with most commercial production. But the QE waves are spectacular. Paula expects to distribute a few CCDs treated in this way for testing soon.

BONANNO: OK. I will read in the future the results of the QE measurements.

JORDEN: On the graph showing QE of the four CCDs over the 2000-10000 Å range - why does the coronene coated thin device show such a high peak (600 nm) response of 80% (relative to 60% on other thin ones)?

BONANNO: I think that the coating acts as an antireflection coating at these wavelengths.

CULLUM: I am not clear on your reference to ESO-type coronene coatings. ESO has not used coronene coatings, only laser dye coatings on an acrylic base. Could you clarify what these coatings consist of?

BONANNO: EEV said that the coating applied on the thinned back illuminated CCD is coronene and the method utilized is similar to the one used from ESO. This is an older technique that is very different from the phosphor sublimation under vacuum conditions.

CCD CONTROLLERS

Robert W. Leach

Dept. of Astronomy, San Diego State University

ABSTRACT: The requirements of current and next generation CCD controllers in the areas of CCD device and system architectures, readout noise, number and speed of readouts are reviewed together with such operational requirements as system flexibility, power consumption, cost and weight. The basic components of a CCD controller are described, including the timing sequencer, clock drivers, video processor and computer interface. The capabilities and implementation of the CCD controller developed at San Diego State are reviewed. An upgraded controller is described to overcome limitations in the area of readout speed and efficient support of multiple readout capability.

1. INTRODUCTION

Improvements in the quality of CCDs developed within the last several years have brought about a corresponding change in the requirements of the controllers used to operate them in ground based optical observatories. The proliferation of different CCD device geometries and readout requirements brought about by the manufacture of CCDs by silicon foundries has led to a much better match of the CCD to its associated optical instrumentation, and has inspired a corresponding improvement in the versatility of their controllers. A parallel decline in the cost of CCD devices has enabled the construction of instruments with more than one CCD, while the increased number of pixels per CCD and the implementation of multiple readout circuits on a single CCD have motivated the development of multiple readout controllers.

Several groups (Reiss 1995, Bonanno 1995, Glass et al. 1995, Leach 1994) have met this challenge by designing controllers containing micro-processors to control the CCD waveforms, clocking voltages and image data flow. This allows considerable operational flexibility since the microprocessor program can be easily changed from the host computer. Different programs can be stored in the host computer for operating different geometry CCDs, for operating the same CCD with different sets of voltages and timing waveforms or readout modes, or to reflect various stages in the developer's understanding of how to optimally operate a given sensor. These controllers are referred to as programmable controllers, as opposed to fixed format or static controllers.

Section 2 below discusses how programmable controllers are implemented, reviewing their general characteristics and architecture as currently implemented in ground based astronomical observatories. Section 3 presents some specific operational parameters of the current programmable controller developed by the author and his colleagues at San Diego State University, and discusses some issues that are motivating a planned upgrade whose design is underway.

2. CURRENT CONTROLLERS

Table 1 lists some of the characteristics of current programmable controllers, and reflects in a general way the four controllers discussed in references mentioned in the second paragraph. Programmability of CCD waveforms and voltages can lead to efficient and reliable system development since electrical components do not have to be removed from the system. Versatile readout reflects not only the diversity of CCDs to be operated but the optimal readout of a given geometry to reflect the astronomical requirements of a particular observation - binning or summing charge from adjacent pixels in either the horizontal, vertical (or both) directions before readout will degrade image resolution but improves the signal-to-noise ratio and can be useful in low signal observations. Reading out only the portion of the CCD that is actually usefully illuminated can reduce the readout time and the image data quantity, while similar gains can be achieved with region-of-interest readouts that select several non-contiguous regions within the CCD field for readout. Region-of-interest readout is particularly useful for observations of small numbers of stars, especially for time variability studies. Multiple readout of several CCDs allows a large increase in image sensor area to accommodate large focal planes, while multiple readouts on a single CCD is useful for reducing the total readout time for the array, an issue increasingly important as CCDs contain more and more pixels. An integrated host computer for loading controller programs, processing user commands, and for storing image can lead to efficient instrument operation and is useful for properly configuring instrument configurations.

TABLE 1

Current Programmable Controller Characteristics

In-situ programmability - CCD waveforms, voltages and readout modes. Versatile readout - binning, sub-image, region-of-interest, multiple readouts. Integrated host computer - for loading controller programs, processing commands, displaying images and image storage.	
Max. readout rate	10 μsec/pixel
Dynamic range	determined by A/D, 16-bits normal 20-bits with special effort
Readout noise	dominated by CCD
Weight	few Kg.
Cost	US $10 - 20k
Power dissipation	< 30 watts
Availability	Some commercially, some not

Quantitative characteristics are harder to discuss since they vary quite a bit amongst the programmable controllers. A maximum readout rate of ten microseconds per pixel to 16-bit accuracy reflects the availability of an inexpensive and compact monolithic CMOS A/D converter manufactured by Crystal Semiconductors, and is sufficient for slow scan readout of currently available CCDs for obtaining readout noise figures down to two electrons rms.

Readout noise is dominated by the CCD even for readout noise values around two electrons rms since they are achieved either with CCDs with high gain on-chip source followers or by long signal integration times. Sixteen bits dynamic range is fixed by the available A/D converters, not by the CCDs, since the dynamic range of some CCDs exceeds 10^5, especially if binning is implemented. Controller dynamic range can be extended to 20 bits by dynamically switching to a low gain analog processing stage ahead of the A/D converter for pixels with high signal levels where the photon shot noise exceeds the digitization error, and then multiplying its A/D counts by the ratio of the gains of the high and low gain stages (Reiss 1995).

Current controllers weigh a few kilograms, including power supply and enclosure, cost somewhere around US $10 - 20 k if a proper accounting of labor charges is made, and dissipate something less than 30 watts, sometimes even less than 10 watts. These properties are entirely suitable for ground-based systems, especially since the cost of the controllers is a small fraction of the cost of CCD sensors typically implemented. Some of these programmable controllers currently in use are available commercially, and some are not.

Fig. 1 shows a block diagram of a typical CCD controller configured for a single readout. A sequencer generates digital waveforms for controlling the clock driver, which translates them into low impedance analog voltages connected to the CCD. A video processor amplifies, filters and digitizes the output of the CCD, which then goes over the communications link to an optional intermediate image memory. This is required in some systems to ensure that images are not damaged by slow host computer response times in multitasking environments. A host computer interfaced to the image memory serves as the user interface, image display, storage medium, and, often, the detector development systems as well. A housekeeping function is included in many system to perform tasks not directly related to CCD control such as shutter timing and control, CCD temperature control, power control and overall system monitoring.

Multiple readout capability is illustrated in Fig. 2 wherein the analog functions of the clock driver and video processor are shown replicated four times for a four readout system. The video processor is replicated because the readout speed for slow scan, low noise astronomical systems is generally limited by the sampling time of the video signal, since the largest noise contribution is usually from white (Johnson) noise of the on-CCD source follower FET. Installing one video processor to each readout allows the video sampling for all readouts to occur simultaneously, so the array readout time is reduced by the number of readouts while still allowing enough time for adequate signal integration time to get low readout noise. The clock driver circuits need to be replicated only as many times as are required to drive the CCD clock driver and DC bias pins in the readout configuration desired. When driving a single CCD with multiple readouts some savings can be realized by driving the parallel and serial transfer registers for each quadrant in parallel, though the readout circuits associated with each on-chip amplifier should be driven separately to minimize crosstalk problems between the readouts. The timing, housekeeping and communication portions of the controller are generally not replicated for modest numbers of readouts since they are constructed mostly from digital circuit components whose bandwidths exceed the video processing bandwidth by enough factors that they can multiplex between the video processors and clock drivers. This can recuce the system cost though it also sets the upper limit to the number of readouts that can be incorporated in a system.

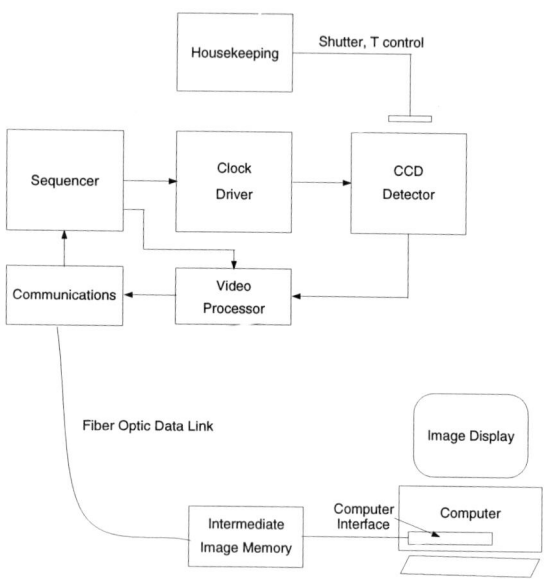

Fig. 1. Block Diagram of a Typical CCD Controller

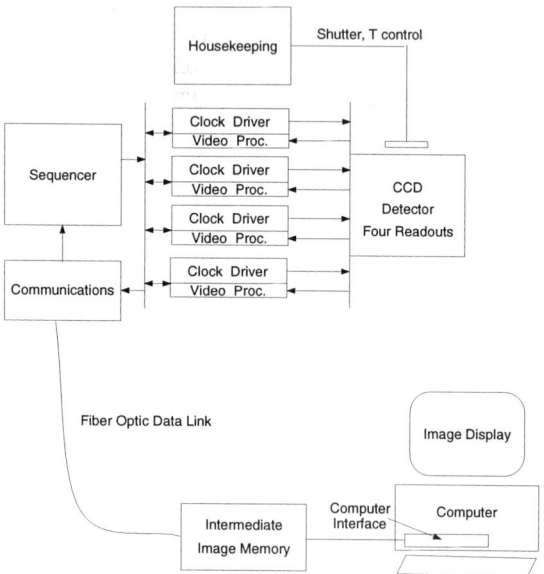

Fig. 2. Block Diagram of a Four-Readout CCD Controller

CCD CONTROLLERS 53

3. SAN DIEGO STATE CONTROLLER

In the interests of specificity, the controller design developed at San Diego State University is described. Table 2 shows a listing of some of its characteristics, listing both the current controller that has been in extensive use, and a planned upgrade.

The current controller has a typical readout time of 21 microseconds per pixel, which includes eight microseconds for sampling the baseline after reset, and eight microseconds for sampling the signal after dumping charge on the output node of the CCD. Both voltages are sampled with an analog integrator in the video processor, and their difference is taken also by the integrator by reversing the polarity of its input between the two integrations. An additional five microseconds is used for clocking the serial charge, resetting the output node and transmitting the 16-bit image data over the fiber optic link. The A/D conversion overlaps with the signal integration. An additional overhead of approximately two microseconds is imposed for each additional readout for updating each set of serial shift register clocks and for the fiber optic data transmission since these operations don't take place simultaneously in the current implementation. If the baseline and signal integration times are each reduced to four microseconds then a faster readout can be had at the expense of larger readout noise, and eight readouts can be accomplished in a time of 26 microseconds. The lowest readout noise obtained with these controllers is reported in this conference as 2.0 electrons rms with four + four microseconds of signal processing time. The system noise is estimated by measuring the rms noise with the input to the video processor grounded at 0.7 A/D units. Most of this is contributed by the A/D converter (0.55 ADU) with the rest evenly divided amongst the video processor stages. At typical system gains this figure corresponds to 1-3 electrons rms readout noise and is not a significant noise source.

The board size is 10 x 26 mm, which is the 3U VME width and a fairly long length. Power dissipation for two readouts is 23 watts, with much of it produced by the analog circuits. Three circuit boards are required for implementing one readout, and one board is required for each additional readout. The practical maximum number of readouts is about eight, with the limit placed by issues such as system size, power dissipation, readout speed inefficiencies and cost. More readouts can be implemented since the digital addressing and backplane transmission accommodates up to 32 readouts, but it becomes a relatively unsuitable system architecture with so many readouts.

The only computer interface currently supported is to VMEbus, although users have developed S-bus, Macintosh and NeXT interfaces as well. The A/D converter is a 16-bit, 10 microsecond conversion time monolithic CMOS device. There are two analog gains software selectable in the video processor. The timing sequencer is based on a Motorola DSP56001 monolithic digital signal processor operating at 100 nanoseconds per instruction, which is also the rate at which timing signals can be generated. A wait mode allows timing resolution down to 50 nanoseconds. There are twelve CCD clock drivers operating over a range of ± 10 volts in programmable steps of 0.1 volts.

A planned upgrade to the controllers focuses around improving the readout speed and the efficiency of supporting a large number of readouts. There are three motivations for improving the readout speed, as follows: (a) Current systems are often used at the telescope or in development labs to acquire images either at high illumination levels or in setup modes where

TABLE 2

San Diego State controller parameters for N readouts

	Current	Planned
Typical readout time	(21 + 2 x N) μsec/pixel	18 μsec/pixel
Fastest readout time for N = 8 readouts	26 μsec	3.2 μsec
Readout noise	0.7 ADU system noise 2.0 e⁻ with LL CCD (Luppino)	
Board size	10 x 26 mm	10 x 23 mm
Power dissipation for 2 readouts	23 watts	18 watts
Boards needed for four readouts	6	4
Practical maximum number of readouts N	N = 8	N = 16
Computer interfaces	VME	VME, S-bus
A/D converter	16-bits, 10 μsec/pixel	16-bits, 2μsec/pixel 19-bit w/ auto-scaling
Selectable gain	2 choices	4 choices
Timing sequencer	DSP56001, 100 nsec/instr.	56005, 40 nsec/instr.
CCD clock drivers	12, in 0.1 volt steps	14, in 0.01 V steps

the readout time for large arrays is burdensome. Such operations as focussing, flat field acquisition, lamp calibrations, field identification, and laboratory optimization and characterization often must be done either quickly or many times, where a reduction in readout time by a factor of five to ten with a corresponding increase in readout noise by a factor of two to three is desirable. Current systems cannot accommodate this. (b) Luppino's (1995) result of obtaining two electrons rms readout noise at eight microseconds total signal processing time suggests that future CCDs can have it both ways - very low readout noise at moderately fast readout speeds. Future controllers will need to read from multiple readouts in times shorter than ten microseconds per pixel to exploit this capability. (c) Infrared arrays, although not specifically targeted by current CCD controllers, have been operated successfully with them by several groups. With infrared arrays becoming larger, incorporating multiple readouts to keep their readout time manageable, and having similar readout requirements as the fast mode being discussed here it becomes feasible to have one controller designed for both optical CCDs and infrared arrays. This can reduce the design and maintenance costs at observatories that operate both types of arrays. Of course, it is a challenge to design controllers that don't compromise too much performance in achieving such versatility.

Features of the upgraded design to accomplish faster and more efficient multiple readout operation are shown in Table 2 in the planned column with the bold items. Faster DSP and A/D converters will be used in an architecture that does not impose a readout time penalty for adding readouts. This will be done by implementing the newer Motorola DSP56005 at a clock speed of 50 MHz and an instruction time of 40 nanoseconds together with hybrid A/D converters with two microsecond conversion times. The clock driver circuit will be changed to the more conventional switched DC arrangement to allow simultaneously updating the clocks going to different CCDs. Surface mount circuit board manufacturing will increase the practical maximum number of readouts to 16 or so by incorporating more than one readout per circuit board.

Other upgrades will be implemented in this process. The clock voltages and DC bias supplies will be specifiable to 0.01 volts rather than 0.1 volts to improve the dispersion from board to board. A 19-bit auto-scaling mode will be implemented to handle large dynamic range readouts as was discussed above. More choices of video processor gain will be provided. A mode to DC couple the video processor to the array readout will be provided for operating infrared arrays. An S-bus computer interface will be implemented to allow direct connection to compact Sun workstations. Expanded memory will be provided by the internal program memory space of the DSP56005 (4608 words) and by external X: and Y: data memory (32k words total). An in-circuit emulator feature of the DSP56005 will be accessible by connection to Motorola supported emulator hardware and software to help users diagnose their DSP code. More attention will be paid to power dissipation issues to offset the increased power consumption of the faster A/D converters.

REFERENCES

Bonanno, G., Bruno, P., Consentino, R., Bortoletto, F., D'Alesandro, M. and Fantinel, D. 1995 in IAU Symposium No. 167, New Developments in Array Technology and Applications, A. G. D. Philip, K. A. Janes and A. R. Upgren, eds, Kluwer Acad. Pub., Dordrecht, p. 319

Glass, I. S., Carter, D. B., Woodhouse, G. F. W., Waltham, N. A. and Newton, G. M. 1995 in IAU Symposium No. 167, New Developments in Array Technology and Applications, A. G. D. Philip, K. A. Janes and A. R. Upgren, eds, Kluwer Acad. Pub., Dordrecht, p. 3221

Leach, R. W. 1994 Optical Astronomy from the Earth and Moon, D. M. Pyper and R. J. Angione, eds., ASP Conf. Series 55, 113

Luppino, G. A. 1995 in IAU Symposium No. 167, New Developments in Array Technology and Applications, A. G. D. Philip, K. A. Janes and A. R. Upgren, eds, Kluwer Acad. Pub., Dordrecht, p. 297

Reiss, R. 1995 in IAU Symposium No. 167, New Developments in Array Technology and Applications, A. G. D. Philip, K. A. Janes and A. R. Upgren, eds, Kluwer Acad. Pub., Dordrecht, p. 317

DISCUSSION

GLASS: Why do you have the image stored between the controller and the computer?

LEACH: To insure that no image data are lost when the computer is a UNIX machine, which has a poor real-time response.

WEAVER: Do you have any problems keeping the DSP clocking noise out of the signal processor chain signal?

LEACH: No. Synchronizing all readout processes is needed.

STOVER: Will your new system have a DSP which is highly synchronized to the readout activity?

LEACH: Yes

STOVER: Will the new generation analog board provide a unique board ID which can be ready by the DSP?

LEACH: It will not be needed because the new analog boards will not need a voltage calibration.

FLORENTIN-NIELSON: Do you find that when performing partial image readout, that the bias will be different from that of a full frame readout?

LEACH: Yes, because the CCD is AC coupled to the video processor.

CCDs IN ACTIVE ACQUISITION SYSTEMS

A. Blecha

Observatoire de Genève

ABSTRACT: The use of CCD detectors as elements of an active acquisition system is reviewed. In such systems, the CCD image acquisition, data-analysis and the instrument and telescope controls are no longer separated elements whose actions are coordinated by an astronomer and/or operator, but are parts of a global system. The interaction between incoming data (nature of the object, registered flux, current PSF and atmospheric transmission), observer's requirement (S/N, spatial and temporal resolution) and forthcoming CCD exposure characteristics (CCD preparation, exposure time, read-out parameters) is examined. The requirements for the CCD electronics, data acquisition system are evaluated and examples of recent application in imaging, spectroscopy, photometry and auxiliary equipment are given. An attempt is made to analyze future technological trends and possible bottlenecks in such systems and to propose simple rules to adopt when designing CCD hardware and software.

1. INTRODUCTION

The CCDs, and electronic imaging devices in general, represent the modern version of photography and as such inherited the overwhelming majority of roles previously assumed by it. The need for a *linear and efficient solid-state integrating detector* was so obvious, that since the very first CCD system, and despite some drawbacks compared to photography, the success was rapid and irreversible. While the immediate availability of data was soon used for rapid visual inspection (and this despite the necessity to resort to very expensive special hardware equipment) it took some time before a more elaborate interaction between recorded data and instrumentation itself was introduced. Several technical, scientific and even psychological factors contributed to slow down the evolution towards an acquisition system with a *direct feed-back* in the acquisition loop. The principal sources of limitations were the hardwired un-programmable electronics, the insufficient and expensive computing resources, the lack of standard basic data-acquisition SW layer and immature data-processing SW.

Each of the limiting factors is subject to evolution, but not all of them are evolving on the same time-scale. While computing resources are growing by a factor of approximately two per year, the requirements driven by the size of CCD chips follow a much slower trend. It took approximately 20 years to go from 400^2 chips to 2048^2 commercially available chips giving a yearly growth of 1.17 and even if we consider a size as large as 4000^2, as available today (mosaic of four two-edge buttable 2000^2 chips), the growth factor would be only 1.25. The actual trend in electronics is towards fully programmable devices, removing another limitation from the above list. As a consequence of this evolution, the technical limitations hindering the use of the CCD as an active element of an acquisition system tend to disappear. The purpose of my paper is to review some existing active applications and evaluate how far we have gone in this direction and list requirements for the design of the control system (CCD controllers and software).

2. ACTIVE ACQUISITION SYSTEMS

2.1 What is an Active Acquisition System?

From the viewpoint of the interaction between data and the acquisition system we could distinguish two types of acquisition:

- The *passive* system merely records the data. No feedback mechanism between the recorded data and observing parameters is built into such a system. The acquisition of a CCD image and data-storing, even with sophisticated pipe-line processing, remains *passive* as long as there is no automatic interaction between what the pipe-line may see and the observing process. Such a system may extract automatically something which is useful for observation, but does not interact directly with the observing process.

- As soon as the result of data-processing directly influences (with or without observers acknowledgment) the observing process (what is done next), the system is *close-looped* or *active*.

A very simple example is the control of CCD saturation. In a *good passive* system, some automatic warning is generated if saturation is reached. Such warning may be provided through image display (saturated pixels displayed in a special color), by a numerical qualifier, special sound or any mixture of these. The warning is the outcome of the operation done on already acquired data and if these are unacceptable, the observation must be repeated. No corrective action is offered. In an *active system*, if saturation control is turned on, the system will take an extra preliminary snapshot of the field, and check for the saturation conditions of the exposure to be made. If saturation is predicted, the system proposes a corrected set of exposure parameters (shorter exposure time, lower gain, ...). The specifications of saturation conditions may be very general (% of the saturated area, with or without removal of cosmic-rays, only central area scanned, ...).

2.2 Existing Realizations

The majority of existing active systems are related to *robotic telescopes*. Automatic Photometric Telescopes (APTs) have been in routine operation for several years (Genet et al. 1987, Bruton et al. 1989, Nelson and Zeilik 1990, Rodono and Cutispoto 1992) but have been built and operated with photoelectric photometers and not with CCDs.

The CCD based APTs applications are still rare. Papers presented during the meeting of the A.S.P. (Filippenko 1991) in a chapter devoted to CCD imaging, describe several interesting applications of active systems. Some systems are devoted to a specific scientific program. The Berkeley Automated SN Search (Perlmutter et al. 1992) uses the result of image processing of the current exposure to decide whether the follow up exposure has to be taken. The Explosive Transient Camera (Vanderspek et al. 1991) detects the new object in the monitored portion of the sky within a time-scale of a few seconds and sends the information to a fast moving photometric telescope to proceed to the measurement of the event. Other systems such as the Australian Automated Patrol Telescope described by Brooks (Brooks 1990) or The Berkeley Automatic Imaging Telescope (BAIT) presented by Treffers (Richmond et al. 1993) are more versatile systems designed more or less to be general-purpose imaging instruments.

Though first to use the active approach, the *robotic telescopes* are not, paradoxically, best examples of active systems. Constraints of unattended operation makes them suitable for routine work, but the requirement that all decisions should be taken automatically (absence of the observer) leads to a very limited number of rather *static* situations where the active system is used. The best laboratory for the active system is a *normally* operated telescope, which, under some conditions, could be run as a *robotic telescope* but remains a general purpose system with active system features. The experience from the operation of the CCD photometric Swiss facility at La Silla is an example of such an approach.

3. SWISS PHOTOMETRIC TELESCOPE AT LA SILLA

3.1 Description of the system

For the last 16 years the Geneva Observatory has been operating at La Silla Observatory a 70-cm Cassegrain telescope equipped with a photoelectric photometer (Burnet 1976) fitted with seven-color Geneva filters (Rufener and Nicolet 1988). In 1991 a CCD photometric camera was permanently attached to the Nasmyth focus to extend the Geneva system to a CCD detector and to serve as the test bench for the project of a new active-system 1.2-m telescope. The CCD camera (Blecha et al. 1991), built around the Astromed 2200 detector fitted with an EEV-GEC p8603 chip is equipped with filters matched to Geneva photometric passbands augmented with R and I filters (Blecha et al. 1994). The CCD camera has been integrated in the general acquisition environment which controls the telescope and instrument functions. Furthermore, the acquisition system has direct access to image and data-processing and display, Guide Star Catalogue (GSC) database (Lasker et al. 1990) and display, the commercial database (/rdb) and the Unix shell.

The acquisition system has been designed to enable easy integration of heterogeneous elements. Existing facilities (telescope) and new commercial hardware (CCD detector) have to be interfaced with a general computer environment and various software services in a way to make the developed system largely independent of the hardware and third party software. The central piece of the control system is the *shell-like* interpreter/pseudo-compiler (Weber 1992, 1993, Blecha and Weber 1992) that ensures coordination between all elements involved. All elements are controlled through server-client architecture and the programming is done through several layers of interpreted language. The user is interacting with the upper layer either via permissive (shortcuts, line completion, parameter defaults ...) line commands or through menus. Such a system allows almost unlimited active functions with an extremely efficient debugging and testing cycle. All active-system features have been implemented during normal observing runs. Since all *logic* of the system is coded in high-level language (plain ASCII files), it is possible to switch instantly from the development environment to the operational one. The software design is based on the following principles:

- All services are implemented as genuine servers. Each server understands and executes a set of commands and returns results to the client but it is a completely independent process(es).

- All hardware and software functions are (also) directly callable from the high-level interpreted language.

- The execution of the shell scripts is preceded by a pseudo compilation during which formal errors are detected. This increases the speed of development and implementation of new features.

- The control of errors and interruptions is supported and handled by the shell script.

The same software system is used for the full on-line reduction of the data from ELODIE, the new compact fiber-fed echelle spectrograph at St. Michel Observatory (Queloz 1995).

At La Silla, several active-system functions are implemented. Some examples are given below.

3.2 Flatfield Acquisition

This is the simplest (though not the less valuable) implementation of active use of the CCD. The everyday regular flatfielding on the sky in several filters is something that requires rather expert manipulation (see for example Tyson and Gal 1993) and in our case (nine filters) it is almost impossible to carry out this task properly without some automatic procedure.

To obtain high quality flatfields the exposure time should be at least ten sec (to avoid shutter modulation), the exposure should be made close to a given position in the local sky (elevation and azimuth) and in a selected area (no stars brighter that m = 15) and the flux per pixel in all filters should be as close as possible to a specified value. To satisfy the above requirements, the order of filters, the exposure scheduling and duration may be adjusted.

Fig. 1 shows the flow chart of the acquisition. After the initial adjustments the system loops in the *flux-meter mode* taking short exposures through reference filter in the reference area (50^2 pixel in central area of the CCD) and estimating the required exposure time until the optimal conditions are reached. Then the FF exposures are made, each time with the net result being a very regular and unattended exposure in all filters. The level of exposure from night to night and filter to filter does not vary more reprogrammed many times, the only genuine *dynamic* parameter is the exposure time.

3.3 Automatic Measurement of a Standard Star

A good distribution of standard stars within the night and within the airmass range is crucial for the quality of the photometry. Usually only one photometric standard star is available within the field and it is neither efficient nor necessary to read the whole CCD. The observing sequence implements the search for the standard star in the database, the telescope pointing, the display of the GSC sky-map, the automatic centering (if required) and the programming of the CCD windowed read-out (window position, and size). The current PSF is used for the determination of the window size. The photometry is done for each filter and

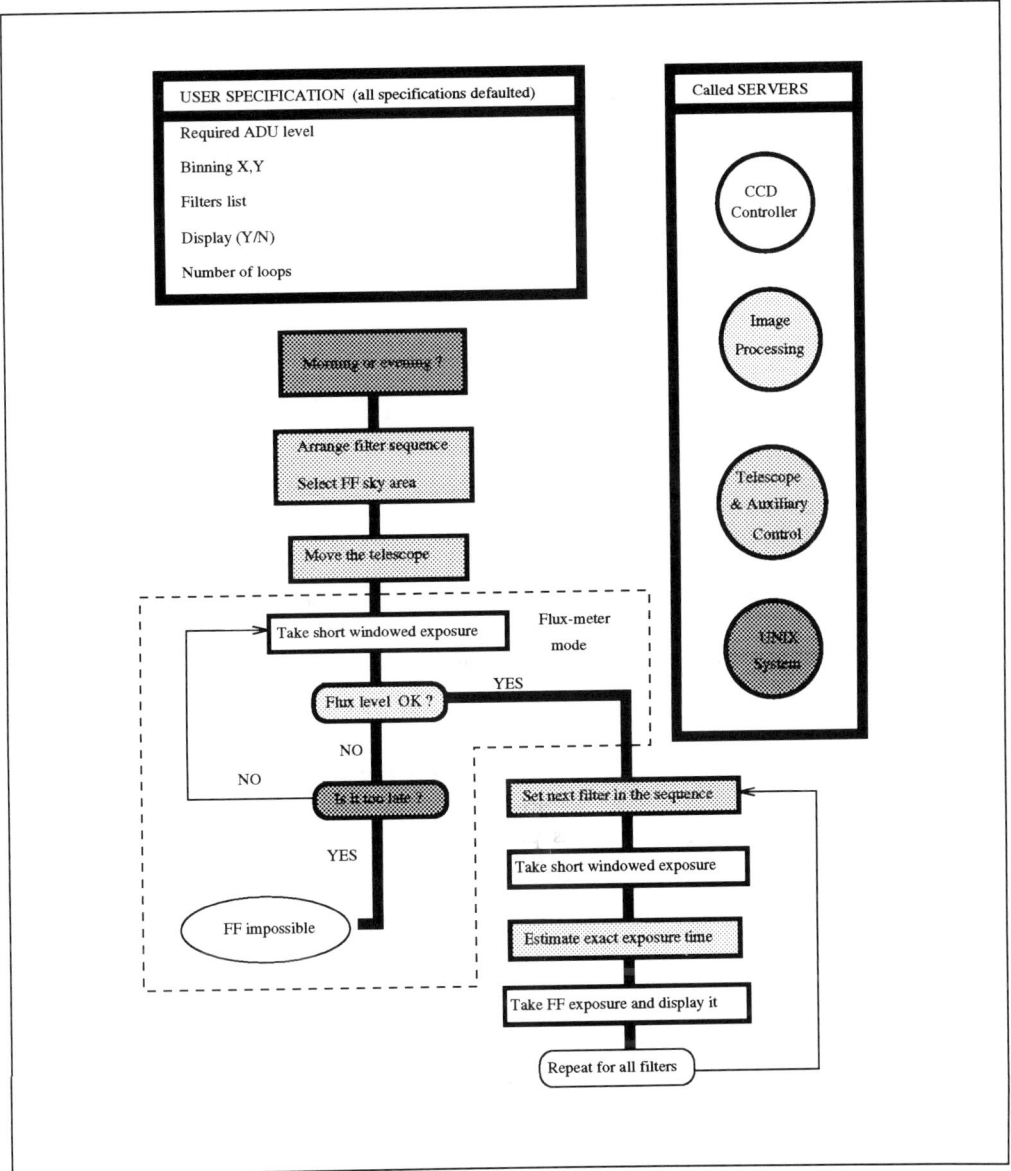

Fig. 1. Automatic flatfielding

repeated if more than one sample is required. Data are stored in a single image-like patched structure similar to that described by Wei (Wei et al. 1990) and during the current exposure, the previous exposure is analyzed. For each exposure the display is updated (patched with the current image), the airmass corrected flux is computed and several checks are made (saturation, position). There is not only considerable interaction between *hardware* elements (results of image processing of CCD data are used to move the telescope, current PSF is used to constrain the exposure time) but also between external SW elements (database, GSC).

The correspondence between the GSC and the sky is established permanently and therefore it is possible to move the telescope by designing the center of the field on the GSC display or inversely to see the GSC sky-map in the currently pointed area.

Several services provided in a classical system by separate elements are assumed by the same CCD camera. The most successful auxiliary role is that of the field camera. When the telescope is pointed, the field is usually shown rapidly (exposure of three seconds without filter, three x three binning and fastest readout) and the telescope position adjustment is made in a most simple way either by designing the object to center or by *dragging* the selected object to a given position in a way similar to CAD software. Each time such centering is done, the database used to establish the pointing model is updated. Several maintenance, auxiliary, and watchdog functions are implemented through the active system. The example of one interesting feature is the *watchdog* comparing the ESO seeing monitor value to a current PSF that issues a warning if the discrepancy between the two is too large (bad focussing).

4. REQUIRED CONDITIONS TO BUILD AN ACTIVE SYSTEM THAT USES A CCD

The CCD hardware and firmware (or system software that cannot be easily modified) should be very flexible. All recently advertised controllers (see session on controllers in these proceedings) are fully programmable and support all standard features that are required (flexible binning, multiples windowed readout, variable gain and speed). What makes the difference is *how many readout parameters may vary* readout parameters may vary and how easily they may be controlled and modified.

4.1 Read-Out Speed

One of the most critical features is the speed. The capability of the system to switch rapidly from one readout mode to another could be characterized by the minimum time required to carry out the programming of the CCD controller for a given mode *including the down-loading of the controller program* if necessary, the readout of the CCD and the transfer of the data into the memory where the control system may use them and finally the data conversion including the formatting and un-scrambling in multi-port or multi-window mode. The time to accomplish this cycle defines a shortest time-constant of any close-loop activity. Several such minimum turnaround times should be examined depending on the application. Typically, we will consider the fastest readout time of the full frame with degraded spatial (binning) and intensity (bits/sample) resolution, the readout time of a small window with the lowest readout noise (this time should be independent of the position of the window) and the time necessary to *flush* the CCD after the exposure of a bright object including the eventual setting time.

It is a widespread belief that the slow CCD read-out is mainly due to the requirement to keep the read-out noise very low. This is true only in a very limited number of applications working with a low S/N image (high resolution spectroscopy of faint objects for example). As soon as the S/N is more than ten, the read-out noise of a modern CCD system becomes negligible compared to other sources of noise (shot-noise, background, ...) and may be increased to some extent if it brings some advantage. **And it does.** The read-out noise obeys approximately the logarithmic law (for ex. Janesick et al. 1988) and the read-out noise (RMS) obtained at a given integration time t_{rd} could be written as RMS $\approx 1/t_{rd}^{0.4}$. Increasing the speed by a factor of six only doubles the read-out noise while a factor of 15 in speed leads to a tripled read-out noise. The above relation is a good approximation in the interval 20 - 0.1 μsec.

It is possible to read some chips (Lincoln Lab.) in correlated-double-sampling mode with the speed of five Mega-samples/sec and RMS of read-out noise 10e. For more classical chips the read-out noise figure has to be multiplied by a factor of two to three.

None of the available controllers is fast enough to handle several-megapixel/sec at 16-bit resolution. The most critical element is the video chain, but probably the whole controller will have to be redesigned.

4.2 Programming of the CCD

Though all recent controllers are programmable, several levels should be distinguished:

- Level 1 - Laboratory programming.

In principle the controller is programmable but this is done almost by *hand-coding*. In such a system almost one-to-one correspondence between the controller state and the instruction exists. It does not make a difference whether the binary or symbolic coding is used.

- Level 2 - High performance parameterization.

The controller understands a high-level control command that specifies all readout parameters. All usual parameters such as binning factor, time-constants, gains or windowing layout are adjustable. Un-scrambling of the data-flow in windowed or multi-port readout is done by the controller and the high-level user has only to define parameters. This is the genuine server. The system may be fast and easy to use. Since only parameters are changed the system is relatively safe, but has some limits. Not all readout modes that the hardware could possibly implement may be supported.

- Level 3 - Programming operational environment.

The controller is designed as an dynamical instrument and efficient development tools are provided. A genuine high-level programming language is provided that enables us to describe waveforms, data flow control and electronics setup into a source file which is compiled into a program executable by a controller. During the execution the Level 2 control may be used. This approach is the most versatile and provides both the laboratory and operational level of programming.

For an active system *level 1* programmability is insufficient. At least *level 2* should be available and for more ambitious projects *level 3* would be preferred.

Starting from 1990 an interesting development has been made by the CTIO CCD group (Smith 1994). A general *Waveform Definition Language* (WDL) has been adopted to program the CTIO Array CCD controller. The symbolic C-syntax language has been used to fully describe the 24-bit sequencer waveform. The source program(s) (ASCII-files) are processed using the standard *cpp* C-preprocessor then compiled. The result is still an ASCII file and is converted to the executable binary data only during the (down)load. Though the compiler produces the hardware specific output the WDL is powerful enough to generate the sequence for virtually any similar electronics. This system is indeed complemented by a *hardware-dependent* run-time system controlling the down-loading, chaining and execution of waveforms. The system does not implement genuine run-time parameters, but provides means to rapidly chain pre-compiled waveform segments.

4.3. Shell Issue and High-Level Programming

The discussion of the general software issue is beyond the scope of this contribution. Let us mention that almost all recent projects of large telescopes use or foresee to some extent an *active control system* with a sort of shell as a central piece of software. Several relevant papers were presented during the recent S.P.I.E. Symposium on Telescopes and Instrumentation for the 21st Century, 13 - 18 March 1994, Kona, Hawaii and many documents have been produced by ESO in the framework of the VLT project.

The specific discussion below is relevant to the UNIX implementation.

4.3.1 Shell Performance

The speed of the CCD controller would be useless if the data are not available rapidly to the processing. In principle any data-processing environment able to supply basic data processing functions may be adopted as a shell. Both, the public domain (various UNIX shells, IRAF, MIDAS) or commercial packages (IDL, Matlab) are possible candidates. Once again, the critical aspect of the choice lies rather in the performance than in the ability to carry out a given task. The most critical aspect is the data input. The system with shared-memory buffers will be the more efficient since the incoming CCD data may be written directly to the working buffer. The performance may nevertheless be poor if the CCD readout and data transfer between the controller and the data analysis system are not done simultaneously. The system with internal working buffers is better than one which relies only on files since often in time-critical operations images are not stored permanently. The data format (encoding) supported by the data-processing system may be relevant. Since *the native* format of the CCD data is usually a 16-bit integer in many operations it is not useful to convert it in the floating format. The *on-the-fly* conversion may be the other alternative.

Nevertheless the time-critical issue is relevant to some limited and foreseeable tasks. The most interesting feature of the active system, *the ability to combine and coordinate the high-level data reduction tasks with the control of data acquisition*, may accommodate with some overhead incurred due to the less efficient, but more *standard* or *easy-to-implement* data communication (pipes, files, message systems). Other aspects may then be more important (available applications, command language, error handling, ...). Since the system should remain open, the two-way (back and forth) communication with the external world should be possible with no loss of information (conservation of keywords, descriptors, ...!).

The shell *programs* should be easy to program, debug and test. All categories of users (astronomer, maintenance staff, developers) should be able to write and modify such programs. Many interesting applications concern technical development and maintenance.

5. CONCLUSIONS

I have described an *active acquisition system* using a CCD. Such a system provides the means for an easy combination of data acquisition and data processing functions controlled from a high-level shell language. The active acquisition approach is valid not only for robotic telescopes, but also for normal observing facilities. It has the virtue of isolating the high-level software from the hardware-dependent layer and offers an efficient development path for new applications. Service observing, remote control and automatic maintenance may be done better through an active system.

With the present technological trends (CCD chips, computer hardware and software) soon only the CCD controller hardware and software will limit the development of active systems.

The CCD hardware should satisfy stringent speed requirements. With improving CCD on-chip amplifier design, the main speed limitation comes from controllers and not from the chip (read-out noise) itself. **There is an urgent need for the megapixel per-second-per-channel class, high precision controllers in the near future.** These are important, not only for *active systems*, but also for all projects with large CCD cameras. Many controllers can handle a mosaic of 16 or even more chips, but cannot process the data at a speed which is already attainable with existing chips in many applications (high S/N observations). Data transmission has to be as direct as possible between the controller and data-processing computer. Intermediate storage has to be avoided or it should be possible to bypass such storage.

The controller has to be integrated in an efficient development and run-time software environment. A genuine high-level programming language should be provided that enables us to describe waveforms, data flow control and electronics setup and store them in a *source* file. Run-time environment should provide control for rapid down-loading and should support real-time parameters which fully define the specific set-up of a current exposure/read-out mode.

REFERENCES

Blecha, A., Weber, L., Simond, G. and Queille, D. 1990 in CCDs in Astronomy, G. H. Jacoby, ed., ASP conference series, vol. 8, p. 192
Blecha, A. and Weber., L 1992 Inter-technical description, Observatoire de Genève
Brooks, P. W. 1990 Proceedings Astronomical Society of Australia, 8, no. 4, p. 377
Bruton, J. R., Hall, D. S., Boyd, L. J., Genet, R. M. and Lines, R. D. 1989 Astrophysics and Space Science, 155, 27
Burnet, M. 1976 Thesis No. 235, Ecole Polytechnique de Lausanne (French)
Filippenko, A. V., ed. 1991 Robotic Telescopes in the 1990s; ASP Conf. Series, Vol 34
Genet, R. M., Boyd, L. J., Kissell, K. E., Crawford, D. L. and Hall, D. S. 1987 PASP 99, 660

Janesick, J., Bredhauter, R., Chandler, Ch. and Burke, B. 1988 in X-Ray Instrumentation in Astronomy. II., Proc. SPIE, 982, R. Golub, ed., p. 70
Lasker, B. M., Sturch, C. R., Mclean, B. J., Russell, J. L. and Jenkner, H. 1990 AJ 99, 2019 and 2173
Nelson, E. R. and Zeilik, M. 1990 ApJ 349, 163
Perlmutter, S., Muller, R. A., Newberg, H. J. M., Pennypacker, C. R., Sasseen, T. P. and Smith, C. K. 1991 Robotic Telescopes in the 1990's, A. V. Filppenko, ed., ASP Conf. Series, Vol 34, p. 67
Crawford, D. L. and Craine, E. R., eds. 1994 Proce. SPIE 2198, Astronomical Telescopes and Instrumentation for the 21st Century, p. 67
Queloz, D. 1995 In IAU Symposium No. 167, Advances in Array Technology and Applications, A. G. D. Philip, K. A. Janes and A. R. Upgren, eds., Kluwer Academic Press, p. 221
Rodono, M. and Cutispoto, G. 1992 A&AS 95, 55
Richmond, M. W., Treffers, R. R. and Filippenko. A. V. 1993 PASP 105, 1164
Rufener, F. and Nicolet, B. 1988 A&A 206, 357
Smith, R. M. 1994 Waveform Definition Language User Manual, beta release, CTIO internal document, available on request.
Tyson, N. D. and Gal, R. R. 1993 AJ 105, 1206
Vanderspek, R., Doty, J. P. and Ricker, G. R. 1991 Robotic telescopes in the 1990s, A. V. Filippenko, ed., ASP Conf. Series, Vol. 34, p. 123
Weber, L. 1993 Inter-Manuel de Reference, Observatoire de Genève (French)
Weber, L. 1992 Inter-Manuel de l'Utilisateur, Observatoire de Genève (French)
Wei, M., Chen, J. and Jiang, Z. 1990 PASP 102, 698

DISCUSSION

D'ODORICO: The approach you have presented is based on the experience you acquired with a dedicated telescope used with a CCD camera. Do you think the same type of active acquisition system could be used in the context of a large multipurpose telescope?

BLECHA: Yes indeed, the large telescope of the future will have to cope with situations where the active approach is the most appropriate. All recent experiences have shown that after the "first light" and "first astronomical result" the telescope system should continuously be upgraded and adjusted. This is exactly the field where the active approach is the most efficient one.

CCD DEVELOPMENT ACTIVITIES AT ESO

O. Iwert

European Southern Observatory

ABSTRACT: Advanced CCD detectors are the backbone of modern astronomical instrumentation working in the 300 - 1000 nm spectral range. The status and results of the different projects to design, manufacture and implement new CCD detectors for the ESO VLT instrumentation are presented. Emphasis is put into ESO's development activities together with industry's, and into first measurement results of thinned arrays in large formats obtained in 1994. In addition some recent cryostat developments are shown to illustrate the different requirements of VLT instruments in this area.

DISCUSSION

JORDEN: Could you discuss the merits of using an IR microscope for mosaic CCD alignment, compared with projecting images and reading out the devices?

IWERT: Projecting images and reading out CCDs: The complete chain of hardware and software (dewar, controller) is needed to read out the devices. It can only be done at room temperature as the devices cannot be easily manipulated when cold. The optical projection must be very accurate, so dedicated optics and mechanics is needed. It can only be tested with working CCDs, whereas it might be interesting to set the mosaic assembly mechanics cold with utilized mechanical samples. Using the infrared microscope: One can see right at the CCDs how the adjustment works and is not stacked by the above listed "obstacles".

FLORENTIN-NIELSON: The impressive QE figures you gave for the Lesser device (in UV) was that after UV flooding? Do you intend to reflood your CCD or keep it cold at all times?

IWERT: Yes, they were obtained after UV flooding. The long term stability of the QE after UV flooding has to be measured. In theory it should be sufficient to keep the CCD cold all the time after flooding.

BESSELL: What anti-reflection coatings have you discussed with Thomson?

IWERT: There is currently no definitive decision about the exact planned anti-reflection coating. However there will be definitely one deposited on the CCD.

INFRARED ARRAY DETECTORS: PERFORMANCE AND PROSPECTS

Ian S. McLean

Department of Physics and Astronomy, UCLA

ABSTRACT: Infrared array detectors, like silicon CCDs a decade before, have revolutionized infrared astronomy. The quality and performance of the current generation of devices has already allowed astronomers to obtain infrared images at wavelengths out to 2.2 microns which are as deep as the best CCD images. High resolution infrared spectroscopy is now a reality and ground-based imaging to 35 microns has been achieved. Several classes of low-noise infrared array detectors with formats of 256 x 256 pixels are now in routine use, and developments are under way which will produce detectors of 1024 x 1024 pixels (for the near IR) within the next year. This paper will briefly review the state-of-the-art and compare and contrast the properties of available arrays. Progress in the field is illustrated with recent near infrared photometry obtained with a new two-channel imaging system developed at UCLA.

1. INTRODUCTION

Infrared wavelengths provide a unique and powerful window on the Universe for the study of many things, from molecular clouds to protogalaxies. Relatively cool objects, such as dwarf stars or giant planets radiate strongly in the IR, and less extinction means that infrared wavelengths allow us to "see" inside dusty, optically-opaque star-forming regions.

Unfortunately, progress at ground-based observatories has been hampered by the nature of the technology available for the detection of IR photons. Until the mid-eighties, only simple "single-element" photovoltaic (PV) or photoconductor (PC) detectors were available. The only way to form an "image" of a scene was to "scan" the image across the detector by pointing the telescope systematically at numerous positions in a grid of coordinates on the sky. The technique was prone to errors and provided little opportunity to "integrate" signal at each position.

In contrast, optical astronomy has benefitted from the Charge-Coupled Device (CCD), which was introduced into astronomy in 1976 only six years after it was invented by Boyle and Smith at Bell Labs (the story of CCDs in astronomy is described in McLean 1989). Because of its semiconductor band-gap, the silicon CCD is not sensitive to wavelengths longer than 1.1 microns. Stimulated by military applications, however, many one- and two-dimensional infrared imaging detector technologies were developed and classified around this time, but none were really suitable for the much lower backgrounds and non-real-time or "staring" applications of astronomy. Fortunately, in the mid-eighties, infrared focal plane "array" detectors developed specifically for astronomy came into operation. The first arrays had formats of 64 x 64 pixels or less, but this was thousands more elements than before!

The impact of the introduction of infrared array detectors has been staggering. I reviewed the field in an article (McLean 1988) following the "turning-point" conference entitled "Infrared Astronomy with Arrays" held in Hilo, Hawaii in 1987 while I was still with the UK Infrared Telescope observatory on Mauna Kea. Only six years later, 300 astronomers and many representatives from the IR detector industry, gathered at my new home, the Departments of Astronomy and Physics at UCLA, from July 18 - 22, 1993 to attend a similar conference entitled "Infrared Astronomy with Arrays: The Next Generation" (McLean 1994). They saw spectacular results from numerous infrared cameras and spectrometers, most using the latest 256 x 256 (65,000) pixel arrays, and heard the plans for the development of 1024 x 1024 pixel arrays within two to three years. This summer, July 1994, Klauss Hodapp at the University of Hawaii, obtained the first images with a 1024 x 1024 HgCdTe array manufactured by Rockwell International. Given the difficulties of this technology, from one to one million pixels in such a short period of time is both amazing and exciting!

2. INFRARED ARRAY DETECTORS

2.1 Current Technology

Several companies have developed IR focal plane arrays which have been used successfully in astronomy applications, including Aerojet Electrosystems, Amber Engineering, Cincinnati Electronics Corp., Hughes-SBRC, Kodak, LETI-LIR, Mitsubishi and Rockwell International; most of these are in the USA.

Infrared-sensitive materials currently in use are summarized in Table 1. The principal materials are PtSi, HgCdTe and InSb for wavelengths less than five microns and doped silicon (e.g. Si:As, Si:Sb) for the 10 - 20 and 10 - 30 micron regions respectively. Work is in progress to develop doped germanium arrays and silicon bolometer arrays for even longer wavelengths. Other papers to be presented at this conference will provide more details on each of these detector options. In general, infrared detectors are formed either from reversed-biased pn junctions (InSb and HgCdTe photodiodes), Schottky-Barrier junctions (PtSi) or Impurity Band Conduction (also known by various trade names such as BIBs, BIBIBs) in doped silicon or germanium.

Currently, the best available arrays for infrared astronomy include the following:

a) 256 x 256 pixel HgCdTe array from Rockwell International Science Center. This device is also known as the "NICMOS 3" array because it was developed for the NASA/University of Arizona Near Infrared Camera and Multi-Object Spectrometer, a second-generation instrument for the Hubble Space Telescope (PI: Rodger Thompson). Dr. Kadri Vural at Rockwell has been a strong supporter of astronomy programs. These devices have quantum efficiencies from 40 - 60% from 1 - 2.5 microns, readout noise of 40 electrons rms, and dark currents of less than one e/s/pixel when operated at 77° K (liquid nitrogen). The pixel size is 40 microns and the charge storage capacity is about 200,000 electrons.

b) 256 x 256 pixel InSb array from the Hughes Santa Barbara Research Center (SBRC). Dr. Alan Hoffman at SBRC was one of the first people to recognize the potential of IR arrays for astronomy and he has pioneered the development of specialized astronomy arrays for many years. A "low background" version of this array is also being developed for the NASA Space

INFRARED ARRAY DETECTORS

TABLE 1

Current detector materials and array formats

Material	Symbol	Cut-off Wavelength	Type	Formats
mercury-cadmium-telluride	HgCdTe (55%Hg)	2.5 microns	pn	256 x 256 1024 x 1024*
indium antimonide	InSb	5.3 microns	pn	256 x 256 1024 x 1024*
platinum silicide	PtSi	~4 microns	Schottky	256 x 256 640 x 480 512 x 512
extrinsic silicon	Si:As	23 microns	PC,IBC	58 x 62, 96 x 96 128 x 128
	Si:Sb	29 microns	PC,IBC	256 x 256
extrinsic germanium	Ge:Ga	113 microns	PC,IBC	*

Notes: * under development

Infrared Telescope Facility (SIRTF), with Giovanni Fazio at the Harvard Smithsonian Center for Astrophysics and Judith Pipher and Bill Forrest of the University of Rochester playing major roles. These devices have quantum efficiencies from 70 - 80% from one to five microns, readout noise of 60 electrons rms, and dark currents of less than one e/s/pixel when operated at about 30 K. The pixel size is 30 microns and the charge storage capacity is about 200,000 electrons.

c) Silicon IBCs are available in up to 256 x 256 formats from Rockwell International (Anaheim, USA) and Hughes-SBRC (Carlsbad, USA), and in smaller formats from LETI-LRI (France). Different versions are possible depending on whether the application is ground-based or space-based, i.e. large well-depth or modest well-depth. To date, the largest format actually used at a telescope is 128 x 128 (see Herter 1994 for a review). These devices are always background limited for imaging.

d) Platinum silicide arrays from Hughes (256 x 256 pixels), Kodak (640 x 480 pixels) and Mitsubishi (512 x 512 pixels) have all been used in astronomy applications. These devices are attractive because they have good uniformity and low noise, but they have inherently very low quantum efficiencies (a few %) and hence their range of applications is limited to wide-field surveys of relatively bright sources. Recent work in Japan is aimed at increasing the QE by almost a factor of ten (see paper by Ueno 1995).

Infrared array detectors are NOT based on the charge-coupling principle of silicon CCDs

(McLean 1993). Instead, the role of detecting IR photons is separated from the role of multiplexing the resultant electronic signal generated in each pixel and relaying it to the outside world. To achieve this, each device is made in two parts, an upper slab and a lower slab. In the upper piece is the IR-sensitive layer which is effectively subdivided into a grid-like pattern of pixels by the construction of tiny junctions. The lower slab is made of silicon and contains a matching grid of "unit cells" in which the infrared detector is connected to the input gate of a FET source-follower and there is also a "reset" FET connected to the same node. Interconnecting the two slabs are tiny columns of indium, called "bumps". Either the upper slab is constructed on an IR-transparent substrate such as sapphire or it is physically thinned and backside illuminated. This construction is known as a "hybrid". The output from each unit cell source follower is connected to an output bus and a final output source follower by a system of shift registers which can address each pixel in turn.

The electronics required to operate IR arrays is very similar to that needed for CCDs, e.g. a small number of clocked lines, level shifters, a few dc bias lines, low noise preamplifiers and an A/D system of typically 14 - 16 bits. Infrared arrays usually have multiple outputs and, because of the much higher background levels, the maximum integration time becomes very short at longer wavelengths, i.e. the bandwidth of IR electronics systems tends to be higher than for CCDs.

2.2 Charge Storage and Collection

In an infrared array detector the absorption of a photon with a wavelength shorter than the cut-off wavelength generates an electron-hole pair. The electrons are collected at the pixel location by electric fields applied either internally or externally or both. For example, in the current generation of InSb and HgCdTe arrays, the field is produced by a reverse-biased pn junction and the depletion region acts like a capacitor which becomes progressively discharged as photogenerated charges accumulate. Since the storage capacitance is at least partially a function of the changing reverse bias, then a small non-linearity is introduced. The effect is always less than 10% worst case and it is smooth and easily corrected. A voltage change applied to the input of the silicon FET is "followed" at the output. The output voltage can be sampled without affecting the input signal, which means that these arrays offer opportunities for noise-improvement using multiple non-destructive sampling. In addition, when the detector is fully de-biased there is no charge "bleeding", but integration ceases and the pixel output is no longer linear with photon flux, i.e. the pixel is saturated. In general there is always a trade-off to be made in selecting the effective pixel capacitance; the larger the capacitance the greater the charge storage or integration capacity ($Q = CV$), but the greater the noise-equivalent charge.

2.3 Detector Properties

In general, there is a very strong, wavelength-dependent, "background" flux of photons comprising thermal emission from the telescope and atmosphere, and non-thermal emission from the night sky - mostly in the form of solar-induced photochemical line emission from the OH radical in the upper atmosphere. Each detector also exhibits an intrinsic "dark current" when not illuminated due simply to thermal excitation of electrons within the semiconductor. This effect can be reduced by many orders of magnitude (to about one electron/s/pixel) by cooling the detector to a very low temperature. For short wave HgCdTe arrays, liquid nitrogen is sufficient (77° K), but for InSb it is necessary to go down to about 30° K (usually with a dual liquid helium/liquid nitrogen system or, in more modern instruments, with a closed-cycle refrigerator). Even lower temperatures are needed for BIB arrays and again liquid helium (4° K) is used. Infrared array detectors also exhibit readout noise (R) like CCDs. A typical value for all the 256 x 256 devices mentioned in this paper is 50 electrons rms, but special multiple non-destructive sampling techniques have achieved noise values as low as 20 electrons.

So far, the InSb detectors have achieved the best quantum efficiencies, followed by the HgCdTe arrays and then the IBC devices. Currently available platinum silicide detectors have very low quantum efficiency (about 2%) at the wavelengths of interest, but they also have the lowest readout noise. Most of the recent devices achieve a good, average non-uniformity over the array of 10% (or better) of the mean. In general, the InSb devices are somewhat more uniform than the HgCdTe arrays and the PtSi arrays are best of all.

2.4 Prospects for Larger Formats

Formats of 256 x 256 pixels are now in routine use for precision astronomical cameras and spectrometers. Astronomy arrays with 1024 x 1024 pixels are currently under development in the USA using both HgCdTe (Rockwell) and InSb (SBRC). The Rockwell project (called HAWAII) is funded by the University of Hawaii and the US Airforce, whereas the SBRC project (called ALADDIN) is funded by the National Optical Astronomy Observatories (NOAO) and the US Naval Observatory (USNO). The current status of these programs is as follows.

The HAWAII 1024 x 1024 array has 18 micron pixels and four outputs configured as four 512 x 512 quadrants. One working device was delivered to Hawaii in time for the comet Shoemaker-Levy collision with Jupiter in July 1994. According to Klauss Hodapp, additional work will be required to establish a stable production line of these chips.

The ALADDIN InSb array has 27 micron pixels and 32 outputs, eight per 512 x 512 quadrant. Al Fowler and Ian Gatley of NOAO report that yields of the new multiplexers have been very good. Hybridization is in progress and results are expected within a few months.

Typical "science grade" devices are costly and range from $60 K to $100 K for current HgCdTe and InSb 256 x 256 devices. This is expensive compared to silicon CCD technology and platinum silicide IR technology. However, both of the 1024 x 1024 devices will be even more costly with estimates ranging from $150 K to $250 K at the present time. Clearly, each detector has advantages and disadvantages, and each must be carefully investigated for its anticipated application.

3. ASTRONOMICAL PERFORMANCE OF IR ARRAYS

Infrared cameras have improved considerably in the past six years. The original 62 x 58 InSb arrays, for instance, had a readout noise of over 400 electrons rms in correlated double sampling mode. Remarkably, this was still sufficiently low to permit background-limited imaging in all broad bands from one to five microns. Since the QE was very high and the actual background per pixel was much less than in old aperture photometers, the result was a huge increase in sensitivity.

Typically, the most recent cameras have a throughput (transmission x quantum efficiency) of 30% or more, readout noise of 40 electrons rms or better, many more pixels, and dark currents which are sufficiently low to allow efficient spectroscopy with resolving powers up to R = 20,000. Many more applications to spectroscopy will be coming in the future which, in itself, is a testament to this technology. Good flatfielding to levels comparable with silicon CCDs, and good stability has led to photometry with IR arrays to levels which are as deep as the best silicon CCDs.

To illustrate progress I will describe some results with a new, unique, twin-channel, cryogenic IR camera system developed at the UCLA Infrared Imaging Detector Lab. The UCLA camera contains a very efficient (96%) dichroic beam-splitter which allows two near IR bands to be observed simultaneously. Light collected by the telescope is first collimated, then separated into two beams by the dichroic and finally re-imaged onto a focal plane array. In the short wave band (1 - 2.5 microns) a 256 x 256 HgCdTe array from Rockwell International is used, while in the long wave band (2 - 5 microns) a 256 x 256 InSb array from Hughes-SBRC is employed. A choice of dichroics allows the so-called atmospheric "K" window, which is a transparent band from 2.0 - 2.4 microns, to be delivered to either channel for cross-calibrations. Each channel is a fully independent, general purpose camera in its own right, with a selection of broad and narrow band filters, grisms (R = 500) and polarizers. On the 3-m UCO telescope at Lick Observatory, the pixel size is 0".7 and the field of view is 180" x 180". Measured backgrounds at K' (1.95 - 2.30 microns) are about 13.1 magnitudes per square arcsecond and throughput is about 30% per channel. (For more details see McLean et al. 1993 and McLean et al. 1994.)

First Light was obtained on the UCO 3-m telescope in June 1993 (McLean 1994). Since the two channels are completely independent, J and H exposures can be "nested" within a longer K exposure. The choice of dichroics means that it is also possible to nest J, H and K exposures inside an L-band exposure. The three sigma, one minute limiting magnitudes for point sources under typical seeing conditions (synthetic aperture of 3.0" diameter) is K' = 18.0 and, simultaneously, J = 19.7. The camera is very powerful for survey work.

Fig. 1 shows an image of the central 7.6 x 7.6 arcminutes of the Local Group spiral galaxy M 33; the image is a composite of J, H and K exposures. Using DAOPHOT, McLean and Liu (1995) have obtained three-color photometry for over 1700 stars, and have separated the bulge and disk components. A (J-K) v K color-magnitude diagram is shown in Fig. 2, and for comparison, typical loci are shown for giant stars in Baade's window and in well-known globular clusters in our own galaxy. The arrays are good enough to allow absolute photometry to a few percent and relative photometry to a few tenths of a percent on sufficiently bright sources. More details are given in McLean and Liu (1994).

Photometry of much fainter objects is also now feasible. Teplitz and McLean have obtained deep JHK photometry of over 150 galaxies in two dense clusters at redshifts of 0.4, Abell 370 and Cl0024+16, both of which exhibit gravitational arcs and arclets. Fig. 3 shows an image of Abell 370 and Fig. 4 gives the derived photometry of the galaxies. More details are given in the poster by McLean, Liu and Teplitz (1995) at this symposium. Only five years ago, good three-color photometry to this depth on so many faint objects would have been a real chore.

As the 1024 x 1024 arrays become available, we will see more and more emphasis placed

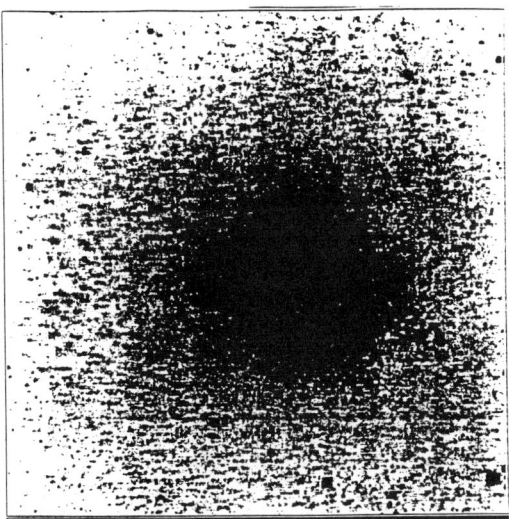

Fig. 1. A JHK composite image of the center of M 33. The field is about 7.6 x 7.6 arcminutes and was obtained by the UCLA twin-channel camera on the 3-m Shane telescope on Mt. Hamilton.

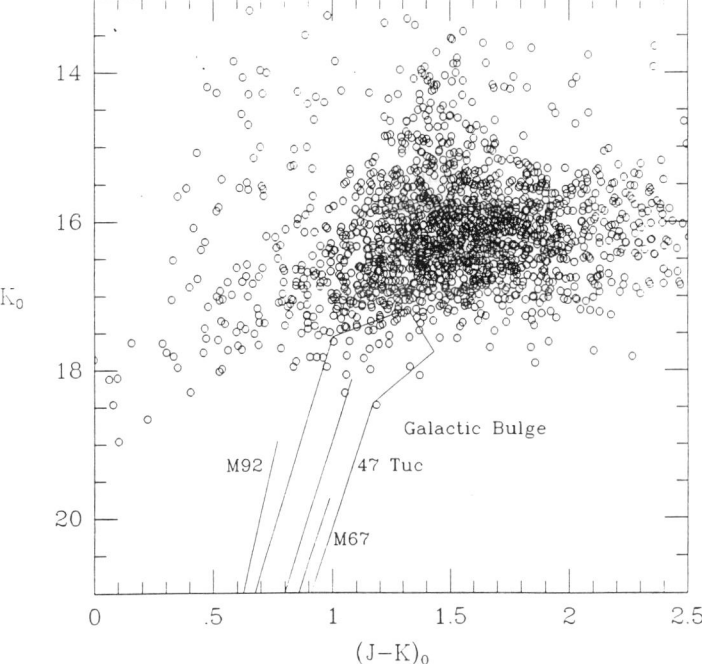

Fig. 2. A (J-K) v K color magnitude diagram of 1700 stars in the central regions of M 33 demonstrating the presence of a population more luminous than any in our galaxy.

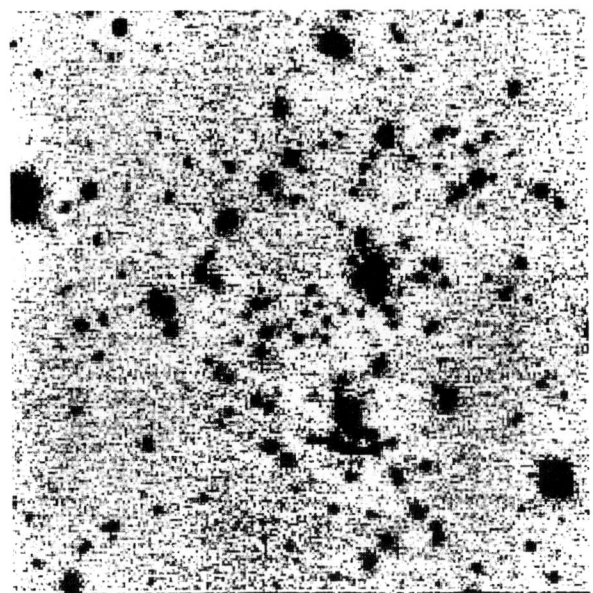

Fig. 3. A JHK composite image of the cluster Abell 370 at a redshift of 0.37. The total exposure time was 90 minutes and the completeness limit is about K = 18.

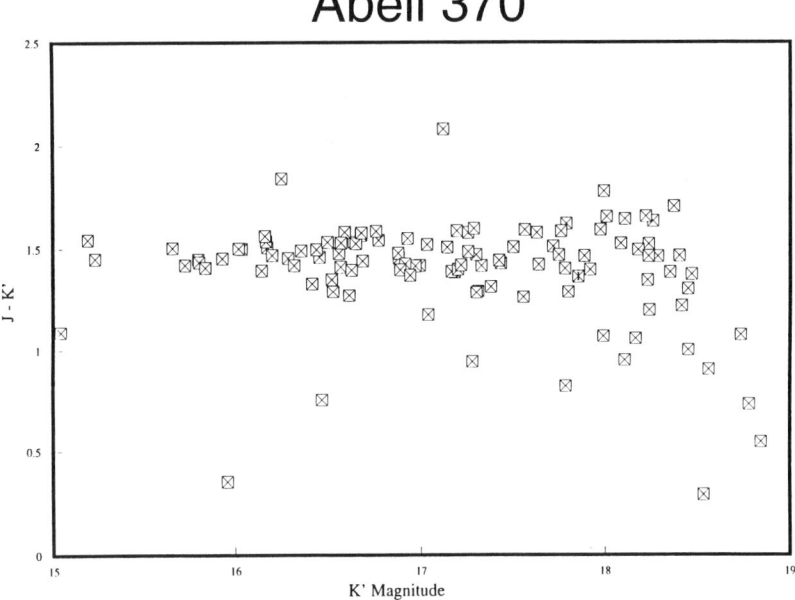

Fig. 4. A color-magnitude diagram for over 150 galaxies in the cluster Abell 370.

on the application of these IR detectors to spectroscopy. These are exciting times for infrared astronomy.

REFERENCES

Herter, T. 1994 in Infrared Astronomy with Arrays, I. McLean, ed., Kluwer Academic Pub., Dordrecht, p. 409
McLean, I. S. 1988 Sky & Telescope 75, No. 3, 254
McLean, I. S. 1989 Electronic and Computer-aided Astronomy: from eyes to electronic sensors, Ellis Horwood Ltd., Chichester, UK
McLean, I. S. 1993 Proceedings of the IV Canary Islands Winter School; Infrared Astronomy, A. Mampaso, M. Prieto and F. Sanchez, eds., Cambridge University Press, p. 336
McLean, I. S. 1994 Infrared Astronomy with Arrays: the Next Generation, I. S. McLean, ed., Kluwer Academic Publishers, Dordrecht
McLean, I. S., Becklin, E. E., Brims, G., Canfield, J., Casement, L. S., Figer, D. F., Henriquez, F., Huang, A., Liu, T., Macintosh, B. and Teplitz, H. 1993 Proc. SPIE, Vol. 1946, Infrared Detectors and Instrumentation, A. M. Fowler, ed., p. 513
McLean, I. S. and Liu, T. 1995, submitted to ApJ.
McLean, I. S., Liu, T. and Teplitz, H. 1995 in IAU Symposium No. 167, New Developments in Array Technology and Applications, A. G. D. Philip, K. A. Janes and A. R. Upgren, eds, Kluwer Academic Pub., Dordrecht, p. 359
McLean, I. S., Macintosh, B., Liu, T., Casement, L. S., Figer, D. F., Teplitz, H., Larson, S., Lacayanga, F., Silverstone, M. and Becklin, E. E. 1994 Proc. SPIE, Vol. 2198, Instrumentation in Astronomy VI, D. L. Crawford and E. R. Craine, eds., p. 457
Ueno, M. 1995 in IAU Symposium No. 167, New Developments in Array Technology and Applications, A. G. D. Philip, K. A. Janes and A. R. Upgren, eds, Kluwer Academic Pub., Dordrecht, p. 117

DISCUSSION

SZECSENYI-NAGY: We have seen some nice results on the IR photometry of intrinsically very bright stars and extragalactic objects. You also have shown us a deep image of the center of the Orion-nebula. As we know this region is extremely rich in red dwarf (dK/dM) stars. Do not you have any interesting and recent results on these absolutely faintest objects?

McLEAN: I have many other results which I could have shown including searches for very low mass stars and brown dwarfs. Others are doing similar work. I recommend to you the Proceedings of the UCLA conference of July, 1993 - Infrared Astronomy with Arrays: The Next Generation published by Kluwer.

SPYROMILIO: How confident are you that the big arrays will survive thermal cycling?

McLEAN: For InSb detectors thinned to about 10 μm one expects good resistance to thermal stresses because the material behaves like a "rubber sheet". For HgCdTe arrays on sapphire, the problem is greater. Rockwell and other companies have proprietary methods which will probably be sufficient - with care.

WATSON: What is the quantum efficiency of a Platinum Silicide detector?

McLEAN: Less than a few percent for commercially available devices.

WEAVER: Is anyone using InAs? Why not (since the QE seems high)?

McLEAN: Not to my knowledge. I think InSb technology is better developed.

CUBY: What about persistence effects with IR detectors?

McLEAN: Some HgCdTe arrays do show charge persistence effects. These can be reduced by resetting and flushing to some extent. Work is proceeding to eliminate the problem during construction.

THE LICK OBSERVATORY TWO MICRON CAMERA

K. Gilmore, D. Rank and P. Temi

Lick Observatory/UCO UCSC

ABSTRACT: Lick Observatory has recently developed a near infrared camera for astronomical imaging. The new camera has been built around a Rockwell NICMOSII 256 x 256 HgCdTe focal plane array. The dewar and optics were manufactured by Infrared Laboratories in Tucson, Arizona while the electronics and data system were designed and fabricated at Lick Observatory.

The new instrument was designed to image faint galaxies and other astronomical objects as well as embedded infrared sources in star formation regions. The camera has cold focal plane scales of 0.2, 0.4 and 0.6 arcseconds per pixel when used at the telescope. Standard H, J and K as well as narrow band special purpose filters are contained in two independent motorized filter wheels.

The camera performance with different plate scales and filters as well as the mechanical, optical and electrical lay-out will be shown. M 42 will be used for characterization of the camera's performance.

DISCUSSION

D'ODORICO: What is the best image quality that you have achieved at the telescope?

GILMORE: This past summer with URC2 at the Lick 3-m telescope, the best image taken was about 0".7.

FINGER: Is the lens close to the entrance window a cold lens?

GILMORE: Yes, the first element of the three element collimator is in contact with the radiation shield.

FINGER: Do you see a reset anomaly? How long does it take for the integration ramp to settle and become linear after reset?

GILMORE: In CDS mode we do not see an anomaly. In multi read mode it is seen. Several μs are required after reset for the ramp to become linear.

FLORENTIN-NIELSON: You have ball bearings inside the liquid nitrogen device. Do you have outgassing from any lubricant in these bearings?

GILMORE: The bearings do not use any lubricant. They are spring loaded into position onto a surface of like material.

INFRARED ARRAYS AT THE EUROPEAN SOUTHERN OBSERVATORY

G. Finger, G. Nicolini, P. Biereichel, M. Meyer and A. F. M. Moorwood

European Southern Observatory

ABSTRACT: This paper gives an overview of infrared array detectors which have been tested and used at ESO. The performance of arrays using Reticon type readouts, CCD readouts and switched FET multiplexers have been evaluated for both InSb and $Hg_{1-x}Cd_xTe$ detectors. Performance limitations specific to the NICMOS3 256 x 256 $Hg_{1-x}Cd_xTe$ detector installed in the ESO infrared array camera IRAC2 are addressed. The first test results with a high well capacity SBRC 256 x 256 InSb array are also presented.

Advanced readout techniques for image sharpening tested on a 2.2-m telescope are discussed briefly. A new generation of instruments being built for the VLT, the very large telescope project of ESO, is designed to house large format 1024 x 1024 IR arrays. A fast data acquisition system is currently being developed at ESO. The system is capable of handling the high data rates generated in the thermal infrared by large format low well capacity arrays. It can also cope with the low read noise required for flux levels of $\leq$ one photon/sec. It will first be installed in ISAAC, the Infrared Array Camera and Spectrometer built for the VLT (Moorwood 1993). The present status of both the detector developments and the data acquisition system is reviewed.

1. INTRODUCTION

In less than ten years infrared arrays evolved from formats of 1 x 32 to 256 x 256 pixels, and their noise performance improved by more than an order of magnitude. In the near future arrays with formats of 1024 x 1024 pixels will be available. A prototype Rockwell 1024 x 1024 $Hg_{1-x}Cd_xTe$ array has already been used to successfully image the SL-9/Jupiter collision in July 1994. In view of this rapid development it is useful to recapitulate the operating principle and the fundamental performance limit of these devices.

2. FUNDAMENTAL PERFORMANCE LIMITS OF PHOTOVOLTAIC DETECTORS OPERATING IN CDM

In the spectral range $\lambda = 1$ to 5 μm infrared arrays are intrinsic photovoltaic devices operating in the capacitive discharge mode (CDM). The absorbed photon excites an electron from the valence band to the conduction band. If the free electron-hole pair is created within a distance of the pn-junction equal to the diffusion length of the minority carriers it is separated by the electric field of the junction. As shown in Fig. 1 the reset switch connects the diode to the external bias voltage. After this switch is opened the detector junction is floating. Each absorbed photon discharges the capacity by one electron. The integrating node capacity C is the combined capacity of the detector and the gate of the source follower belonging to each pixel. The integration ramp is sampled at the beginning and end of the exposure. The variance of this voltage difference $\{DV\}^2$ is given in equation 1:

$$\{\Delta V(\tau)^2\} = \frac{2KT}{C}\left(1 - e^{-\frac{\tau}{\tau_C}}\right) \qquad 1$$

τC is the autocorrelation time RC corresponding to the diode impedance R and the capacity C of the integrating node

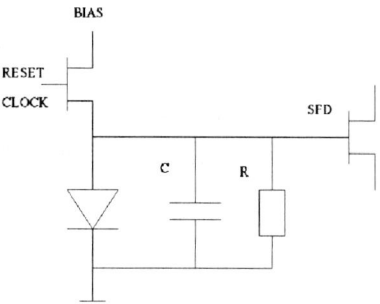

Fig. 1. Photovoltaic detector operating in capacitive discharge mode.

and τ is the time interval between the two voltage readings. By multiplication with C^2 and taking the square root of both sides of equation we obtain the noise $\Delta Nrms(\tau)$ expressed in rms noise electrons as given in equation 2:

$$\Delta Nrms(\tau) = \sqrt{\frac{2KTC}{q^2}\left(1 - e^{-\frac{\tau}{\tau_C}}\right)} \qquad 2$$

Two limiting cases are of interest assuming typical parameters for state of the art detectors e.g. C = 30 fF, darkcurrent of one e/sec at one V reverse bias, τ = 600 sec yielding τ_C = 1.8 10^5 sec. For the uncorrelated case, when $\tau \gg \tau_C$, equation 2 simplifies to the well known KTC noise $\sqrt{2KTC}/q$ and yields 51 e⁻ rms. For the correlated case, when $\tau \ll \tau_C$, the two voltage readings are completely correlated and equation 2 reduces to $\sqrt{2KTC\tau/R}/q$ yielding 2.9 e⁻ rms. The photon shot noise of photon flux Φ_{phot} is equal to $\sqrt{\Phi_{phot}\tau}$. If the smallest detectable photon flux Φ_{limit} is defined to have an associated photon shot noise equal to the correlated case of equation 2, we obtain:

$$\Phi_{limit} = \frac{2KT}{q^2 R} = 0.014 \quad \text{photons/sec} \qquad 3$$

For the idealized case of a noise free detector multiplexer and acquisition system the smallest detectable photon flux Φ_{limit} is proportional to the detector temperature T and inversely

INFRARED ARRAYS

proportional to the detector impedance R and becomes independent of the integration time τ. The HAWAII array (see Table 4) achieves read noise values of 8.6 e rms which is of the same order of magnitude as the fundamental noise limit derived in equation 3 for the correlated case and long integration times.

The pixel transfer function which gives the signal in volts per electron is inversely proportional to the pixel capacity. To reduce the read noise, the capacity of the detector must be as small as possible. However, in many applications the detector has to handle the high flux levels of the thermal infrared as well. The available readout time depends on the full well of the detector and is proportional to the pixel capacity. To reduce the required read speed for thermal broad band imaging, a trade-off between full well capacity and readout noise has to be considered.

Detectors operating in CDM mode are intrinsically nonlinear devices for two reasons. First, the pixel capacity is a function of voltage which is changing continuously during the integration, since the detector voltage is floating in CDM operation. Second, for small flux levels, the nonlinear I-V characteristic of the diode becomes important. The linear regime of the detector is confined to the following two conditions: First, only a small fraction of the full well capacity is used. Second, the photon generated current dominates the dark current of the detector diode.

3. OVERVIEW OF INFRARED ARRAYS AT ESO

In Table 1 the detector arrays tested at ESO and installed in various instruments are listed. The first array introduced at ESO was a linear 1 x 32 InSb array bonded to a Reticon multiplexer with a pixel size of 200 μm. The limitation of the Reticon readout is due to the fact that all pixels must share a common video line. The read noise was reduced to 1200 e rms by a special weighted double correlated clamp (Finger et al. 1987a). This detector was installed in the cryogenic infrared spectrometer IRSPEC in 1985 and was in service until 1991.

The first two-dimensional array tested at ESO was a 64 x 64 $Hg_{1-x}Cd_xTe$ array read out by a CCD. The cutoff wavelength of the detector is 4.2 μm but can be set between 2.5 μm and 10 mm by varying the composition x of the $Hg_{1-x}Cd_xTe$ alloy. The detectors consist of cylindrically shaped p-n junctions which are connected to a buried channel CCD by a special loophole interconnect technology (Finger et al. 1987b). The input stage of the CCD consists of a direct injection circuit. At low flux levels the charge injection efficiency drops, since the input impedance of the direct injection circuit is inversely proportional to the current. Another problem is the threshold uniformity of the injection circuit which requires very uniform doping concentration of the buried channel CCD. The MCT CCD was installed in the infrared array camera IRAC1 in 1988.

Advances in the large scale integration of silicon made it possible to place a source follower in the unit cell of each detector. Multiplexers having one source follower per detector (SFD) have become the generally accepted way of read out. A 64 x 64 InSb array was tested in the lab and the famous 58 x 62 InSb SBRC array (Fowler et al. 1987) replaced the linear array in IRSPEC in 1991. The possibility of tuning the band gap of the alloy $Hg_{1-x}Cd_xTe$ (MCT) to the energy of a photon of λ = 2.5 mm allows the subdivision of the λ = one to five mm spectral range in to the non-thermal infrared shortward of λ = 2.5 mm which is covered by MCT

TABLE 1

Infrared Arrays at ESO

Company	Device	Format	λ_c [μm]	Instrument	Read out	Pixel pitch [μm]	Well e	Noise e rms	QE
CINCINNATI	InSb	1x32	5	IRSPEC	Reticon	200	$2.4\,10^7$	1200	0.85
MULLARD	MCT	32x32, 64x64	4.2	IRAC1	CCD	48	$3.1\,10^6$	200	0.28
CINCINNATI	InSb	64x64	5	LAB	SFD	100	$2\,10^6$	180	0.75
SBRC	InSb	58x62	5	IRSPEC IRAC1	SFD	75	$3.2\,10^5$	200	0.8
ROCKWELL	MCT	256x256	2.5	IRAC2A	SFD	40	$2\,10^5$	25	0.55
CINCINNATI	InSb	256x256	5	LAB	SFD	30	$8.4\,10^5$	?	0.8
SBRC	InSb	256x256	5	IRAC2B	SFD	30	$5\,10^5$	120	0.65
ROCKWELL	MCT	1024x1024	2.5	ISAAC	SFD	18.5	$6\,10^4$	<10	0.6
SBRC	InSb	1024x1024	5	ISAAC	SFD	27	$3\,10^5$	25	0.8
LIR	Si:Ga	64x64 128x192	18	TIMMI	SFD	100 75	$3\,10^7$	2500	0.25 0.3

detectors. Longward of λ = 2.5 mm, in the regime of the thermal infrared, the 1 - 5 mm response of InSb detectors is utilized. The big step both in performance and array format came with the NICMOS3 256 x 256 $Hg_{1-x}Cd_xTe$ array having a cutoff wavelength of 2.5 μm. This array has been producing science in the large infrared camera IRAC2A since 1992 (Moorwood et al. 1992). At present two high well InSb 256 x 256 arrays are being tested at ESO. The Cincinnati array uses an extra capacity of 0.17 pF at the gate of the source follower which augments the detector capacitance of 0.1 pF, resulting in an integrating node capacity of 0.27pF (Blessinger et al. 1993). The SBRC array uses the standard CRC-463 multiplexer and will be discussed in more detail below.

Both ROCKWELL and SBRC are developing 1024 x 1024 arrays which will be complementary with respect to wavelength response, readout speed and noise performance. With growing array format there is a general tendency towards smaller pixels and lower pixel capacity. The λ = 8 - 13 μm region is served by a 64 x 64 Si:Ga array manufactured by LETI/LIR. In 1993 this array was installed in TIMMI, the ESO 10 μm camera built by the Service d'Astrophysique at Saclay. A new 128 x 192 successor of this array is under development (Lucas et al. 1994).

4. NICMOS3 MCT 256 x 256 ARRAY

The NICMOS3 array is an excellent device in most respects and has been evaluated extensively. Results are discussed in more detail elsewhere (Finger et al. 1993). The characteristics of NICMOS3 are given in Table 2. The array has a very low dark current of 0.4 e⁻/sec due to its short cutoff wavelength of λ_c = 2.5 mm and an excellent noise performance of 20 e⁻ rms.

TABLE 2

NICMOS3 MCT 256 x 256 Array Characteristics

Detector	$Hg_{1-x}Cd_xTe$ LPE on Al_2O_3 substrate buffered by CdTe (PACE 1)
Wavelength range	1 - 2.5 µm
Number of pixels	256x256
Pixel Pitch	40 µm
Read-out	CMOS shift registers organized in 4 independent quadrants, SFD unit cell
Output amplifiers	4 (1 per quadrant)
Package	68 pin LCC
Full Well	$2 \cdot 10^5$ e⁻
Noise	20 e⁻ rms with multiple sampling
QE	52% at 2.2 µm, 29% at 1.25 µm
Darkcurrent	< 0.4 e⁻/sec
Operating Temperature	70 K
Problems	reset anomaly, amplifier glow, persistence
Status	fully tested, operating in IRAC2 since May 1992

However, the array suffers from some problems relevant to low flux applications. One problem is the persistence effect. If the array is exposed to ambient temperature K-band radiation a subsequent dark exposure displays a dark-current of 9 e⁻/sec, which decays exponentially and decreases to 2.7 e⁻/sec after 700 sec. Subelectron darkcurrent can only be achieved after the detector has been kept dark for many hours. The reset anomaly is another problem with which to cope. After the reset strobe has been applied to the detector diodes to recharge the integrating node capacitance, the integration ramp is strongly nonlinear during the first 100 msec. For this reason the first nondestructive readout in the multiple nondestructive read scheme has to be discarded. Only the linear part of the integration ramp should be used. Multiple nondestructive sampling is adversely affected by the luminescence of the multiplexer which is strongest in the corners of the array close to the output amplifiers. In the center of the array, where this effect is smallest, a 300 sec dark exposure with 656 nondestructive readouts gives a signal of 7582 electrons. The signal consists of the following components: 3400 electrons may be assigned to the amplifier glow, 3900 electrons are due to the shift registers and 270 electrons are due to the diode dark current. The photon shot noise generated by the multiplexer glow limits the reduction of read noise by multiple sampling to 20 e⁻ rms.

5. HIGH WELL CAPACITY SBRC 256 x 256 InSb ARRAY

Since the one to five micron wavelength region covers both the non-thermal and the thermal infrared, the detector arrays used for multimode instruments at large 8-m class telescopes have to cope with flux levels varying from less than one photon/sec for high resolution spectroscopy to 10^8 photons/sec for M band imaging. If all instrument modes must be served by a single detector, a trade-off between read noise and full well has to be considered. Small pixel capacities result in excellent noise figures but require prohibitively high read speeds for broad band imaging in the thermal infrared. At an eight-m telescope broad band imaging in L of a 46"x 46" field with the standard SBRC 256 x 256 InSb array requires that one pixel is read out in less than 435 nsec if all four quadrants are read in parallel.

Two possibilities have been explored to increase the capacity of the integrating node. An extra capacity can be added to the gate of the source follower in the unit cell of the multiplexer (Blessinger et al. 1993). This requires a special multiplexer design. Alternatively, the pixel capacity can be increased by increasing the doping concentration of the detector diode. The capacity of the pn junction is proportional to the square root of the doping concentration N_{doping} as shown in equation 4:

$$C(V) = A_{Diode} \sqrt{\frac{\varepsilon_{InSb} N_{doping} e}{2(V_{builtin} + V)}} + C_{Gate} \qquad 4$$

SBRC has developed a special high doped 256 x 256 InSb array hybridized to a CRC 4 63 multiplexer which is compared to the low doped standard InSb array in Table 3. The integrating capacitance is approximately 0.19 pF, much higher than the 0.06 pF for the standard low-doped arrays. With a 400 mV bias, the well capacity becomes 4.8 10^5 electrons. Typical noise on the high-doped arrays is 100 electrons rms, compared with 55 electrons for low-doped arrays. Preliminary dark current measurements which may be distorted by thermal background photons due to imperfect shielding gave 15 e/sec at 350 mV reverse bias. The cosmetic quality of the array is good. It has 270 bad pixels 170 of which are clustered at the edge of the array. The expected readout time for one pixel is increased to more than one μsec for broad band imaging in L, which is feasible with state of the art electronics. The quantum efficiency specified in Table 3 was measured by the manufacturer.

6. SPECIAL READOUT TECHNIQUES

To reduce the read noise we implemented the nondestructive multiple sampling scheme. After applying the reset strobe the array is read out many times, but without resetting the integrating node capacitance to the external bias voltage, which is illustrated in Fig. 2. During the stare time the integration ramp is sampled nondestructively as often as the bandwidth of the data acquisition system allows. The images read nondestructively are used to update the

TABLE 3

Comparison of Standard and High Doped SBRC 256 x 256 InSb Array Detector

Detector	C [pF]	Well [e]	Noise [e rms]	QE [%]	pixel time [nsec]
standard	0.06	1.85 10^5	55	80	435
high Well	0.19	4.8 10^5	100	65	1377

error equations of a regressional fit to the integration ramp. The read noise $\Delta Nrms_n$ for n nondestructive readouts per detector integration is given by equation (5), $\Delta Nrms$ being the read noise of double correlated sampling:

$$\Delta Nrms_n = \Delta Nrms \sqrt{\frac{6(n-1)}{n(n+1)}} \approx \Delta Nrms \sqrt{\frac{6}{n}} \quad \text{for} \quad n \gg 1 \qquad 5$$

In principle nondestructive readout can reduce the read noise to negligible amounts by increasing the detector integration time and the number of possible readouts. In practice other limitations arise, e.g. the NICMOS detector becomes shot noise limited by the luminescence of the multiplexer for integrations longer than three seconds.

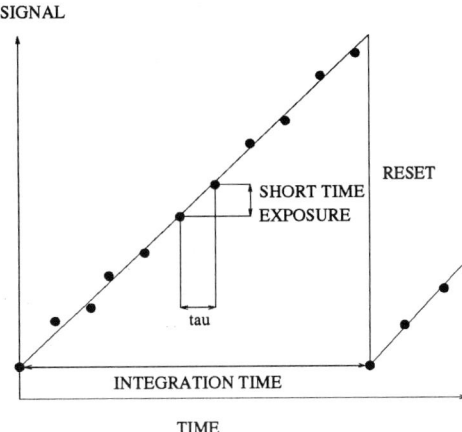

Fig. 2. Illustration of multiple nondestructive sampling readout and on-chip tracking algorithm. The difference of two nondestructive readouts corresponds to a short time exposure which can be recentered to compensate image motion due to atmospheric seeing.

Ground based astronomical observations suffer from seeing effects which result from distortions of the plane wavefront of a star image by atmospheric turbulence (Roddier et al. 1992). Within the nondestructive readout scheme a powerful technique of image sharpening can be implemented. This technique is called on-chip tracking and allows correction of image motion in real time which is equivalent to correction of first order wavefront perturbations (Finger et al. 1993, Finger et al. 1994a, b). The information required to compensate the image motion is contained in the difference of two consecutive readouts. The difference is equivalent to a short time exposure taken during the time interval between two readouts. Positional information obtained from a guide star within the field seen by the array is used to shift the whole data array representing non-destructive readouts. The shifted differences are used to reconstruct the integration ramp. A threshold set for the peak intensity of the guide star can be defined to discard moments of bad seeing.

On chip tracking has the same effect as a tip tilt correction by an active optical element (Close and McCarthty 1994) - but without the extra complexity of adding such a device. A comparison of on chip tracking with simple shift and add using destructive readout and double correlated sampling shows that the readout speed can be increased by a factor of three without any hardware modifications. Furthermore, under-sampled images profit from the noise advantage of multiple sampling.

The tracking mode has been tested at a 2.2-m telescope in the K band with the infrared array camera IRAC2 equipped with the NICMOS3 array. Under good seeing conditions the spectral bandpass is well tuned to the telescope diameter to perform first order wavefront corrections since the ratio of the telescope diameter D to the Fried coherence parameter r_0 is expected to be $D/r_0 \sim 3$ at $\lambda = 2.2$ mm (Roddier et al. 1992). Fig. 3 shows a subframe of a three second exposure containing the pre-main-sequence binary star S CrA. The left image is untracked while the right image is tracked. The data are untreated raw data. The pixel scale is 0.14 arcseconds which means that the diameter of the airy disk is sampled by 3.7 pixels. The detector was read out nondestructively every 100 μsec. Tracking improves the Strehl ratio by a factor of 2.7 and the binary star is well resolved.

Fig. 3. K' image of the pre main sequence binary star S CrA. Left: untracked. Right: tracked. Integration time three sec. Pixel scale: 0.14 arcseconds.

THE IMPACT OF INFRARED ARRAY TECHNOLOGY ON ASTRONOMY

Giovanni G. Fazio

Harvard-Smithsonian Center for Astrophysics

ABSTRACT: Over the past ten years a technological revolution has occurred in the development of two-dimensional infrared array detectors for astronomical observations. The wide application of these arrays for both ground-based and space observations has resulted in a profound change in the capabilities and perspective for infrared astronomy, resulting in new views of the infrared sky. A review will be presented describing these detectors, the numerous advantages they provide for astronomical observations, the present state of array technology, and the potential for future growth.

MONOLITHIC Si BOLOMETER ARRAYS: DETECTORS FOR FAR INFRARED AND SUBMILLIMETER DETECTION

Harvey Moseley

Laboratory for Astronomy and Solar Physics
Goddard Space Flight Center

ABSTRACT: The improvement of photoconductors and photovoltaic detectors for $\lambda < 200$ μm has displaced bolometers as detectors of choice for many applications requiring high sensitivity. Continued development of bolometers for operation at low temperatures (T $<$ 0.1 K) has resulted in significant improvements in their sensitivity, making them excellent choices for many broad band applications at $\lambda > 200$ μm, cryogenic spectrometer applications in the submillimeter, and applications requiring extreme stability and ease of calibration.

I will describe the development of bolometers over the past decade, with particular emphasis on the detectors we have developed at GSFC. Detectors have been constructed in 36 element arrays with NEP $\sim 5 \times 10^{-18}$ W/$\sqrt{\text{Hz}}$ and a response time of ten ms. Such detectors are very useful for many current problems in Cosmic Microwave Background studies and submillimeter spectroscopy. We will describe the current state of development of the detectors and the improvements we are pursuing.

There has been significant progress in the development of superconducting tunnel junction detectors for operation in the far infrared and submillimeter spectral range. Though at an early stage, these detectors have great promise of excellent sensitivity and ease of array fabrication. I will discuss the current developments in this area.

DISCUSSION

WAMPLER: Can you compare hetrodyne interferometry with imaging interferometry using bolometer arrays?

MOSELEY: Bolometer arrays probably have a sensitivity advantage for interferometry over current hetrodyne systems for continuum sources, but the hetrodyne systems may be technically easier to implement.

OBSERVATIONAL CONCERNS AND TECHNIQUES FOR HIGH-BACKGROUND
MID-INFRARED (5 - 20 micron) ARRAY IMAGING

Daniel Y. Gezari

NASA/Goddard Space Flight Center
Infrared Astrophysics Branch

1. INTRODUCTION

Broadband observations at 5 - 20 micron with large ground-based telescopes are dominated by thermal background radiation from the telescope optics, instrument and sky. Astronomical sources typically contribute less than one percent of the total detected flux. In most cases the direct, unprocessed image of a bright source (before background subtraction) is indistinguishable from an image of blank sky (Fig. 1). This paper gives an overview of the fundamental concepts, conditions and limitations of high-background mid-infrared imaging. The techniques used to acquire and process the image data are summarized, illustrated with examples from our 58 x 62 pixel mid-infrared camera astronomy program.

Bright, mid-infrared astronomical sources are few and far between. If you point a telescope equipped with a near-infrared imaging camera randomly at a thousand places in the sky you are likely to detect some sources (stars) at every position. But if you observe with a mid-infrared camera at a thousand places in the sky you will probably see nothing in a ~1 arcmin field at every position. There are only a few hundred mid-infrared sources which are bright enough to detect easily, and many of these are well-known standard stars. Altogether in the Catalog of Infrared Observations (Gezari et al. 1993) there are about 2000 mid-infrared sources which are brighter than one Jansky or four mag (our practical detection threshold). But this is a tiny population compared to the enormous number of anonymous and uncataloged near-infrared sources which can easily be imaged from the ground.

The infrared background flux level is determined by the optical throughput (focal ratio of the detector input beam), and by the temperature and emissivity of the telescope optics, the instrument optics, and the sky. Most of the total background comes from the telescope mirrors at a good observatory site. Small changes in the background flux during the observations can contribute to degraded images and dramatically reduced sensitivity. The largest background variations are spatially uniform in the focal plane (due primarily to atmospheric opacity changes) but they can result in second-order spatial defects in combination with subtle linearity and gain effects (primarily in the detector system) or optical efficiency differences (primarily in the telescope) which can become significant after the background is subtracted.

In a typical observatory environment, the thermal background flux in the Cassegrain focal plane of a large (~three-meter) conventional telescope from the T ~ 270° Kelvin night sky and telescope mirrors is about 10^9 photons $sec^{-1}m^{-2}micron^{-1}arcsec^{-2}$ at ten microns. But the detector "well" capacity of infrared photoconductor arrays is typically only 10^5 - 10^6 electrons. Thus it would seem that exceptionally short exposures (sampling rates faster than available A/D converters and the time constants array multiplexer circuitry) would be required to avoid saturating the detector. However, the small detector pixels, actual optical efficiency, low photoconductive gain at which our detector can be operated, and reasonably fast sampling rate (30 Hz) combine to reduce the number of electrons which actually accumulate in each detector well during a 30 msec exposure by about four orders of magnitude, making broadband operation of the large-format array feasible under high-background conditions.

2. LIMITING NOISE SOURCES

The sensitivity of broadband mid-infrared observations is limited (ideally) by statistical fluctuations (roughly 1000 electrons per exposure) in the background level (roughly 10^6 electrons per exposure), as compared to detector read noise ($\sim$ 100 electrons per read) and dark current noise ($\sim$ ten noise electrons per second). Evaluating or specifying signal/noise in an image is little different than doing so for a single detector. But since the individual detector pixels behave very much alike (especially after gain correction of the data), measuring the scatter among adjacent pixels in a featureless area of the image approximates the noise performance of any one of the individual detectors if it were monitored during the total integration interval.

The term "sky noise" is not generally used to describe the background shot noise in the incident photon flux (statistical fluctuations proportional to $n^{1/2}$), but rather to describe random variations in the sky emission due to opacity and temperature structure in the atmosphere moving overhead. If the astronomical source of interest and the blank sky could be imaged simultaneously, a nearly perfect sky background subtraction could be made, and effects due to sky opacity structure would be eliminated. Chopping comes close to approximating simultaneous observations, with the chopper frequency determining the degree of temporal coherence.

The frequency spectrum of "sky noise" is another concern. Noise due to opacity differences can have a characteristic 1/f dependence, the tendency toward long term drifting of the sky emission level, and can be reduced by chopping and sky-subtracting at a sufficiently high frequency (determined by sky conditions). As the chopper beam separation is increased, spatial coherence decreases. But since atmospheric effects occur in the extreme near field, chopper throws would have to become extreme (more than a few arcminutes) for appreciable differences in sky emission between the two chopper positions to become a concern. These effects become aggravated under conditions of poor atmospheric transmission.

Twenty years ago, when single bolometers were commonly used for infrared astronomical observations, fast ($\sim$ 30 Hz) chopping was standard operating procedure. Changes in the temperature of a bolometer caused by the incident photon flux are measured as changes in bolometer resistance. But bolometers were not easy to use as DC detectors since high resolution A/D converters were not then available. Therefore, the difference between two DC voltages (corresponding to the detector temperature at the two chopper beam positions on the sky) was measured using a lock-in amplifier synchronized with the chopper frequency. And since both the bolometers and the sky exhibited noise with 1/f characteristics, fast chopping was commonly employed with bolometer detector systems. Integrating photoconductor detectors and modern A/D converters now make DC measurements with arrays straightforward. The 1/f current noise of bolometers is not a concern in photoconductor arrays. However, noise is associated with the stability of the electronic system bias and reset levels, and noise performance can be optimized by choice of operating voltages. Appropriate detector operating temperature can minimize dark current noise due to thermal ionization of carriers when the temperature is too high (> 15° Kelvin) or loss of responsivity (and consequently degraded signal/noise) due to charge trapping when the temperature is too low (< 7° Kelvin).

3. BACKGROUND SUBTRACTION AND FLATFIELDING

A standard set of data reduction procedures is applied to high-background image data. For background subtraction, "reference" images of blank sky (observed simultaneously using the telescope chopping secondary mirror) are subtracted from "source" images. The resulting difference images are flatfielded by dividing each by a normalized blank sky image (a measure of the relative instrumental gain pattern of camera, or the "gain matrix"). Any residual sky signal (background offset level after subtraction) can be removed by subtracting another blank sky image from each difference image (an alternative to nodding the telescope while taking data). After being aligned spatially the flatfielded images are averaged together to improve signal-to-noise and produce a final mosaic.

A single 30 msec camera exposure is referred to as a "frame", and an a co-added series of frames is an "image". A source image $S(x,y,t)$ contains the intensity distribution of the astronomical object $O(\alpha,\delta)$, an atmospheric background flux component $B_a(t)$ which is generally featureless but can vary rapidly in intensity, and large background contributions from the telescope $B_t(t)$ and instrument $B_i(t)$ which change more slowly. A "reference" image $R(x,y,t)$ taken on adjacent blank sky contains everything but the source contribution, although the detected background level could have changed slightly from its value during the "source" observation. The net photon flux is multiplied by the instrument "gain matrix" $g(x,y,t)$, which combines the detector response of each pixel and the optical efficiency of telescope and camera.

$$S(x,y,t) = g(x,y,t) [B_a(t) + B_t(t) + B_i(t) + O(\alpha,\delta)]$$

$$R(x,y,t) = g(x,y,t) [B_a(t) + B_t(t) + B_i(t)]$$

The basic astronomical image $I(x,y,t)$ is the difference between a source and reference image. If the background contributions are equal in the source and reference images, subtracting them yields the object intensity distribution $O(\alpha,\delta)$ in the sky, modified by the instrument gain distribution

$$I = S - R = g(x,y,t) O(\alpha,\delta)$$

The basic flatfielding operation actually occurs in the initial sky subtraction procedure. If any image defects persist after this step, they can be further corrected for by dividing the image by a normalized reference image R_o of nearby blank sky, the observationally generated map of net camera response described by $g(x,y,t)$.

$$O(\alpha,\delta) = (S - R)/R_o$$

The reference image used for the division can be the direct sky image obtained simultaneously with the chopper, or a sky frame taken separately under the gain and background conditions as similar as possible. A more rigorous flatfield can sometimes be obtained by dividing by an image composed of the difference between two sky images taken at high and low airmass, which removes the instrument and telescope background contributions from the denominator (in the same way that the corresponding factors in the signal and reference images subtract out in the numerator). But spatial variations in the instrument and telescope backgrounds are usually negligible over time intervals comparable to the integration time, so dividing by a single normalized reference frame usually produces equally good results.

In the raw images shown in Fig. 1, a gradient of about 30% can be seen across the field due to gain differences among the array pixels. The dark spots are array defects of up to 25% of the mean resulting from defective bump-bonds between the detector and multiplexer wafers of the hybrid array chip. Despite these gain nonuniformities, the sky-subtracted images can be flatfielded to better than 0.01% (1s) of the raw image mean. Flatfielding of high-background array images can be the most demanding instrumental calibration requirement. Flatfielding techniques have also been discussed by Hoffmann et al. (1987) and McCaughrean (1987).

Unfortunately, differences in the telescope background, DBt, between the two positions of the secondary mirror can cause a different kind of problem, a spatial offset (or "residual sky") which persists after the sky subtraction and flatfielding process. This offset arises from small differences in the optical path through the telescope optics between the two chopper positions, and is often due to asymmetrical views of the telescope structure, vignetting, etc. This offset can present itself as a constant background level or as an intensity gradient (wedge) across the image. This residual sky background can be removed from an image by nodding the telescope so that the astronomical object appears alternately in the S and R images, which are subsequently subtracted from each other in pairs. Or, a single image of nearby blank sky can be subtracted from a number of different object images. The final object image which is co-added from several such corrected images would not limited by the noise level of the single blank sky image, provided that the telescope is repositioned slightly for each integration (to put the object at a slightly different position on the array). When the images are aligned on the common object for averaging, the position of the common sky image becomes shifted in the aligned stack, and the noise is randomized in the average.

Fixed-pattern image defects commonly appear in the image when processing is attempted with a reference image obtained at a different background intensity (Fig. 2). If the source and sky images are not too different, a simple level matching process (adjusting offset and gain factors) can be used to generate a matching reference image. However, large background differences combined with small non-linear detector effects (discussed below) can result in two images which can not be successfully matched. In principle, the same reference image can be used to correct many different data images. However, the background flux and electronic gain of the detector system have to be quite stable on the time scale of the observation.

4. DETECTOR LINEARITY

The linearity of the detector response affects more than just the first-order photometric accuracy in imaging array applications. Since most of the charge in the well is generated by background photons, we are always operating at about the same level in the detector well,

CONCERNS AND TECHNIQUES FOR MID-INFRARED ARRAY IMAGING

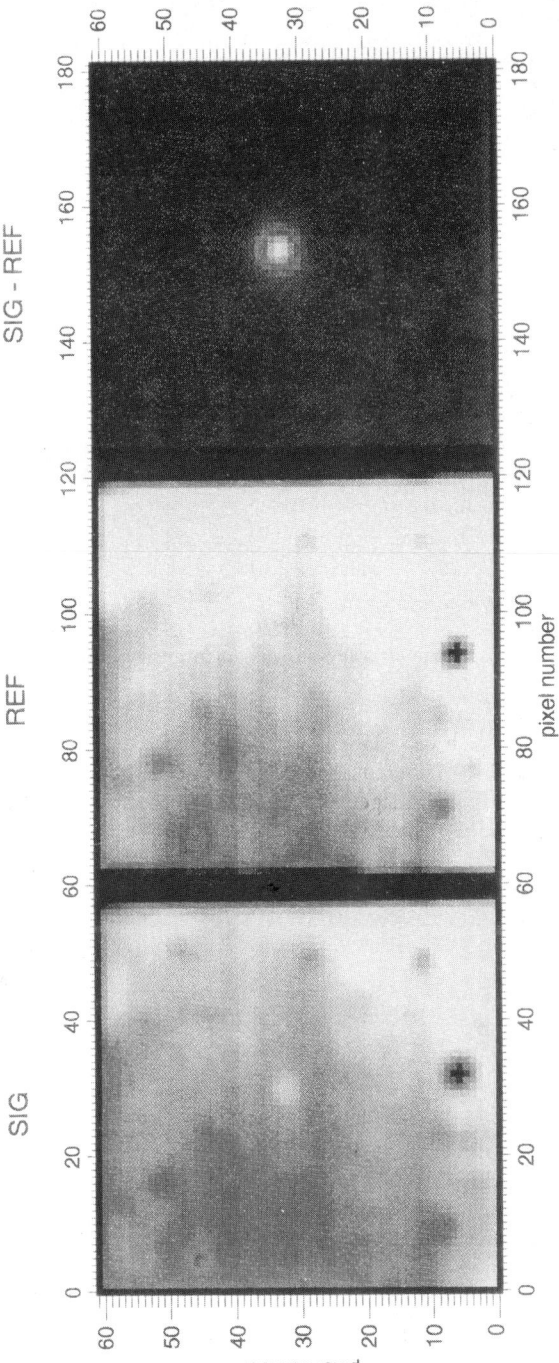

Fig. 1. Individual source (SIG) and reference (REF) images of the bright star α Boo, and the final sky-subtracted image (SIG - REF) obtained by taking the difference between the two. α Boo can barely be seen above the large background level in the SIG image. The REF image represents an observational gain matrix of the system response.

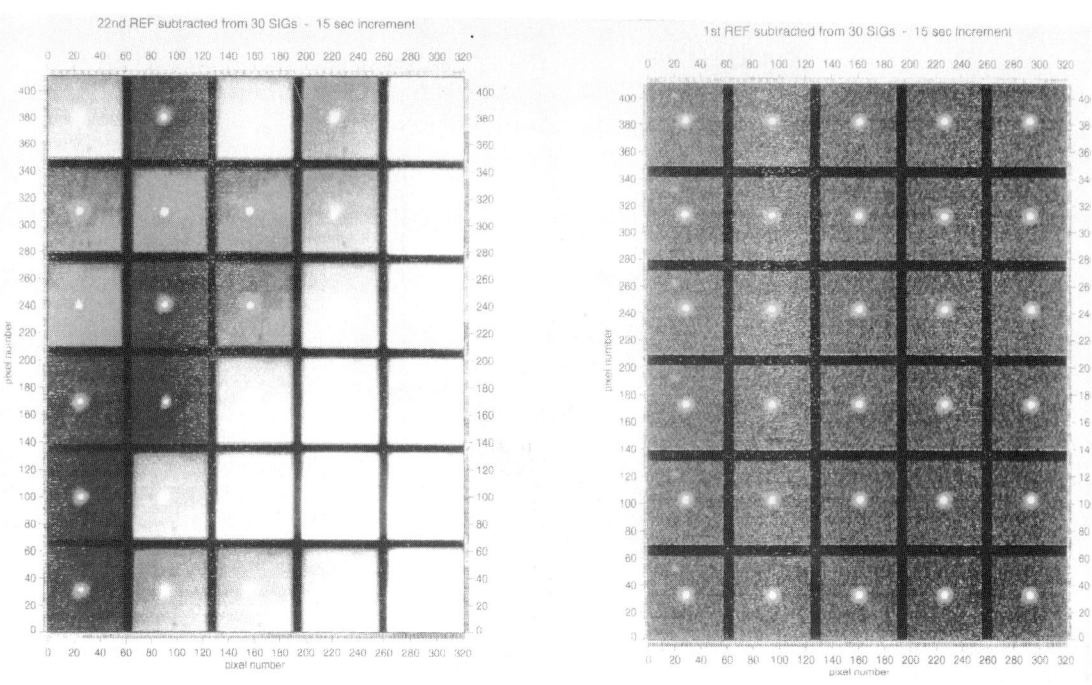

Fig. 2. A comparison of chopped (3 Hz) and unchopped observations of NGC 1068 on the IRTF with good weather (each with one minute of integration time). The unchopped image shows defects due to the gain matrix response to different background levels, and has about a factor of four worse signal/noise than the chopped image. The unchopped SIG and REF images were taken over an elapsed time of approximately six minutes (a three minutes exposure on the source followed by a three minute exposure on nearby blank sky). The bright point-like nucleus of NGC 1068 has a flux density of about 30 Janskys at 12.4 micron.

regardless of source strength for all but the brightest sources, and detector linearity would not seem to be a major concern. But very subtle detector problems (latency effects, defective bump-bonds, cross-talk, etc.) become magnified when the background is subtracted and the < 1% of the well which represents the source flux is expanded to full contrast range in the final image or photometry. In this sense high background imaging is much less forgiving than low-background (where most of the detected signal is due to source photons). Linearity difficulties are aggravated by any changes in system gain or bias levels. Therefore, considerable attention must be paid to the design of a stable detector electronics system.

5. ARRAY CAMERA SYSTEM

The infrared array camera system described here was developed for broadband, diffraction-limited 5 - 20 micron astronomical imaging with large observatory telescopes. The camera uses a 58 x 62 pixel gallium doped silicon (Si:Ga) photoconductor array detector manufactured by Hughes/Santa Barbara Research Center (SBRC). The detector array is a hybrid device, assembled from a wafer of Si:Ga detector material (nominally sensitive between 5 - 17 micron), bump-bonded to a Hughes CRC-228 direct readout (DRO) integrated circuit multiplexer chip (Hoffman 1987). The array pixels are read out serially, although the switched FET multiplexer design allows them to be sampled in any order, or polled non-destructively to determine the fullness of the well in low signal/low background applications (with the penalty only of added read noise). A detailed description of the array camera optical and electronic design, detector characteristics, and operating procedures has been presented by Gezari et al. (1992).

One of the very desirable characteristics of this SBRC array is that the photoconductive gain, G_{pc}, of the detectors can be adjusted as a function of net detector bias (G_{pc} can be reduced to about 0.1 by operating at a net detector bias of four volts). This characteristic of the device permits broad-band operation of the array at higher backgrounds, with no compromise in detected photon noise statistics. Since G_{pc} is a post-detection gain factor, reducing the photoconductive gain reduces the number of electrons generated per incident photon, but does not change the incident photon statistics or signal/noise of the observation.

Under typical high-background conditions the Si:Ga array is operated at about 1/2 full-well capacity (full well ~ 7×10^5 electrons) using a ~ 30 Hz frame rate (30 msec integration time per pixel), and with photoconductive gain set at about 0.1. This results in noise equivalent flux density NEFD = 0.03 Jy/min$^{-1/2}$pixel^{-1} (1s) with broadband ($\Delta\lambda/\lambda$ = 0.1, transmission ~ 80%) interference filters (1 Jy = 1 Jansky = 10^{-26} Wm^{-2}Hz^{-1}). The NEFD expressed as noise equivalent brightness is NEB = 0.45 Jy min$^{-1/2}$ arcsec^{-2} (1σ). The point source NEFD (1s) is about ~ 0.5 Jy min$^{-1/2}$, that is, a 0.5 Jy point source would produce a detector signal equal to the (1s) noise in an integration of one minute, since the point source flux is spread over 20 - 30 pixels by diffraction and, to a lesser extent, atmospheric seeing. A point source of ~ 0.1 Jy could be detected in one minute (1σ) if five x five pixels are binned together in subsequent data analysis. But these performance numbers can be deceiving. Consider an infrared galaxy which has a ten micron flux density of 1 Jy in a point-like (one arcsec) nucleus, and another one Jy distributed in a surrounding uniform disk about ten arcsec in diameter (not an unrealistic model for the brighter infrared galaxies, but a faint source by our standards). The nucleus would be detected with reasonable confidence in about one minute of integration time. But the surface brightness of the disk is 100 times lower, and would require ~ 10,000 minutes (~ 150 hours) of integration time to achieve the same signal-to-noise. Binning up five x five array pixels to an effective 1.3 x 1.3 arcsec synthetic aperture would reduce the integration time by a factor of five to ~30 hours, a more feasible but still sobering amount of observing time. A low-background, liquid nitrogen-cooled infrared telescope has been proposed for 5 - 20 micron imaging at the South Pole (Gezari 1994), which could greatly improve the efficiency of such observations.

6. DATA ACQUISITION AND IMAGE ANALYSIS

The array data frames collected from the source and reference positions of the telescope chopping secondary mirror are sorted and co-added into two algebraic arrays by the data acquisition computer, synchronized with the chopper drive signal. Data are ignored while the chopper mirror is moving between end positions. Two final images with total integration times of typically one minute (1800 30 msec frames) are down-loaded to the host computer and stored as image pairs.

We have developed an image analysis software package called MOSAIC to process large-format array image data (Varosi and Gezari 1992). The MOSAIC software is now available for use by interested researchers on a limited basis. Fig. 3 shows the Orion BN/KL infrared source complex at 20.0 micron (Gezari and Backman 1995), a 40 x 40 arcsec mosaic image made up of 23 individual overlapping images, to illustrate how large mosaic images of complex fields can be assembled successfully from many individual overlapping array images.

7. CALIBRATION PROCEDURES

Array detectors provide intrinsically high relative astrometric accuracy. Astrometric calibration requirements include determination of the array pixel angular size (plate scale) in both dimensions, field rotation and distortion. However, there are several practical factors which complicate astrometric calibration and limit astrometric precision, including telescope tracking, chopper stability, telescope encoder accuracy and mechanical flexure. The telescope magnification also changes slightly with focus setting.

There is a real scarcity of cases where two or more stars fall within a single array field of view (15 arcseconds) and are also easily detected in both the infrared and visible, much less which have relative positions which are measured with high accuracy. Telescope encoder displays are generally not sufficiently precise for sub-arcsecond telescope plate scale calibration measurements. We ultimately determined the array plate scale by programming the telescope to offset reproducibly between the four visible Trapezium stars (which have accurately determined relative positions but which are not bright mid-infrared sources), repeating the programmed motion on a strong, nearby ten micron star, recording multiple exposure infrared images with the array, and measuring positions of the four infrared star images in the infrared data.

For the purposes of absolute flux calibration, images of a standard star must be obtained at least once during the observation of an infrared source, or at least each hour to determine an atmospheric extinction correction at each wavelength. The calibration star images must have signal/noise comparable to the final source images to be calibrated. A dark frame (cold shutter) is useful for instrumental calibration purposes but is not required for reduction of differential (chopped) data if the image contains blank sky.

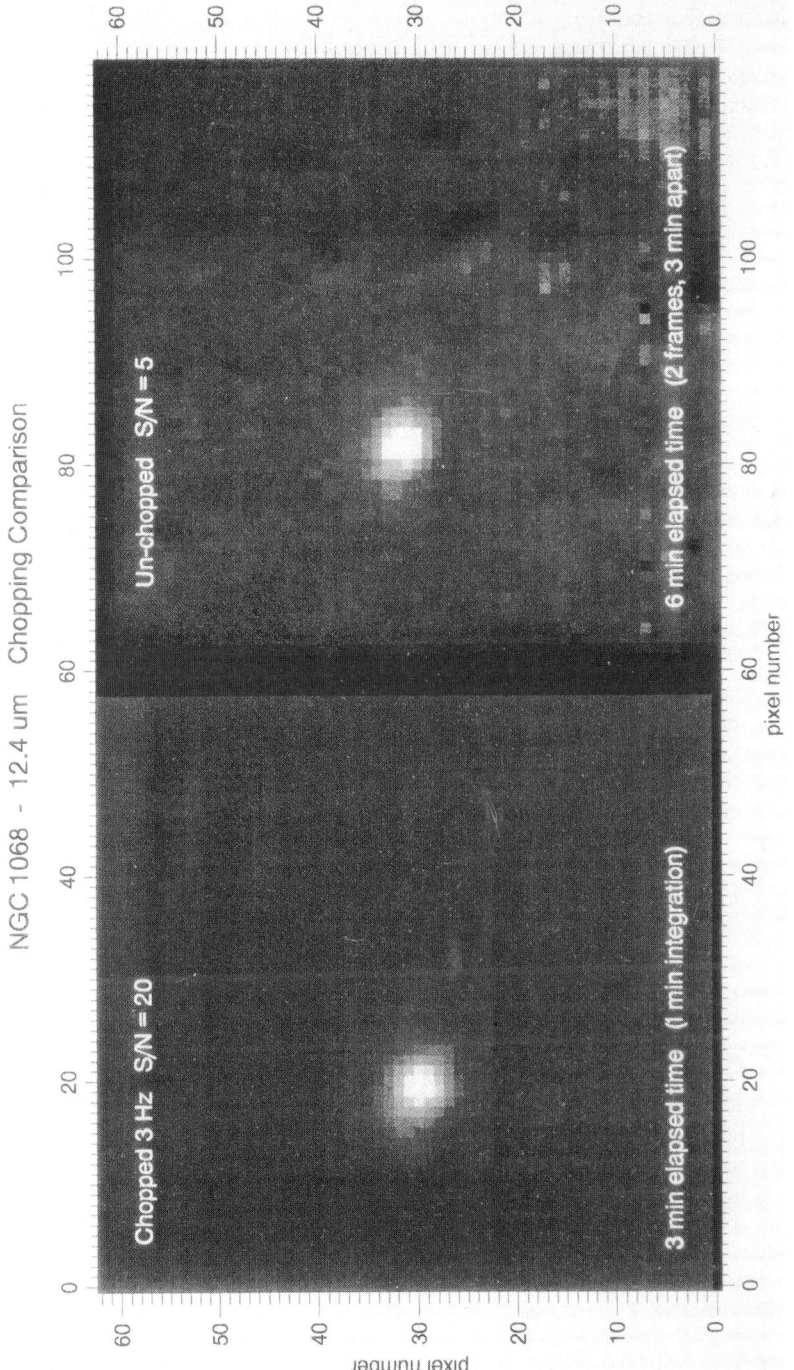

Fig. 3 (Left): A large-scale 20.0 micron image of the Orion BN/KL infrared source complex (Gezari and Backman 1995) assembled from 23 individual images (1 min integrations) covering an area of about 40 x 40 arcsec, showing the borders of the individual array image fields co-added for the final mosaic. Right: Final 20.0 micron image of BN/KL produced with our MOSAIC software package (Varosi and Gezari 1992). The brightness scale displayed is about 1 - 170 Jy arcsec^{-2}.

Standard star calibration fluxes have not yet been adopted for all of our mid-infrared (5 - 30 micron) filter wavelengths. The calibration star flux can be extrapolated from observational photometric data for that star found in the list of Bright Infrared Standard Stars compiled at the NASA/Infrared Telescope Facility, or from observations listed in the Catalog of Infrared Observations (Gezari et al. 1993). Since mid-infrared standard star data are generally expressed in relative units (magnitudes), flux densities corresponding to 0.0 mag have to be extrapolated to our filter wavelengths from the mid-infrared absolute flux calibration established by Rieke, Lebofsky and Low (1985), i.e.: at 10.6 micron, 0.0 mag = 36.0 + 1.2 Janskys; at 21.0 micron, 0.0 mag = 9.4 + 0.5 Janskys. For additional discussions of mid-infrared calibration standards, see Gillett, Merrill and Stein (1971), Gehrz and Woolf (1971), Tokunaga (1984), and Rieke, Lebofsky and Low (1985).

8. CONCLUSION

One of the most frequently asked questions is whether chopping is necessary with mid-infrared array detectors. Sky subtraction and flatfielding operations seem to break down on time scales of roughly one minute when making staring (unchopped) observations under average weather conditions. Of course, the ability to observe without chopping would provide significant practical advantages. High speed secondary mirror chopper mechanisms would be not be needed (a concern in the design of larger telescopes) and image quality and astrometric accuracy would not be limited by chopper positional stability.

If some part of an image is known to contain blank sky, and an accurate gain matrix exists for that image, the two can be used to create a synthetic sky frame for background subtraction without chopping. Sky background subtraction can also done by "dithering" (shifting the position of the source on the array in two images), then subtracting the shifted images, and deconvolving the source structure (a process similar akin to nodding). In practice, sky subtraction using these alternate methods can be rather tedious, and does not treat the high frequency noise components.

Nodding the telescope (moving alternately between two positions in the sky) can be effective for sky subtraction but has some limitations since nodding is slow, and may not allow for sky subtraction to minimize 1/f sky background variations. Low frequency changes in the sky background are generally largest, and this 1/f behavior of the detected sky flux level means that sky noise components below the chopping frequency can be significantly reduced. At some point chopping faster, at the high cost of mechanical complexity, stability, heat dissipation, vibration, and image quality, will not result in significant noise reduction.

At very good, dry observatory sites the 1/f noise component can become negligible at low frequencies, and 1 Hz chopping seems to be adequate for minimizing noise associated with atmospheric emission variations. However, at sites with poor atmospheric conditions, faster (ten Hz) chopping could provide a significant advantage. Chopping against blank sky provides real-time sky background subtraction and data for gain matrix correction at a reasonable cost in observing efficiency. Depending on weather conditions, chopping may not always be required for routine high-background array observations. But chopping can always improve the limiting sensitivity achievable for imaging faint sources.

We are grateful to Mary Hewitt of Hughes/Santa Barbara Research Center (SBRC) for her expert guidance during the development of the array camera system. We acknowledge stimulating discussions with Dick Joyce, Harvey Moseley and Frank Varosi. We thank Michael Hauser at NASA/Goddard for his on-going support of this program. This research is funded by NASA/OSSA (RTOP 188-44-23-08).

REFERENCES

Gehrz, R. D. and Woolf, N. J. 1971 Ap J 165, 185
Gezari, D. Y. 1994 Proposal for Directors Discretionary Fund Support, NASA/Goddard Space Flight Center, (preprint)
Gezari, D. Y. and Backman, D. E. 1995 (submitted to Ap.J)
Gezari, D. Y., Folz, W. C., Woods, L. A., Wooldridge, J. and Varosi, F. 1992 Proc. SPIE, 973, 287
Gezari, D. Y., Schmitz, M., Pitts, P. and Mead, J. L. 1993 Catalog of Infrared Observations, NASA RP-1294
Gillett, F. C., Merrill, K. M. and Stein, W. A. 1971 Ap J 164, 83
Hoffman, A. W. 1987 Proc. Hilo Detector Workshop, Univ. of Hawaii, p. 29
Hoffmann, W., Fazio, G. G., Gezari, D. Y., Lamb, J., Shu, P. and McCreight, C. 1987 Proc. Hilo Detector Workshop, G. Wynn-Williams and E. E. Becklin, eds., Univ. of Hawaii
McCaughrean, M. J. 1988 Ph.D. Thesis, University of Edinburgh
Rieke, G. H., Lebofsky M. J. and Low, F. J. 1985 AJ 90, 900
Tokunaga, A. T. 1984 AJ 89, 172
Varosi, F. and Gezari, D. Y. 1992 Astronomical Data Analysis Software and Systems. II., R. J. Hanisch, R. J. V. Brissenders and J. Barnes, eds., ASP Conf. Series 52, 393

DISCUSSION

D'ODORICO: Whether chopping is needed for infrared arrays at large telescopes and if so, at which frequency is a relevant question for the engineering of the secondary mirror units of these telescopes. Do you know at which chopping frequency it is foreseen to operate the mid-infrared camera at the Keck telescope?

McLEAN: The mid-infrared camera and the long wavelength spectrometer for Keck have not yet been commissioned. Nevertheless, the f/25 infrared secondary is a chopping secondary and it has been used with the near infrared camera. I do not know what frequency will be used ultimately at 10-m, but the secondary is capable of fast chopping.

FINGER: Measurements in the mid-infrared consistently show that noise is decreasing with increasing chopping frequency.

GEZARI: It is not clear that going to chopping frequencies faster than about 1 Hz results in an improvement in noise associated with the sky emission. Longer term transparency variations seem to dominate, suggesting that fast nodding may be a viable alternative to chopping in large telescopes. But the question is open and more work needs to be done on this.

MOSELEY: What chopping frequency was required to reduce noise to background limit?

GEZARI: In La Silla at the 2.2-m telescope (IRAC1) and the 3.6-m telescope (TIMMI) noise measurements have been performed in the L and N bands. The noise is decreasing with increasing chopping frequency. Depending on the atmospheric conditions background limited performance is achieved at frequencies between a few Hz and 10 Hz.

AN INFRARED CAMERA BASED ON A LARGE PtSi ARRAY

I. S. Glass, K. Sekiguchi and

South African Astronomical Observatory

Y. Nakada

Kiso Observatory, Inst. of Astronomy, Univ. of Tokyo

ABSTRACT: A large-format PtSi array (effectively 1k x 0.5k pixels) has been incorporated into an infrared camera intended for survey work using the 0.75-m telescope at Sutherland. The array is very uniform and almost free of cosmetic defects. The camera, its electronics and the operational procedures that we use are described, together with the methods of data reduction. The limiting magnitudes that can be observed in the JHK broad bands are given.

1. INTRODUCTION

This instrument is a joint project between the South African Astronomical Observatory, the Institute of Astronomy of the University of Tokyo and the National Astronomical Observatory of Japan. It has provision for up to six filters, normally including the J(1.25), H(1.65), K(2.2) and K'(2.15) bands. The K' filter is one which omits the long-wave part of the regular K band, i.e. that which contributes most to the thermal background arising from the telescope and sky.

Because of its large format but low quantum efficiency, the camera is expected to be employed mainly for survey projects. For example, it is being used to survey a number of fields close to the galactic center ($A_v \sim 30$ mag) to verify the apparent build-up of luminous stars towards the center noted by Catchpole, Whitelock and Glass (1990). This requires exposures in two or three colors to estimate the interstellar reddening along the line of sight. While at K the extinction will be about 2.5 mags, at J it is about 7.5 mags and only foreground objects are visible.

A further project concerning the galactic center is to monitor fields at a single wavelength reasonably frequently to pick out variable stars such as miras which are expected to be particularly numerous. The K band is again expected to be the most useful for this purpose in spite of the high foreground extinction.

The same techniques will be applied to survey certain fields in the Magellanic Clouds for luminous AGB stars and variables, though these appear to be close to the limit achievable with the 0.75-m telescope.

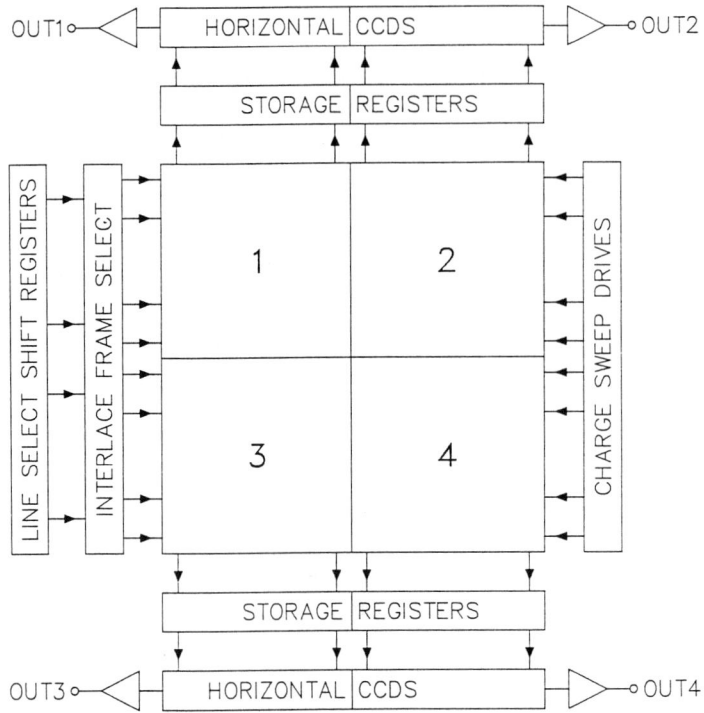

Fig. 1. Functional diagram of the chip. It has four independent sectors.

The camera is also being used to examine the late-type stellar content of globular clusters. Here again, the large format of the detector outweighs its low quantum efficiency.

In addition, the Comet Shoemaker-Levy impact with Jupiter during July 1994 turned out to be an ideal subject for the PtSi camera. Some of the observations were made in late afternoon more than an hour before sunset.

2. CHARACTERISTICS OF THE CHIP

The chip is from the family described by Kimata et al. (1992). It is of monolithic construction, using proven silicon techniques, so that it has few defective pixels. The detectors are PtSi Schottky-barrier type. The device has four independent sectors (see Fig. 1), each of which may be regarded as a separate 520 x 260-pixel detector. There are no dead spaces between the sectors. The readout commences simultaneously in the four outer corners of the chip and progresses line by line until the center is reached. In each quarter, the charges from pairs of lines are switched one pair at a time into four-phase vertical CCDs which "sweep" them into storage registers. Because only one pair is being swept down at a time, the operation can be carried out many times to overcome any charge transfer inefficiency effects due to the narrowness of the channels. Following the charge sweep operation, the storage registers are emptied into conventional four-phase horizontal CCDs for readout via floating-gate amplifiers.

The pixel spacing is 17 microns in each direction, with 53% fill factor. Because the lines are read out in pairs, the pixel size is effectively 17 microns in the horizontal direction by 34 in the vertical. The quantum efficiency of the device varies from about 6% at J to about 2% at K. About 1% drop in response (non-linearity) is noticed when the wells contain about 600,000 electrons.

The problem mentioned by Glass et al. (1994), in an earlier report on this device, namely that the background varied slightly from line to line, has been eliminated by a change in the waveforms applied. In effect, the vertical CCDs are now operated separately from the horizontal ones (they were previously interleaved). The unevenness arose due to very slight irregularities in the waveform timing, induced by the assynchronous nature of the Transputer link which transmits the data after each A/D conversion.

3. CONTROLLER

The chip is driven by one of the SAAO/RAL Transputer-based controllers described by Glass et al. (1994). This consists of a number of circuit cards in a Eurocrate. The clock waveforms are controlled directly from software and their timing is governed by the Transputer which can output words by DMA (direct memory access) every 200 ns. D/A converters on two "clock" cards generate the actual pulses to be applied as well as a number of DC supplies. The outputs from the chip first go to differential preamplifiers mounted on the dewar and then to four "video cards" which contain the 16-bit A/D converters. The digitized data are sent in Transputer link format (eight bytes or four 16-bit words at a time) over optical fibers to the data acquisition computer from which the observer works and where the data are stored. Other modules in the controller are used to operate the filter wheel and to monitor the temperature of the chip. The Transputer is programmed in OCCAM and assembler and the program is stored on an EPROM.

4. DATA ACQUISITION COMPUTER

The observer's computer acts as a terminal to the controller. Simple one-character messages sent to the controller are used to initiate readouts of various kinds, change filters, measure temperatures etc. The composition of an exposure sequence in terms of pre-wipes and actual readout is controlled from the data acquisition computer. The exposure sequence proper is a pre-wipe followed by an exposure time measured in the Transputer and, finally, the readout with A/D conversion. The data are sorted into their correct order in a 1040 x 520 matrix as they arrive. After the readout they are recorded with their FITS header onto disk. They are then scaled and displayed according to their maximum and minimum values and a histogram of pixel values is also placed on the screen.

The processor is a 486/50MHz PC; there are 8 MB of memory and 340 MB of disk space. Pictures are displayed on a super-VGA monitor in 1028 x 768 256-color mode. An auxiliary black and white monitor is used for alphanumeric data. A set of programs similar to PC-VISTA

(Treffers and Richmond 1989) enable some on-line reduction to be performed. In particular, files can be displayed and images analyzed for focussing and crude photometry. The data are saved on a DAT tape at the end of the night. The PC uses Gnu C for 32-bit addressing.

A special data acquisition program was written for the Comet Shoemaker-Levy/Jupiter event to allow continuous exposures with readouts as frequently as every ten seconds. Because there were of necessity no pre-exposure wipes, a (mainly) bias background frame had to be subtracted from each exposure to reveal the planet. Only one quarter of each frame was recorded on disk so that up to three hrs coverage could be obtained without interruption.

5. OPTICAL MATCHING

The camera is used on the 0.75-m telescope at Sutherland at an effective focal ratio of f/16 and a scale of 17 arcsec per mm. The diffraction limit of this telescope at 2.2 microns is 0″.7. There is some lateral spherical aberration due to the displacement of the Cassegrain focus below

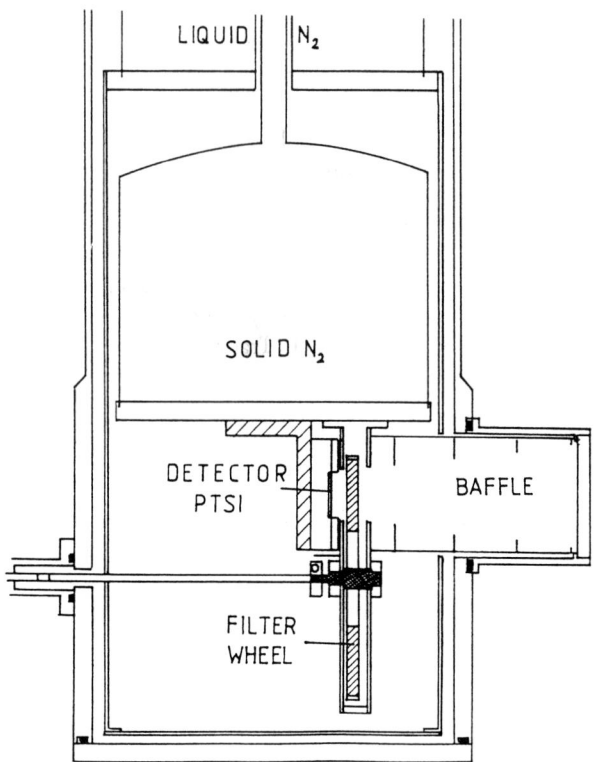

Fig. 2. Cross-section of lower part of cryostat. The filter wheel has space for six filters of 30 mm diameter, but one position is normally blanked off for use as a shutter.

AN INFRARED CAMERA BASED ON A LARGE PTSI ARRAY

its design position. The best achieved FWHM of the image is 0".8 to 0".9 in very good seeing. The scale of the image amounts to 3.3 pixels per arcsec. Such oversampling was felt to be necessary because of the 53% fill factor of the chip.

There is no field imaging, and hence no position in which to place a cold "Lyot" stop. This was at least in part due to a lack of design information for a lens having the necessary resolution and wavelength range. With the present set of cold baffles, the detector "sees" around 16 times more solid angle of telescope and sky than desirable. This is not a very serious problem at J and H, but it restricts K exposures to little more than five minutes during the warm part of the year. A longer cold baffle, giving a reduction of two to three times in the background, is to be installed soon. The K signal-to-noise ratio should then improve by a few tenths of a magnitude.

6. TEMPERATURE

The detector needs to be maintained at $60° \pm 2$ K, representing a compromise between excessive dark current and charge transfer inefficiency. Using a simple micrometer valve on a large vacuum pump, the temperature stays within $0.5°$ K during the night.

7. SIGNAL ANALYSIS

There is no mechanical shutter since it would have to be cold as well as fast in order to be useful. Instead, a "dummy" readout (taking 1.6 s) defines the start of an exposure. The real readout (6.25 s) defines the end of the exposure. This approach requires that the data from each frame must be corrected to allow for the change of exposure time with position.

The signal from the detector results from (a) photons truly from the field (b) photons from sky and telescope surrounding telescope exit pupil (c) dark current ($\sim$ seven electrons/sec at $60°$ K) (d) Spurious radiation from the amplifiers (e) "Trapped" electrons depending on the recent "history" of the chip.

The amplifier radiation is the most serious of the spurious components. The amplifiers are situated in the corners of the chip and, if driven in the manner recommended for normal high-speed readout, they would saturate the scene immediately. A number of methods are used to reduce this problem.

a) The output amplifiers are turned on only when the readout commences.
b) The readout time is minimized - at 6.25 s its lower limit is mainly determined by the speed of the Transputer link and its protocol.
c) The load resistors of the output emitter followers, across which the signals are measured, are kept as high as practicable to reduce the current in the device.
d) The bias of the first stage of the output amplifiers (each amplifier is a three-transistor device) is kept as low as possible consistent with reliable operation.

With these precautions, the effective read noise towards the center of the chip is equivalent to about 30 electrons. The first pixels to be read out are those nearest the output amplifiers and the spurious charge does not have enough time to build up. The worst remaining effects of amplifier emission are visible on the outer ends of the last lines to be read and affect the baseline noise level, causing difficulties with the DoPHOT (Schecter, Mateo and Saha 1993) reduction procedure which normally assumes a uniform noise level.

The chip is given ten dummy readouts before an exposure to clear it of charge from the previous exposure. However, it is found that small numbers of pixels in the extreme corners are not cleared consistently. It is believed that this problem is caused by the saturation of parts of the chip close to the output amplifiers during the preceding readouts. (Note added later: Dr M. Ueno has made some suggestions about changes to the "dummy" readout waveforms which should help to solve this problem.)

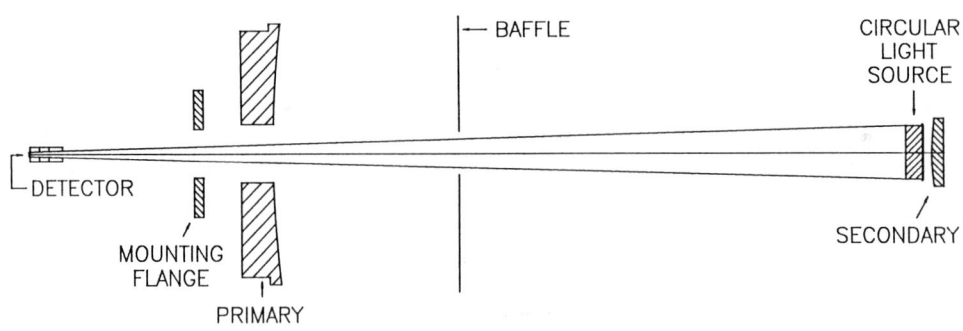

Fig. 3. Flatfield arrangement on the 0.75-m telescope.

8. FLATFIELDS

The over-wide pupil denies the use of the sky or screens within the dome for flatfielding purposes. The solution that we have adopted is to place a flat circular diffuse light source in front of the secondary (close to the position of, and having the same apparent size as the exit pupil). A baffle along the tube cuts of light from outside the exit pupil. The response of the detector is measured with the light on and off to determine the difference which is then taken as representative of light from the telescope pupil.

9. REDUCTION PROCEDURE

The basic reduction procedure is to: (a) Correct the fields for uneven exposure time, (b) Divide by the appropriate flatfield, (c) Subtract the appropriate sky frame, (d) Measure the images, for example by DoPHOT (Schecter et al. 1993) and (e) Standardize.

9.1 Median Averaging of Frames

Because of the high background at K and to some extent at the other wavelengths, it is generally necessary to subtract sky frames taken under the same conditions as the object exposures. This not only removes the extraneous background, but also that due to the dark current and the output amplifiers. Unfortunately, the survey fields are large and often crowded, so that it is usually impossible to find one devoid of stellar objects nearby. In addition, the time taken for a survey is doubled. However, it is possible to construct sky frames from a number of survey frames by median averaging (pixel-by-pixel). Usually, at least five frames are necessary to avoid chance coincidences of stellar positions.

The formation of median averaged frames is not completely straightforward. The sky background in the near-infrared is primarily due to OH⁻ emissions originating high in the atmosphere and is variable by several percent on time scales of a few minutes (Ramsay, Mountain and Geballe 1992). At wavelengths longer than about 2.3 microns, the thermal background from the atmosphere and telescope become dominant. These also show slow fluctuations on time scales comparable to those used for exposures.

To remove the effect of the fluctuations we treat them at present as if they are completely uniform over the entire frame. A group of exposures is first averaged pixel-by-pixel in the ordinary way. The average frame is subtracted in turn from each individual on a histogram is made of each difference frame and the commonest pixel value is used to calculate an offset to correct the original frames to a standard background level. The median average is then obtained and used to subtract the background from each corrected frame, which is then cleaned to remove bad pixels etc. and passed to DoPHOT for further analysis.

It is intended to use a procedure similar to that described by Balona (1994) for obtaining accurate photometry and surveying for variable stars.

9.2 Accuracy Achieved

Calculations of the number of photons that should reach the detector from a star of a given magnitude agree reasonably well with the actual numbers measured. From the observed backgrounds due to thermal radiation and the read-out amplifiers it is immediately obvious that they control the limiting magnitudes that can be achieved.

So far, very limited quantities of data have been analyzed in detail. The same quarter of two five-minute galactic center K pictures taken at different times with fairly good seeing conditions (~ 1.5 s FWHM) have been processed using DoPHOT and compared. It should be noted, however, that our implementation of DoPHOT has not yet been fully optimized for this task. Four stars brighter than $K = 9$ have photometry differing by $<= 1\%$. By $K \sim 11$, the magnitude differences are around 10%; the exact figure is uncertain as the data are somewhat contaminated by misidentifications of objects between the two frames. Theoretically, under similar seeing conditions, an error of about 0.03 mag would be expected at $K = 11$. The larger apparent error may be due to the fact that the flatfields were each derived from single exposures and the median sky frames from only five exposures each, including one near the galactic center which can be expected to be excessively crowded.

The limiting magnitudes in the J and H bands for similar errors are controlled by the noise due to amplifier radiation. Calculations indicate that at J we should reach 14.7 and at H 13.5 at $\pm$ 10% error.

10. FUTURE WORK

The limiting magnitude at K will be made a few tenths of a magnitude fainter by the addition of a longer cold baffle to reduce the extraneous radiation from the telescope and it is expected that this will be installed during September 1994. The possibility of introducing relay optics and a cold pupil stop are also being investigated. General improvements will be obtained by taking more flatfield frames during each observing run. It appears that the flatfield characteristics of the array are quite stable.

Additional improvement may be possible in the suppression of amplifier radiation and experiments are planned in this area. The effects of changes in the clocking waveforms remain to be investigated.

ACKNOWLEDGMENTS

We wish to thank Mr D. B. Carter for his work on the electronics of the camera; Dr L. A. Balona and Dr J. W. Menzies for their patient help with programming queries; Dr M. Ueno, University of Tokyo, for information concerning the properties of the chip; Mr W. Pearson, Mr D. Weir and other members of the SAAO technical staff for their help at various stages of this project.

REFERENCES

Balona, L. A. 1995 in IAU Symposium No. 167,New Developments in Array Technology and Applications, A. G. D. Philip, K. A. Janes and A. R. Upgren, eds., Kluwer Academic Pub. Dordrecht, p. 187
Catchpole R. M., Whitelock P. A. and Glass I. S. 1990 MNRAS 247, 479
Glass, I. S., Carter, D. B., Sekiguchi, K. and Nakada, Y. 1994 in Infrared Astronomy with Arrays: The Next Generation, I. S. McLean, eds., Kluwer Academic Pub., Dordrecht, p. 285
Kimata, M., Yutani, N., Tsubouchi, N. and Seto, T. 1992 SPIE 1762, Infrared Technology XVIII
Ramsay S. K., Mountain C. M. and Geballe T. R. 1992 MNRAS 259, 751
Schecter, P. L., Mateo, M. and Saha, A. 1993 PASP 105, 1342
Treffers, R. R. and Richmond, M. W. 1989 PASP 101, 725

DISCUSSION

FAZIO: Are the 1040 x 1040 PtSi arrays commercially available?

GLASS: I believe they are developmental types released only to the Japanese astronomers.

D'ODORICO: In view of the dramatic development in the size and capabilities of InSb and HgCdTe array, do you think that PtSi arrays will still be competitive in the future?

GLASS: If the QE of PtSi chips can be improved dramatically I think they will remain competitive, especially because of their high cosmetic quality and relative uniformity.

PtSi IR ARRAY IN MOSAIC CONFIGURATION

Munetaka Ueno

Department of Earth Science and Astronomy
The University of Tokyo

Fumiaki Tsumuraya and Yoshihiro Chikada

National Astronomical Observatory Japan

ABSTRACT: The rapid progress in the focal plane technology enables us to use large format infrared sensors, such as 256 x 256 InSb/HgCdTe and 1040 x 1040 PtSi arrays. Infrared two-dimensional sensors make possible not only the imaging observations but a deep detection limit. The development of a large format infrared array is one of the most important breakthroughs in observational astronomy.

We propose to build a mosaic infrared camera for the SUBARU 8-m telescope. The SUBARU telescope is designed to reach a diffraction limited image at infrared wavelengths with a wide field of view (six arcsec at the Cassegrain focus). The camera is designed to cover the entire field of view with PtSi infrared sensors and to employ a weighted shift-and-add operation and a real-time image processing. The efficiency of the mosaic infrared camera and power of the 8-m telescope have a strong potential to meet challenging problems. Most of the regions of the near infrared sky are not covered with enough sensitivity. It is essential to conduct infrared deep and wide surveys.

1. INTRODUCTION

It was only ten years ago that the first infrared camera was applied for astronomical observations (Forest et al. 1985). It was a very drastic change in infrared astronomy since the infrared array was the first imaging tool in the infrared region. The infrared array was a very large quantum jump in observational astronomy, similar to that of the photographic plate. This decade has been the epoch in which a number of infrared cameras have been developed for astronomical uses and the format size of infrared array has been getting larger and larger. We have been developing PtSi infrared arrays under collaboration with the semiconductor research and development laboratory, Mitsubishi Electric Co. in Japan (Ueno et al. 1992), and we are building a prototype model of the mosaic infrared camera.

2. THE SUBARU TELESCOPE

The SUBARU Japanese National Large Telescope project is being promoted by the National Astronomical Observatory, Japan. The SUBARU project aims to construct an 8-m infrared/optical telescope with excellent image quality atop Mauna Kea in Hawaii. The SUBARU telescope employs an active support system of the primary mirror to maintain a precise optical surface and to reduce gravitational distortion, a dome flushing system to reduce atmospheric turbulence inside the dome and precise driving of the telescope with direct driving motors. Using these new technologies, the SUBARU telescope is designed to realize 0.2-0.3 arcsec near infrared seeing and diffraction-limited imaging with an adaptive optics/speckle interferometer. The field of view of the primary focus and the Cassegrain focus of the SUBARU telescope are 30 arcmin and six arcmin respectively. The prime focus of the SUBARU telescope contains optical and infrared corrector lenses.

3. THE PtSi ARRAYS

The PtSi infrared array detectors have been developed under collaboration with Mitsubishi Electric Co. (Kimata et al. 1987, Ueno et al. 1992). The current performance and specifications are tabulated in the following table.

TABLE 1

Parameters for Infrared Arrays

512 x 512 PtSi Infrared CSD

Number of pixels	512 x 512
Pixel size	20 micron x 26 micron
Read-out scheme	CSD (Charge Swept Device)
Quantum efficiency	6% @ 2 micron
Filling factor	71%
Read noise	< 30 e^-
Dark current	< 3 e^-/sec @ 60 K
Full well capacity	1.5×10^6 e^-
Uniformity	> 98%
Linearity (gain error)	< 1%
Defective pixels	< 0.01%

1040 x 1040 PtSi Infrared CSD; developed in 1992. 1040 PtSi array has four output amplifiers and is manufactured by the Mitsubishi Electric Co.

Number of pixels	1040 x 1040
Pixel size	17 micron x 17 micron
Read-out scheme	CSD
Quantum efficiency	6% @ 2 micron
Filling factor	51%
Read noise	< 30 electrons
Dark current	< 1 electron/sec @60K
Full well capacity	1.5×10^6 electrons
Uniformity	> 98%
Linearity (gain error)	< 1%
Defective pixels	< 0.01%

The quantum efficiency of the PtSi array is about ten times lower than those of InSb and HgCdTe detectors. However the PtSi array has excellent uniformity, large format size, good stability, low read noise and an inexpensive cost. If the format size of the PtSi array is ten times larger than other's, the total efficiency for a wide field survey with the PtSi array must be quite comparable to that with other systems. These features of large format size, mass productivity and cheap costs suggest that we should build a mosaic infrared camera with PtSi arrays.

In 1991, we conducted a galactic center survey at H and K bands using the 512 x 512 PtSi camera and a Newtonian telescope with 25-cm aperture (Ueno et al. 1993, Fig. 1). In our survey, 12 square degrees of the galactic center region are covered with 35 frames of H and K band images. The stable response of flatfielding makes possible complete background limited observations and wide field mosaic image.

Recent progress in PtSi technology enables us to improve quantum efficiency. In addition to optimizations of the thickness of the PtSi layer and the cavity structure of the detector, the strong electric field of the contact layer of the PtSi is very effective in improving the quantum efficiency of PtSi arrays (Konuma et al. 1995). According to the new theory of the PtSi detector, the quantum efficiency is estimated to be more than 10% at two microns. The PtSi array uses CCD architecture for the readout system and the CCD has enough capability to get low readout noise under rapid operation. A low-noise HDTV CCD achieves 11 electron/pixel read noise for 70 Mpixel/sec readout speed, that is, two electron/pixel read noise for 512 x 512 pixel/100 m sec. The possible improvement in the quantum efficiency and readout noise shows the future capability of PtSi arrays. Improved PtSi arrays seem very suitable for low background uses such as spectroscopic applications and use as a speckle interferometer.

4. BEYOND THE CONVENTIONAL SPECKLE TECHNIQUE

Two approaches are considered to bring about high spatial resolution imaging, adaptive optics and use of an array as a speckle interferometer. Adaptive optics employs mechanical corrections using deformable mirrors to maintain the wave front, disturbed by atmospheric turbulence, while a speckle interferometer uses computational correction of disturbed images. The adaptive optics technique is a real-time operation and is very effective for spectroscopic applications as well as high spatial resolution imaging. The speckle interferometer is generally an off-line operation and is not suitable for spectroscopic observations and visible imaging because the read noise of the detector has a harmful effect on the sensitivity of the observations. Adaptive optics works only within an isoplaneatic angle since the mechanical compensator of the wave front is very difficult to expand beyond the isoplaneatic angle while the speckle interferometer can correct the entire field of view.

The conventional speckle interferometer has faced several barriers in applications for practical observations. The conventional technique uses independent observations of objects and calibration stars lying beyond the isoplaneatic angle using a self-calibration technique because the field of view of the camera system must be small at visible bands and also the isoplaneatic angle is typically ten arcsec in the visible region. In addition to this problem, the strong effect of the read noise reduces the visible sensitivity because a typical speckle life time is less than 10 m sec.

On the other hand, the isoplaneatic angle at two microns is typically one arcmin and its speckle life time is 100 m sec. We can get a sufficient number of guide stars in the near infrared region, that is, at least one guide star within an isoplaneatic angle. Freedman's parameter (D/R0) for the SUBARU telescope at two microns is five to ten. Under such a small Freedman's parameter condition, a tip-and-tilt operation as well as shift-and-add operation are usable for the high resolution imaging. The shift-and-add operation can align the peak of images and drive 5 to 10% flux into the diffraction angle.

The weighted shift-and-add operation seems a much more efficient technique under these conditions. The weighted shift-and-add operation aligns several peaks, for example first to seventh largest peaks, in a frame using the calibration stars in the same frame, then integrates the frames. More than 60% of the flux is estimated to fall within the diffraction angle after the weighted shift-and-add operations (Tsumuraya 1994). The weighted shift-and-add operation seems very suitable for a large aperture telescope with a good seeing condition.

5. DEVELOPMENT OF PtSi ARRAYS FOR THE SUBARU TELESCOPE

We have started developing new PtSi arrays under collaboration with Mitsubishi Electric Co. and Nikon K. K. in Japan. The goal of the new array is shown in the table below.

TABLE 2

Goal of New Array

Number of pixels	1024 x 1024
Output amplifiers	4
Pixel size	TBD
Read-out scheme	CSD
Quantum efficiency	> 10 - 15% @ 2 micron
Filling factor	> 75%
Read noise	< 3 e$^-$/pixel for 3 M pixel/sec
Full well capacity	> 2 x 10^5 e$^-$
Device architecture	monolithic structure

The new PtSi array is a very powerful detector for not only the mosaic applications but also for usual observations. We plan to develop a very cheap infrared camera system and make collaborations with small and middle size telescopes. In the infrared region, we have a chance to conduct good science even with a small telescope.

REFERENCES

Forrest, W., J., Moneti, A., Woodward, C. E., Pipher, J. L. and Hoffman, A. 1985 PASP 97, 183

Kimata, M., Denda, M., Yutani, N., Iwade, S., Tsubouchi, N. 1987 IEEE Journal of Solid-State Circuits, sc-22, No. 6

Konuma, K., Asano, Y. and Hirose, K. 1995 Physical Review, in press

Tsumuraya, F. 1994 Ph. D. thesis
Ueno, M., Ito, M., Kasaba, Y. and Sato, S. 1992 Proc. of SPIE, conf. vol 1762, p. 423
Ueno, M., Ichikawa, T., Sato, S., Kasaba, Y. and Ito, M. 1993 AIP Conf. Proc. 278, 64

DISCUSSION

FINGER: What is the limit of QE one can achieve with PtSi at the K-band and how can it be achieved?

UENO: According to the new theory in the Schottky-barrier mechanism, the QE of the SBD detector is determined not only by the work function of a metal material and thickness of the metal layer but also by the electric field behind the Schottky-barrier contact layer. A strong electric field improves the QE of SBD. This technique requires a small modification in the array structure.

IWERT: Large thinned CCDs are so far not produced by Japanese companies. Also the Subaru Project does not intend to enforce the contacts to Japanese industry to initiate these developments. However we see on your talk developments by Mitsubish and possibly Nikon for the development of large PtSi arrays. Where are the differences, and why does the Subaru Project not favor the development of CCDs from Japanese industry?

UENO: The Subaru Project still makes much effort to produce a large format CCD in Japan. However most of the CCDs are produced in commercial divisions. A typical commercial division of Japanese industry has no interest about scientific uses. The PtSi arrays are developed in R & D divisions and their engineers are strongly interested in the scientific instruments.

MOSELEY: Could you describe the optical system to be used with your mosaic array?

UENO: I plan to use 1:1 optics for the Cassegrain focus. (During lunch time discussion, Dr. Finger of ESO gave me a very helpful idea for the optics.) The Subaru telescope employs a primary focus corrector for 1 - 2.5 μm. We can put our mosaic camera simply at the prime focus if we would like to make wide field imagings. The pixel size of the PtSi array is designed to fit a diffraction limit scale at the F/13 Cassegrain focus and fit a seeing size at the F/2 prime focus.

JORDEN: For what reasons do you hope to achieve a readout noise of two to three e^- at 3 megapixels/sec? Is this for a single, or multiple output device?

UENO: The PtSi array is designed to have four output amplifiers. The read noise of a CCD detector is determined by the node capacitance and gate noise density of the output source follower amplifier. The PtSi array will have 10 fF of node capacitance and very low noise MOS FET. These technologies are well realized in the HDTV CCDs in Japan.

CCD PHOTOMETRY - PRESENT AND FUTURE

Alistair R. Walker

Cerro Tololo Inter-American Observatory

ABSTRACT: I will briefly treat some of the developments relating to CCD photometry which have taken place over the last two to three years, and speculate on those which can be anticipated to take place in the near future. CCDs have been in widespread use as astronomical detectors for slightly more than a decade. For the majority of visual-light projects they have displaced the photographic plate and the photomultiplier as the detector-of-choice, except in applications requiring large area coverage or high time resolution, but even here the development of arrays and the use of more sophisticated electronics has permitted encroachment into these domains. The use of large, low-noise detectors for CCD photometry places stringent demands on telescope optics, instruments, controllers, calibration procedures, data reduction methods and data storage, and forces a holistic approach to managing the data flow from detector to final storage medium.

For an earlier review on this field see Walker (1993); other papers presented at IAU Colloquium No. 136 and IAU Symposium 167 (these proceedings) should be consulted for different topics, alternative viewpoints, and more in-depth discussions of specific subjects.

1. THE PRESENT

Changes in the last two to three years include:

1.1 More General Availability of High-Performance CCDs

For an over-all summary see P. Jorden (1995). Concentrating on devices used for direct imaging, SITe (previously Tektronix) are delivering, albeit at a rather slow rate, thinned CCDs of high cosmetic quality with format 2048 x 2048 and 24 micron pixels. A significant fraction of these CCDs have four working amplifiers, and perhaps their only problem is that the CCD surface has a bow such that the center is some 250 microns higher than the corners. This is significant when these CCDs are used in fast beams. For a dedicated dewar the window could be replaced by a suitable "field-deflattener" lens, and this approach has been successful when using these CCDs in very fast spectrograph cameras. Together with their smaller brethren (1024 x 1024, 512 x 512), the SITe CCDs are near-ideal for direct imaging. They do have quantum efficiency response that falls off below 4000 Å, but are much improved in this respect when compared to Tektronix CCDs produced four to five years ago. Typical QE figures are 25% at 3000 Å and 60% at 4000 Å. CCDs with the same number of pixels have been obtained via foundry runs at Loral (previously Ford Aerospace), but the number of thinned versions that have been produced is extremely small. The same is true of other producers and it is clear that without Tektronix CCDs photometrists would be very much disadvantaged.

Another important development is that manufacturers are now usually fabricating CCDs with MPP (multi-pin phase) implants. When a CCD is clocked in MPP mode the dark rate at a given temperature is lower than in non-MPP mode, at the expense of lowering full-well capacity. Although to achieve dark rate floors of a few e$^-$/hour it is still necessary to cool to typically -100° C, MPP CCDs achieve dark rates at temperatures -40° C to -60° C that are still low enough that dark noise is not the major noise source for exposures of up to a few minutes. At these temperatures Peltier coolers rather that liquid cryogens can be used, and several systems based on this technology are on the market. Using relatively inexpensive front-illuminated CCDs, they seem particularly suitable for those not requiring an expensive state-of-the art system and indeed there are many scientifically valuable programs (e.g. variable stars, supernovae searches) for which such a system is entirely adequate. It is perhaps unfortunate that the emphasis in the amateur realm, particularly in the USA, seems mostly aimed at producing pretty pictures.

1.2 Higher Performance CCD Controllers

A new generation of CCD controllers is appearing (R. Leach 1995 and poster papers, these proceedings). These offer many advantages. They are more reliable due to lower component count and the use of modern packaging techniques such as multi-layer boards and surface mounting of components. Performance has improved due to better op-amps, faster, more accurate, and self-calibrating Analog-to-Digital convertors (ADCs) and the utilization of fiber optics to aid isolation. Power consumption need only be a few watts. Typically, each unit contains four separate video paths to accommodate CCDs with multiple amplifiers, and include the large number of clock and bias levels needed to be able to accommodate the possible readout schemes. In general, these controllers incorporate powerful processors (DSP's, transputers) which can provide significant computing power in the controller itself. This can be used for pre-processing the data (essential for IR arrays) or, for instance, to control a flexible real-time display (Walker 1995). Making these controllers compact and reliable as well as delivering state-of-the-art noise performance is not a trivial task and, so as not to degrade the effectiveness of the hardware, the software tends to be both complex and extensive. This can have distinct advantages, for instance sophisticated software tools can aid development (eg CCD waveform generation tools), and can easily control DACs (Digital to Analog convertors) to provide programmable voltages which, apart from initially setting the bias and clock levels, can monitor these levels to make sure they are always correct. It is relatively simple to incorporate a DAC to provide controllable video gain, and so on. The user can take advantage of this flexibility in that it is possible to tailor the CCD data acquisition system to be optimal for a specific program. A corollary of this flexibility is that the detailed characteristics of such a CCD system tend to be less well-known. Photometry is difficult enough already without adding a poorly-characterized detector system to the list of the observer's woes, and a wiser approach would seem to be to identify a small number of options that are adequate for the great majority of programs, and characterize these modes of operation in detail.

1.3 Better Data Acquisition Software, More Powerful Computers

Over the past decade the improvement in computer processing power has more than kept pace with the increase in numbers of pixels of a typical CCD (Blecha 1995). Computer memory technology, together with disk and tape storage, has also advanced in similar fashion. The efficient processing and storage of data from detectors such as the SITe 2048's is, if not something to be dismissed as a minor problem, certainly soluble in terms of hardware that costs very much less than the detector itself. At the same time, the data acquisition process has become more efficient for the user, by provision of easily-learned user interfaces and the integration of observation and reduction processes. High band-width networks allow multiple observers to share, possibly remotely, observation and reduction tasks. It is feasible to ascertain the quality of the incoming data in near real-time by examining image focus, measuring signal/noise, doing aperture photometry of standard stars, etc. This ability has a real effect on the quality of data, for too long have many CCD photometrists tolerated data acquisition processes that have given them far less feed-back on their data than typically enjoyed by those doing photo-electric photometry.

1.4 HST

The resolution of better than 0.1 arcsec, as a result of the highly successful 1993 refurbishment mission, is resulting in deeper photometry than ever achieved from the ground, and is allowing work in crowded fields impossible to study effectively even with the best ground-based telescopes. However some care must be taken. HST images are poorly- or under-sampled, even on the high-resolution PC CCD where images with fwhm = 1.8 pixels are routinely achieved. the star-cosmic ray differentiation must be done with care (Windhorst et al. 1994). Since the Loral CCDs used on HST WFPC-2 are front-illuminated, there will be considerable intra-pixel response variations (Jorden et al. 1994). Thus for best results exposures should be split and the telescope jogged slightly between each individual exposure to randomize the position of any given star with respect to its distance form a pixel center. Given these precautions, and by applying standard reduction techniques, excellent photometry can be obtained by conventional aperture or psf photometry. Although the STScI provides absolute calibration figures for the CCD-filter combinations, a safer and, if properly done, an inherently more accurate technique is to perform the absolute calibration via ground-based observations.

This need not require a large telescope. Another aspect to be considered is that the WFPC broad-band filters are not particularly good reproductions of UBVRI. Ground-based comparisons for many of the filters appear in Harris et al. (1993), see also Harris et al. (1991) and Walker (1994), and from these data it is in principle possible to transform HST measurements to the standard system. An alternative procedure is to calibrate via ground-based observations using reproductions of the WFPC filters, and then derive the requisite astrophysical parameters via isochrones or by theoretical fluxes obtained by convolving model atmospheres with the filter passbands.

1.5 Standard Stars, and Photometric Calibrations

As far as broad-band UBVRI observers are concerned, the work by Landolt (1992) is of prime importance. He provides photometry for many stars in the (equatorial) Selected Areas and also in a number of convenient, CCD-sized fields. The latter generally consist of fields chosen to contain a blue star and a number of redder stars in order to provide stars of similar magnitude but with a wide range in color. The number of stars of magnitudes 10 - 15 with accurate photometry is extensive, and this work is of the utmost value, since it allows many more standard stars to be observed for a given integration plus read-time.

For standard star work the read-time overhead is often dominant when compared to integration time, particularly for large CCDs on moderate to large telescopes. It is has certainly been the case in the past that CCD imaging projects done on large telescopes have been inadequately calibrated due to a perceived large overhead of time required to do the calibration work. This is no longer the case. Many of the stars in the Landolt (1992) lists have rather few observations, and the extensive CCD observations that have been made of these fields can in principle be collated and improved magnitudes derived. Work of this nature is in progress (P. Stetson, private communication). Menzies (1992) has emphasized that the differing implementations of the Johnson UBV system is worrying, and differences can be large, especially for stars of extreme colors, and for (U-B) in general. We should also mention the useful bibliographic list of UBVRI sequences (Ritzmann 1992). An important minority of CCD photometry uses filters other than UBVRI. An important subset of observations are made using narrow-band filters to study emission-line sources such as supernovae remnants and H II regions. Hamuy et al. (1992, 1994) provide a new, accurate set of spectrophotometry for the southern secondary and tertiary standards.

Specialized systems such as Gunn, Washington, Strömgren, DDO, Geneva and Vilnius are used by a minority of CCD observers, in general to provide sharper discrimination of astrophysical quantities (e.g., temperature, metallicity, gravity) than is possible with the UBVRI system. The conference references mentioned above provide several papers devoted to work using these systems, an example of their usefulness can be gauged from the paper by Geisler et al. (1991) which provides the metal abundance calibration from Washington photometry, while examples of precision work on standard stars are (for DDO) provided by Cousins (1993) and (for Strömgren) by Kilkenny and Laing (1992).

1.6 Larger Telescopes, Improved Image Quality

The great majority of CCD imaging projects would benefit from improved image quality, with the few exceptions including surface photometry of large clusters or galaxies, programs where only bright, isolated stars are measured, and cases where the pixel size is such that the images are already under-sampled in moderate seeing. The problem is two-fold. Firstly, seeing monitor measurements show that many mountain-top observatories enjoy median seeing of typically ~0.7 arcsec and this is then often ruined by some combination of the telescope optics and the telescope environment. Examples such as the ESO NTT telescope at La Silla and the WIYN telescope at Kitt Peak show that telescopes designed with stringent imaging error-budgets can indeed achieve close to the site seeing, and this approach is being carried out even more carefully for the new generation of very large telescopes. The example of the CFHT shows that improvements to a previous generation of telescopes are indeed possible and can produce similar performance. Various papers in Stepp (1994) are a useful introduction to the literature, as well as describing in detail both the performance goals of the new generation of very large 8 to 10-m class telescopes and programs for improving already existing telescopes. Smaller (1 to 2-m) class telescopes are often used for CCD imaging programs and can also benefit from such upgrades. For these, replacement of optics, often made in an age where seeing was measured to one significant figure only, can be advantageous and need not be financially crippling.

2. THE NEAR FUTURE

2.1 Large Multi-Amplifier CCDs

There will be more general availability of large multi-amplifier CCDs. Manufacturers such as EEV, Reticon, SITe, and Thomson are actively pursuing the scientific market, while custom devices can be from procured from Loral, Orbit and some of the manufacturers just mentioned. The successful thinning of CCDs is still a black art which few have mastered, and for which the user pays a substantial premium; however gains of two to three in blue QE and 1.5 in the red are very substantial. Anti-reflection coating technology improvements will allow the option of optimizing for higher QE at the ends of this optical range, with only minor trade-offs elsewhere. Amplifier design improvements should allow lower read times without compromising read noise. A more general worry for the future is that most of the CCD manufacturers rely on outdated four to five inch diameter wafer fabrication lines for which it is becoming increasingly difficult to buy high-quality silicon. The size of the scientific CCD market alone precludes the substantial investment needed to set up a modern fabrication line using larger wafers.

2.2 Mosaics of CCDs

Mosaics of CCDs will provide fields of view comparable to those that have been obtainable using photographic plates. This will lead to a renaissance of wide-field imaging projects, of two general types. The first is the study of large structures, such as distant clusters of galaxies, and near-by galaxies. The second type is deep surveys for rare objects, such as high red-shift quasars, field-star studies in our galaxy and in the Magellanic Clouds. CCD controllers able to operate mosaics will be constructed from the new generation of extensible controllers which are just now appearing. The advance of computer technology will mean that the problems of processing and storing gigabytes of data per night will be surmountable. Projects such as MACHO, which uses a mosaic of four 2048 x 2048 CCDs and pipelines the data-taking process through to the photometry of the stars observed, are already showing the feasibility of wide-field imaging projects. Projects using much larger arrays (e.g., the Sloan Digital Sky Survey) are in the planning or construction phase. These large CCDs place critical demands on the quality of telescope optics, and telescopes optimized for wide-field imaging are a pre-requisite. The use of Schmidt telescopes with relatively coarse pixels scales will allow surveys to cover fields of several square degrees at a time to greater depths than has been possible photographically. For extended objects the detection limit for a given pixel size is dependent only on f/ratio, and small Schmidt telescopes can be surprisingly competitive.

2.3 Improved Quality of Photometry

Improved quality of photometry will result from attention to bandpass-matching and the provision of more (and fainter) standards. Further work on modeling the behavior of photometric systems by convolution of theoretical fluxes with the filter-detector bandpasses (e.g., Bell et al. 1994) will help in the astrophysical interpretation of photometry. However the over-subscription of the larger telescopes at national facilities which has resulted in small time allocations per investigator has often, judging from the published literature, caused data to be published where the attention to standardization has been less than adequate.

This is a great pity since the CCD has shown itself to be capable of highly accurate results. It is suggested that investigators always propose to carry out a supplementary calibration program on a smaller telescope to support observations made on a larger one. It is, unfortunately, necessary to reiterate that one cannot expect to do good CCD photometry in non-photometric conditions, and the widely practiced technique of exposing through thin cloud has dubious worth if accurate photometry is desired. Thin cloud is rarely uniform over the field of a large CCD, even when integrated for many minutes.

2.4 Reduction Software

Reduction software will need to keep pace with mosaics, and in particular will need to run in memory to minimize slow input/output steps. The initial bias subtraction, flatfielding and geometric transformations will require powerful data acquisition systems with lots of memory, disk storage, and efficient tape or optical disk storage. Similar machines will be required to make operation of programs such as DoPhot and DAOPHOT run efficiently. Such systems (e.g., SparcStation 20, 512 MB memory, 30 GB disk system, fast display, double exabyte) cost under $100,000, and can typically process (trim, bias subtract, flatfield, register) a typical night's worth of data (~ 25 GB, 100 images) from an 8k x 8k mosaic in just a few hours (S. Grandi, 1994). Certainly even more capable machines will be available for a similar price in the near future, however it is very clear that the sheer quantity of data produced will require careful attention to data management aspects. The software effort needed to minimize the time between acquisition and processed data will be considerable. Schemes such as the KPNO "Save-the-bits" (Pilachowski et al. 1993) automatically back-up raw data frames immediately following read-out, which apart from removing the onus of storing raw data from the observer serves as a data archive.

2.5 Adaptive Optics

Adaptive optics is a new and exciting way to improve the resolution of ground-based observations, over a small field. Development of the technology is rapid (Ealey and Merkle 1994), following the military declassification of the basic techniques. A challenge for the future is to develop the means for making photometric measurements on CCD frames for which adaptive techniques have been used to sharpen the star images. Variable-PSF methods have been very successful at modeling star images on CCD frames where aberrations cause star profile variations as a function of field position. The need to be able to do photometry on aberrated HST CCD frames spawned many new and modified techniques, with varying degrees of success. The challenge for the future will be to refine these methods in order that reliable photometry can be achieved on frames where the star profiles are two-component (sharp peak, plus low intensity wings) and vary with field position.

REFERENCES

Bell, R. A., Paltaglou, G. and Tripicco, M. J. 1994 MNRAS 268, 771
Blecha, A. 1995 in IAU Symposium No. 167, New Developments in Array Technology and Applications, A. G. D. Philip, K. A. Janes and A. R. Upgren, eds, Kluwer Academic Press, Dordrecht, p. 57
Cousins, A. W. J. 1993 SAAO Circ. 15, 30
Ealey, M. and Merkle, F., editors 1994 Adaptive Optics in Astronomy, Proc. SPIE 2201

Geisler, D., Claria, J. J. and Minniti, D. 1991 AJ 102, 1936
Grandi, S. 1994, NOAO internal report
Hamuy, M., Walker, A. R., Suntzeff, N. B., Gigoux, P., Heathcote, S. R. and Phillips, M. M. 1992 PASP 104, 533
Hamuy, M., Suntzeff, N. B, Heathcote, S. R., Walker, A. R., Gigoux, P. and Phillips, M. M. 1994 PASP 106, 566
Harris, H. C., Baum, W. A., Hunter, D. A. and Kreidl, T. J. 1991 AJ 101, 677
Harris, H. C., Hunter, D. A., Baum, W. A. and Jones, J. H. 1993 AJ 105, 1196
Jorden, P. R. 1995 in IAU Symposium No. 167, New Developments in Array Technology and Applications, A. G. D. Philip, K. A. Janes and A. R. Upgren, eds, Kluwer Academic Press, Dordrecht, p. 27
Jorden, P. R., Deltorn, J. -M. and Oates, A. P. 1994 Instrumentation in Astronomy VIII, D. L. Crawford and E. R. Craine, eds., Proc. SPIE 2198, p. 836
Kilkenny, D. and Laing, J. D. 1992 MNRAS 255, 308
Landolt, A. U. 1992 AJ 104, 340
Leach, R. 1995 in IAU Symposium No. 167, New Developments in Array Technology and Applications, A. G. D. Philip, K. A. Janes and A. R. Upgren, eds, Kluwer Academic Press, Dordrecht, p. 49
Menzies, J. W. 1992 in IAU Colloquium No. 136, Stellar Photometry - Current Techniques and Future Developments, C. J. Butler and I. Elliot, eds., Cambridge Univ. Press, Cambridge, p. 40
Pilachowski, C., Seaman, R. and Bohannan, B. 1993 NOAO Newsletter 35, 32
Ritzmann, B. M. 1992 Bull. Inf. CDS, 40, 41
Stepp, L. M., editor 1994 Advanced Technology Optical Telescopes V, Proc. SPIE 2199
Walker, A. R. 1993 in IAU Colloquium No. 136, Stellar Photometry - Current Techniques and Future Developments, C. J. Butler and I. Elliot, eds., Cambridge Univ. Press, Cambridge, p. 278
Walker, A. R. 1994 PASP 106, 828
Walker, A. R. 1995 in IAU Symposium No. 167, New Developments in Array Technology and Applications, A. G. D. Philip, K. A. Janes and A. R. Upgren, eds, Kluwer Academic Pub., Dordrecht, p. 339
Windhorst, R. A., Franklin, B. E. and Neuschaefer, L. W. 1994 PASP 106, 798

DISCUSSION

HOG: I often miss the information on QE curves whether the CCD is thin or thick. Why not write it?

WALKER: Thick CCDs are the ones with peak QE less than 50%.

PEL: Do you know anything about the prospects for interference filter layers deposited directly onto the chip?

WALKER: Most thinned CCDs are coated with single layer anti-reflection coatings. Two layer coatings (by Mike Lesser, U. of Arizona) show excellent QE throughout the whole optical range, for thinned UV-floated CCDs.

WATSON: Do you use a field-flattener in the Schmidt telescope with 60 arc-min field coverage?

WALKER: No, but certainly for any CCD or mosaic larger than this curvature of field is a problem that will need to be solved either with a field-flattener or by constructing the mosaic such that it conforms to the focal surface.

COOK: What is the opening/closing time for the shutter of the proposed 8k x 8k mosaic?

WALKER: A curtain shutter will be used. This type of shutter has, by design, equal exposure lines everywhere on the mosaic even for very short (eg. second) exposure times.

PENNY: The faintest stars in the HST CMD you showed seemed to have quite small errors. I presume that there were fainter stars with the expected increase in errors as you go fainter. Was there a reason why these stars were not shown?

WALKER: The CMD was a preliminary one and the processing is not yet complete. The faintest stars were not shown.

O'DONOGHUE: Does intra-Pixel QE variability discovered by the RGO group apply to thinned, back-illuminated CCDs or did Paul Jorden say the problem applied to thinned chips but (perhaps puzzlingly) front-illuminated. I know I should be asking this to Paul Jorden.

WALKER: Inter-Pixel QE variations are more significant for front-illuminated (thick) CCDs than they are for back-illuminated (thinned) CCDS.

JORDEN: To clarify - we measured a thick, frontside, coated EEV device and a thin, backside TEK device. Contact me, or see references for more details (Jorden et al. 1974, Proc SPIE 2198, 836).

CCD PHOTOMETRY: SOME BASIC CONCERNS

C. Sterken†

University of Brussels (VUB)

ABSTRACT: We review the status of CCD photometry with emphasis on applications in projects that require considerable amounts of observing time at medium-aperture telescopes and at tasks complementing projects carried out at large telescopes. Associated problems of CCDs are discussed, together with unsuspected difficulties affecting millimagnitude accuracy mainly caused by pixel-size mismatch, cooling, data acquisition, non-linearity of response and improper standardization.

1. INTRODUCTION

When the photomultiplier (PMT) became in widespread use, its advantages compared to previous detectors, viz. the eye and the photographic plate, led to an explosion of the number of telescopes equipped with photometers for replacing and complementing photographic photometry. Its advantages were clear: high dynamic range, high quantum efficiency (QE), and - if properly applied - a linear response. The principal drawback was its inability to be used for two-dimensional photometry. Now, CCDs have been in widespread use for more than a decade; they essentially offer the same qualitative and quantitative improvements as did the PMT compared to the photographic plate, but CCDs enable us to carry out photometry in two dimensions as well. So, one may expect soon that CCD-based instruments will be the only photometers available at most observatories.

The advantages and problems of CCDs are elucidated in many papers presented at this meeting, and we hear a lot on recent developments in CCD detection. The present paper is not meant to be an exhaustive list of all aspects involved; it is in the first place intended to remind us of some important points which are relevant to those who use, or intend to use, CCDs for photometry. The paper adresses a broad spectrum of people and overviews the field, but with a restricted approach, namely that of a *photometrist* trying to determine the *magnitude* of an object and the *color* of an object, however, at milli-magnitude accuracy level. This approach is in a very particular framework, namely the framework of an astronomer who does not have large blocks of observing time at a very large telescope, but who belongs to a University Observatory that lacks a staffed software department but has a telescope of the one-meter class or smaller, and is obtaining a one-time grant for acquiring that large, efficient, all-lambda, low dark-current, low-RON four-million pixel chip we hear about at this meeting. That astronomer wishes to complement work done at a large telescope, carry out projects that require considerable amounts of observing time and which cannot be done at a large instrument, do calibrated direct imaging and also time-series photometry at millimag accuracy, and drive the telescope to the edge, to the limit of its possibilities.

---

† Belgian Fund for Scientific Research (NFWO)

2. PRIME ADVANTAGES of CCDs

Accurate time-resolved photometry is difficult to obtain for objects fainter than about 15th magnitude. It is very true that - even taking into account efficient multi-channel photometers - the high QE of CCDs leads to an important gain in observing time. Photoelectric photometers are almost never scheduled today on telescopes much larger than approximately one meter (Howell 1990). A CCD, moreover, permits simultaneous differential stellar photometry (though it may be difficult to find a good comparison star within the small field of view). Therefore, a major advantage of the CCD is that useful data can be obtained in less-photometric conditions. Also, stars with close companions - for which normal single-channel photometry is difficult - can be measured. Another strong side of the CCD is that it leads to accurate *relative* magnitudes from point-spread function (PSF) fitting.

For an 11th mag star, a 50-cm telescope like the SAT (Strömgren Automatic Telescope, Florentin Nielsen et al. 1987) barely records 100 counts per second in uvby (half of that amount in Hbeta N or W, see Fig. 1), and long sequences of integrations are needed to achieve a favorable S/N ratio. A CCD camera on the same telescope would do the job in a minute. But the real advantage of the CCD compared to the PMT is in the error budget through the simultaneity of star and sky-background measurement, as seen in Fig. 1, which was constructed with data taken at the moment of strongly varying sky brightness preceding moonset. Note that the data are raw counts per second, not cleaned for any effect of centering (u, v, b and y data were obtained simultaneously, and so were N and W in an alternating star - sky sequence with repetitive recentering of the stellar image). The scatter leaves room for doubt, as does the gradient of the descending sky brightness. Careful and well-planned PMT photometry will allow the extraction of high-precision magnitudes and colors for such a moonlit night, but it is clear that a long integration on the star, followed by a short (or even long) integration on the sky will introduce a systematic error. Such problems will not arise in CCD photometry, not even for fainter stars. This advantage of the CCD versus the PMT in fact extends to any application in which a variable sky background might interfere: not only cometary photometry which (by virtue of comets' brightest phases when they approach the Sun) is done during twilight, but also faint-star photometry in external galaxies (LMC/SMC), and even any study of **local** variations of foreground interstellar extinction.

3. PRIME DISADVANTAGES OF CCDs

A first drawback in CCDs is that the integration time of all objects on a single CCD frame is the same, thus the S/N for each source is in general not the same all over a single frame. This is a severe limitation for close-binary star programs, where the exposure time must be appropriately short for not overexposing the primary, forcing exposures to be repeated as many times as needed for receiving enough photons from the secondary to obtain an acceptable S/N ratio, a situation that is a severe restriction on the potential range of magnitude differences between the components (Van Dessel et al. 1992). A second drawback is the "dead" time caused by reading out the chip and storing the data (about 5 min for a 2048 x 2048 chip). This dead time is the reason why many observers adopt the approach of observing standards during only one very good night (Sinachopoulos 1994, Jønch-Sørensen 1994) and spend little time on extinction measurements. But the most troublesome property of CCDs for accurate photometry is likely to be their high reflectivity - the reflected light getting back to the detector a second time, often in a position where it does not belong (red and blue leaks). That reflection is

CCD PHOTOMETRY: SOME BASIC CONCERNS   133

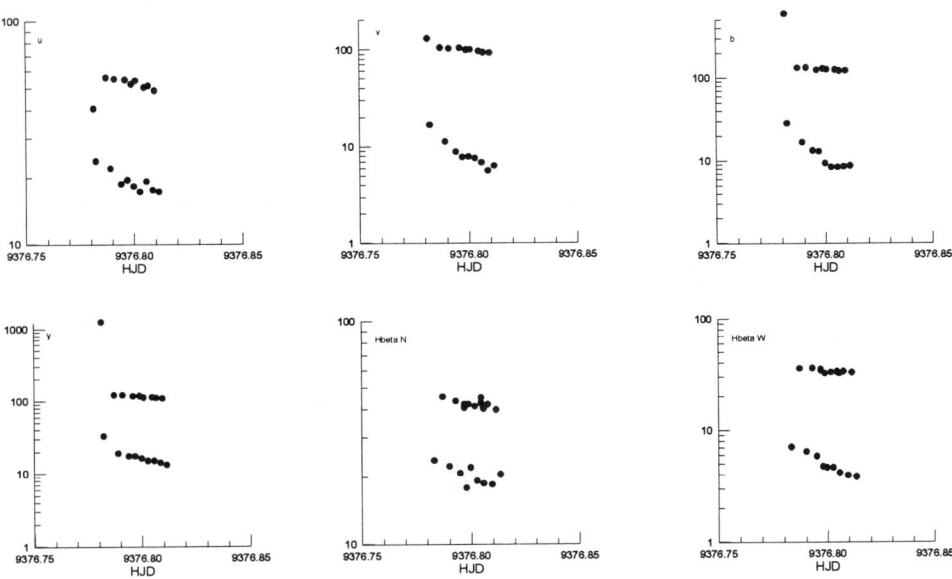

Fig. 1. uvby and Hβ, NW count rates for an 11th magnitude star observed with SAT shortly before moonset. Note the strong decrease in sky brightness during the period of measurement, and also some deviating points

---

specular, leading to structured ghosts in instruments equipped with CCDs, thus to highly local photometric errors (Tinbergen 1993).

The mayor disadvantage, however, is the huge amount of work involved with data reduction. A general photometric reduction program - that is, to convert PMT and CCD raw instrumental magnitudes to a consistent standard scale - has been distributed by ESO as the PEPSYS context of the MIDAS data-reduction system. It also includes broad introductory information on photometry (Young 1994). But it is the rectification of the raw data frames (shutter-time corrections, removal of cosmic-ray events, bias and dark, flatfielding) and the extraction of instrumental magnitudes that is so time consuming. Walker (1986) states that for a one week observing run one can expect to take at least a same amount of time reducing the data (for extracting a few [~ five] stars per frame in uncrowded fields). This agrees with Sinachopoulos (1994), who consecrated about 25 hours of work to obtain the instrumental magnitudes of one night to an internal accuracy of 0.003. But it is possible to obtain a hundred CCD frames in a night, each frame potentially containing of order 10,000 stars and of order 1,000,000 individual data-numbers. Stetson (1993) has developed software to carry out these tasks with a minimum of human effort. He can reduce the entire body of data from a typical three- or four-night observing run (from raw data frames through the profile-fitting to final, calibrated publishable photometry, see Stetson 1990) in a period of order of three to four forty-hour weeks - still a considerably long time. The real stumbling block thus is the astronomer's time, and we must devote some serious thought very quickly to increasing the efficiency of the data reduction techniques, probably with a view to the automation of as many as the tasks as possible (Sneden 1990). Stetson (1990) advocates the elaboration of comprehensive automatic software for carrying out the numerous reduction steps.

4. PROBLEMS USING CCDs

Apart from the particular problems on standardization, there are several problems at millimag accuracy level, viz. how to deal with a varying PSF, and how to deal with the fact that the response inside a pixel is not uniform (non-uniformities of 10% can arise inside one pixel). It is not unusual to find that about 1% of the pixels in an array are faulty - they may be of low sensitivity, noisy, dead or have high dark current (Glass 1993).

Besides, observers have to overcome two major obstacles: shutter timing, and twilight flatfielding. There are two shutter timing effects, viz. the (spacially non-uniform) difference between actual and presumed exposure time (a problem that ruins any calibration at millimag level), and the pixel-to-pixel differences (with very adverse effects for short exposures, such as for standard stars). The flatfielding of pixel-to-pixel variations in sensitivity and chip illumination is a fundamental step in reducing CCD frames: its failure will ruin every following step (the pixel-to-pixel variations in sensitivity can amount to a factor of 1.5). Short exposures are unavoidable when observing frames with standard stars. Hainaut et al. (1994) avoid this by exposing defocused standard-star images for 10 - 20 sec, and estimate overexposure due to shutter delay to be 0.5 ± 0.3 sec (the real value could, however, not be determined).

Twilight flatfielding poses two problems:

a) Short exposures are inevitable - and the two-dimensional flatfield structure will be significantly altered by the shutter function at exposure times of a few seconds and less.[1] Thus, with a shutter delay of the order of 0.5 s, also a flatfield exposure shorter than 5 sec will be systematically affected on the level of more than 10%.

b) Only very little time is available for taking the frames - the readout overhead is not less than for science exposures, and this leads to only a fraction of effectively useful twilight time (too dark skies bring stellar images on the CCD, too bright skies saturate the detector). Though the short-exposure problem decreases when using narrow-band filters, Jønch-Sørensen (1994) points out that getting enough sky flatfields with appropriate intensity levels in six filters requires many nights. Note that the importance of correcting for the shutter effects in twilight flats will increase with the larger CCDs because of their long read-out times.

Fig. 2 illustrates the rate of change of sky surface-brightness level, as measured on two nights using the SAT (uvby simultaneously, Hβ, NW simultaneously). As the insert shows, there is a slight wavelength dependence in the slope of the change (the deviating N,W data are

---

[1] but also by the difficulty of discriminating cosmic-ray events.

from a different night, both yield exactly the same slope). The discontinuity in uvby is due to the removal of a neutral-density filter, and it illustrates that the time available for taking twilight flatfields can be considerably extended using stacked or wedged neutral-density filters. There is a substantial curvature during the "bright" part of twilight, an effect that is partly due to the dead time of the system (such curvature, however, can also have a physical cause, such as the presence of volcanic dust in the atmosphere). However, the non-linearity seen in these curves is not a problem of PMT-based photometry alone! Saturation effects in CCDs may produce similar symptoms: Hodapp et al. (1992) report an excess dark current which follows the detected pattern of a previous strong exposure. Although its effects are negligible after a few complete reads of the device for normal background-limited exposures, it can remain a nuisance for up to an hour for exposures with low backgrounds. There is also a recently-discovered nasty problem (Fosbury 1994), viz. the fact that charge bleeding in columns seems to alter the bias level, and this change depends on the intensity of the brightest star that causes the charge bleeding. That problem, of course, not only affects the flatfield frames, but also the science exposures.

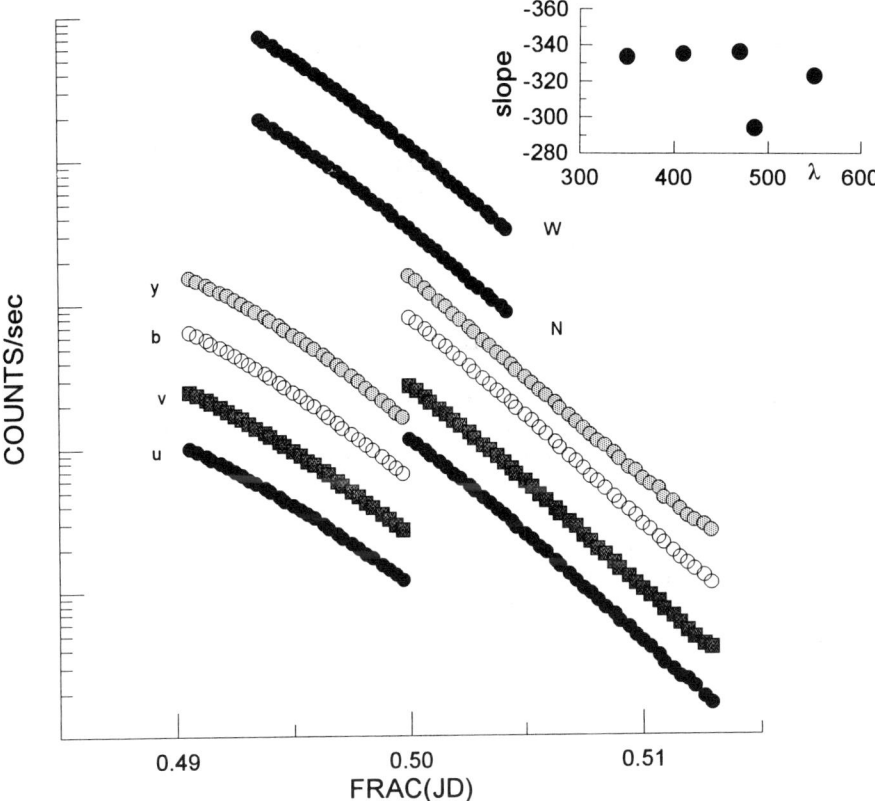

Fig. 2. Log of the twilight sky brightness for uvby and NW (the latter were obtained the night after the uvby measurements. All count rates were shifted along the vertical axis. The upward jump in uvby is due to the removal of a neutral-density filter. The insert shows the wavelength dependence of the slope of the sky brightness variation (log counts per day)

Tyson and Gal (1993) have derived a very useful expression for calculating an optimal time series of flatfield exposures adjustable to any given CCD read-out time. They give tables that allow all frames to record similar count levels - their data are taken at the zenith, which is the case for most flatfields. Surma (1993) develops a simple method to deconvolve the intrinsic flatfield and the two-dimensional shutter function from a series of flats of different exposure times. Using these intrinsic calibration functions it is possible to flatfield long-exposed object frames accurately and to construct flatfields which include a specified shutter contribution and thus are valid for any given short exposure time in an object exposure.

Tobin (1993) discusses CCD flatfielding using an illuminated dome screen: despite uniform screen illumination, non-flat response results from telescope flexure and by light scattered off by telescope baffles: a 0.3% accuracy is achievable. An extremely important matter is that beams from lamps should resemble beams from the telescope, and they should therefore have the same focal ratio, and proceed from a pupil at the same distance and with the same central obstruction (Tinbergen 1993).

5. PITFALLS EMPLOYING CCDs

CCD cameras, sometimes cheap commercial models, are often installed at small (Schmidt-Cassegrain) telescopes at university (and private) observatories on account of the availibility of these telescopes, and because of the fact that on these telescopes a very large amount of observing time is obtainable. Such simple systems, often with standard imaging and reduction software, can lead to deterioration of precision in some cases. But even high-tech professional implementations are prone to systematic disturbances: one is, for example, parasite light caused by LEDs. Other disorders are as well due to misconceived construction details as to mistaken views of application.

5.1 Pixel Size

The small size of some chips tempts the use of a scale which maximises the sky coverage (for variable stars, where one seeks to observe a comparison star on the same frame) but which leads to undersampling. This is particularly serious when there is dead space between pixels or uneven sensitivity within pixels. To cover star images with several pixels and to fully utilise any good seeing, a pixel size no greater than $0\farcs5$ is required. With pixel sizes ranging from 20 to 30 μ that implies a focal length of 10 - 15 meters. Photometric accuracy falls off rapidly as the star images become undersampled; for work that does not require the small pixel size (e.g. galaxy surface photometry) shorter focal lengths can be used (Walker 1986). Another point is that CCDs are now approaching the size of small photographic plates, and since the FWHM can be 10% larger at the corner of a 13 arcmin square field (CTIO 0.9-m f/13.5 telescope, Walker 1993), a field flattener is needed to avoid a strongly varying PSF as a function of field position. The same problem can occur if the CCD chip itself is warped.

5.2 Cooling

At large observatories, detector cooling is achieved by liquid gas cryogenics or by Peltier cooling. At smaller installations, only the latter - or no cooling at all - is the rule (note that these Peltier elements can be small in size, but are inadequate for installing inside the tube of a Schmidt telescope). As is the case with PM'l's, the dark current is a function of the detector's working temperature. So, Jønch-Sørensen (1994) found nightly variation of the bias level when the temperature control was occasionally defective.

It is well known that the temperature in a dome fluctuates rather irregularly during the night, and - though the fluctuations are an order of magnitude smaller - the temperature inside an apparatus often varies in a way uncorrelated with dome temperature. This is illustrated in Fig. 3, which gives the temperature measured in the SAT dome and inside the SAT photometer during four nights: (a) shows the trend of the temperature variation at the detector and at two different measuring points in the dome, (b), (c) and (d) compare dome temperature and detector temperature at the same time resolution as the scientific measurements. Whereas (a) and (b) exhibit a fairly linear behavior for the detector temperature, this is not so in the other cases, where a non-linear trend with superposed strong fluctuations (with a gradient of 0°2/min) is seen. Our data show that the simple approach of measuring dome temperature at the beginning and at the end of a night and applying some kind of correction linearly with time is a totally unfounded procedure. Let us note that dome temperature usually undergoes a steep drop during the first hour of the night, and that the pattern of variability of dome temperature is strongly dependent on the type of observations carried out: time-series photometry, with dome orientation that only slowly changes in time yields a different temperature pattern than does all-sky photometry with its higher frequency of dome rotation. A cooling system must be able to correct for such fast changes. Tyson (1990) points out that stability is crucial if imaging at 100 parts in 10^6 of the night-sky background is required (not only for what concerns temperature in particular, but for CCD stability in general).

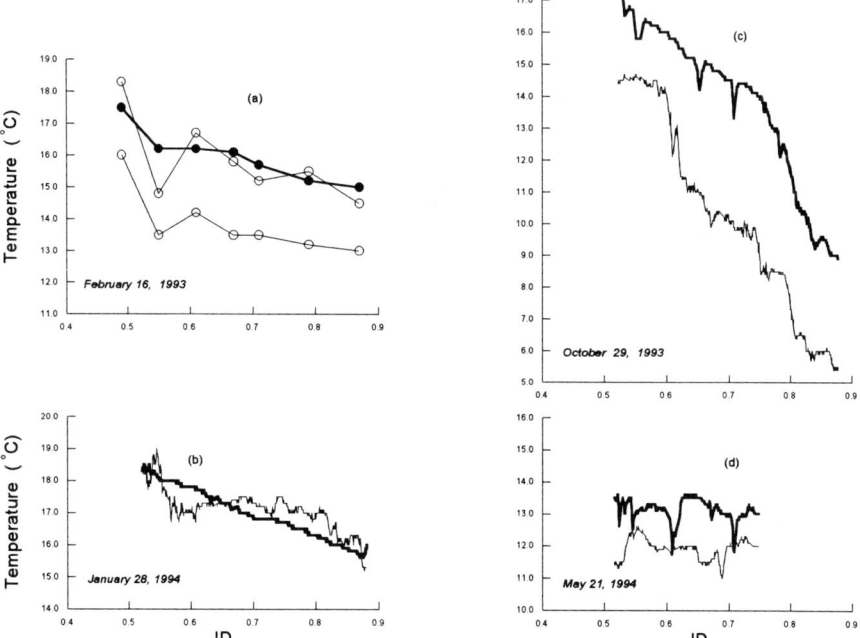

Fig. 3. (a) depicts the run of dome temperature (open circles) measured at two different locations in the dome, and detector temperature (full circles). (b), (c) and (d) give dome temperature (thin line) and detector temperature (thick line) for three other nights

5.3 Data Acquisition at the Telescope

Many papers deal with data reduction software, but what about bug-like errors that are generated at the time of observation, and that may accumulate to large effects? One such problem already occurs when the telescope control system (TCS) is directly linked to the data acquisition system (DAS). In such cases, the telescope coordinates are recorded in the headers of the data files, and the person who carries out the data reduction (often a student who is not familiar with the instrumental configuration) is tempted to use these recorded coordinates as absolute coordinates. Such procedure imports errors due to the more or less frequent application of the "common correction" (CC) routine - that is, an instruction that corrects for the difference between the actual telescope pointing and corrected catalogue coordinates - -leading to repetitive introduction of a non-rigid coordinate system for the objects measured, especially when a CC is done during every cycle of measurement of program and comparison stars when monitoring with an automatic or semi-automatic telescope. The effect is even more dramatic for TCS systems that apply approximate formulae for precession corrections and, as such, introduce large errors for objects with large positive or negative declination (the error will also appear when small mechanical shifts or oscillations in α or in δ occur). The consequence is illustrated in Fig. 4, where the differences $X - X_{tel}$ of the airmass calculated with catalogue coordinates and telescope coordinates (data obtained with the SAT) is plotted in function of airmass. The top figure gives the result for two low-declination stars a few arcminutes apart (as could appear on a same CCD frame) that have been measured sequentially for several hours. For airmasses below 1.7, the difference $X - X_{tel}$ amounts to not more than 0^m001 mag, and does not influence the accuracy at millimag level. The other points all refer to two stars at declination $\sim -70°$ and separated by several degrees. They were observed sequentially, and every four to five minutes a CC command was given. Besides the two curved sequences, there are a number of dispersed points which correspond to moments when the autocentering failed for some reason, or when a reboot of the TCS occurred. In differential photometry, as illustrated in Fig. 4, pseudo airmass-differences amounting to 0.15 may be generated, a situation that may lead to breakdown of any photometry even at the *centi*-mag accuracy level.

5.4 Non-linearity of Response

Most CCDs appear to be linear to at least 0.1 - 0.5% from very low signals to at least 80% of the full well capacity (Walker 1993). A surprising number of CCD instruments turn out to have non-linearity and other problems of the sort long familiar to photographic photometrists (Young 1994). Schwartz and Abbott (1993) describe non-linearity of response amounting to 4% due to parasite noise in the analog-to-digital converter board. However, linearity is generally assumed on the basis of absence of adverse effects, but no vigorous linearity tests seem to be carried out, and it would be highly valuable if all observing sites provided a robust linearity-check routine at the telescope (Jønch-Sørensen 1994).

6. EXTINCTION AND STANDARDIZATION

Extinction determination is, in principle, in no way different for CCD data than for PMT measurements. Building transformations to a standard system is a similar job. Still, virtually no CCD photometrists derive extinction directly from their own data. They spend a minimum of time on technical observations (remember, frequent observations of standards all over the sky

CCD PHOTOMETRY: SOME BASIC CONCERNS 139

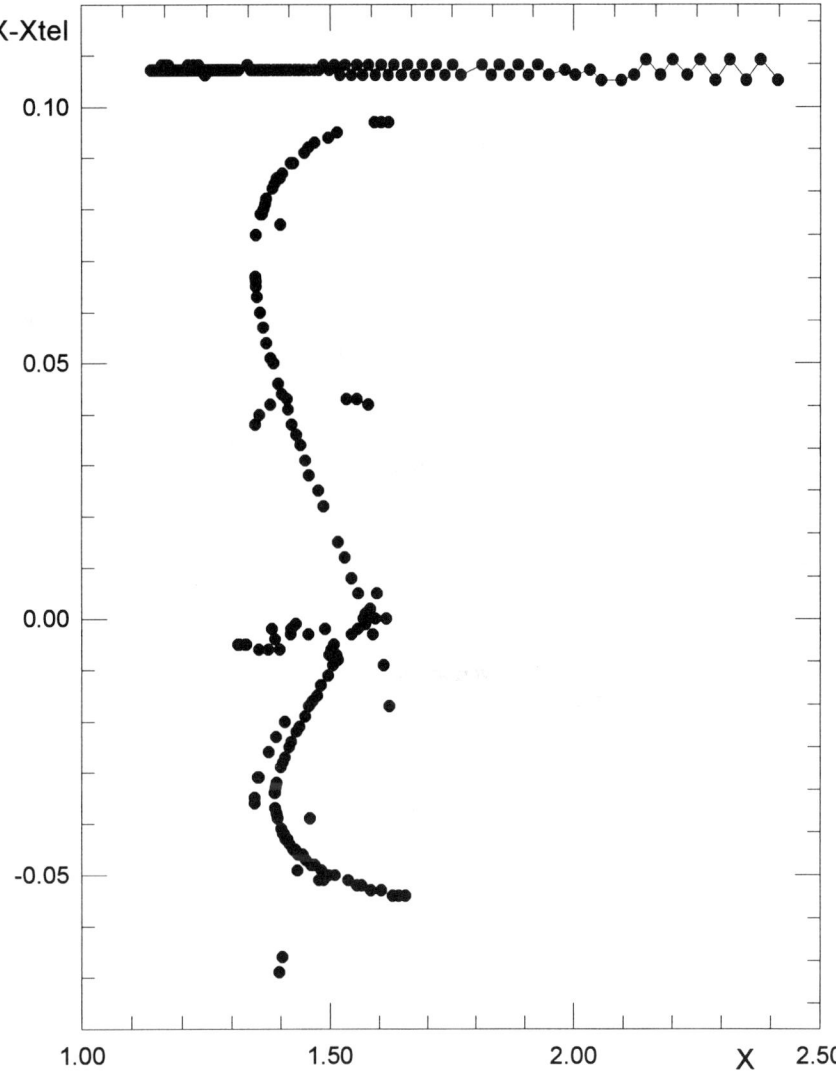

Fig. 4. Difference in airmass using (accurately precessed) catalogue coordinates, and coordinates recorded by the telescope control system. The top curve is for two very close low-declination stars, bottom curves are for two stars several degrees apart at $\delta \sim -70°$ (see text)

also demand a fast-slewing and fast-setting telescope), and supplementary information on extinction from a nearby small (50-cm) telescope is often sought (Hainaut et al. 1994, Jønch-Sørensen 1994). Taming atmospheric extinction will require frequent short observations of standards all over the sky and a means of monitoring extinction *variations* near the object being observed (Tinbergen 1993). The response curves of CCDs vary much more than those of PMTs (also, CCDs have an extended red response, and their blue response falls steeply). Thus, CCD photometric passbands poorly match existing standard systems[2] and, in principle, each CCD chip would need its own custom-designed filters, a solution that cannot be straightforwardly considered for every CCD camera.

Fig. 5 illustrates acquired accuracies from two sources of photometry. For faint stars the (internal) errors are not dominated by photon statistics but by the accuracy of the flatfielding, the quality of removing cosmic ray events and CCD defects and the quality of the profile fitting. When it comes to calibration, and transformation to a standard system, several additional systematic errors may be introduced. First of all, exposure times may strongly differ with passband (see Fig. 4), so that for a same star a blue frame is shutter-time correct, while a red frame may be, at the same time, affected with errors of several per cent. Such situation ruins the accuracy of the associated color index; the color index, in turn, introduces the damage directly in the transformation equations. Furthermore, standards as faint as the objects under study are not available (standards with a good range in color should be on a single CCD frame), so one must either use short exposures, or recur to defocused images. Defocusing not only ruins surface - or crowded-field photometry, it also changes the light path, as is stressed by Walker (1986): " ... the defocusing of bright stars is not in general to be recommended since one of the major advantages of the CCD is that the standard star frames and the program star frames can be measured in precisely the same way. Any departure from this must introduce a systematic error at some level". Young (1994) underlines that there must be angular effects in CCDs, which should show up in defocused images: the response depends on angle as well on position. A pixel in an out-of-focus image receives light from only a small part of the pupil. But in flatfields, it sees the whole pupil. These effects, at least, should vary according to the Fresnel

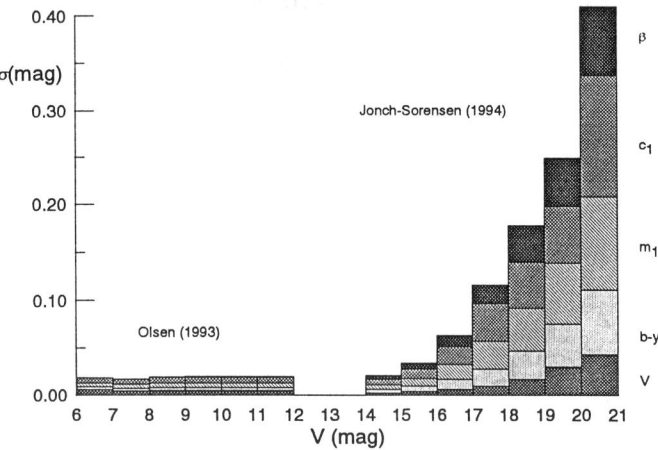

Fig. 5. Standard deviations for uvby (Olsen 1993, PMT, SAT telescope) and uvbyβ (Jønch-Sørensen 1994, CCD, Danish 1.5-m telescope)

[2] see Young's paper in this conference

reflection coefficients in back-illuminated CCDs, and should be larger due to shadowing by electrodes in front-side illumination.

Let us stress, however, that most of the standard star work at SAAO has been carried out with the 0.5-m telescope, and much of the success of the programs was due to the large amount of observing time available at the telescope (Menzies 1993). It is clear that a repetition of such enterprise at a larger telescope is impossible today. Menzies suggests that new CCD-sized fields in the E regions would be the best option as far as accurate transfer of zero point and color scales is concerned, but it will be a very time-consuming exercise to find suitable fields containing a reasonable number of stars per CCD frame.

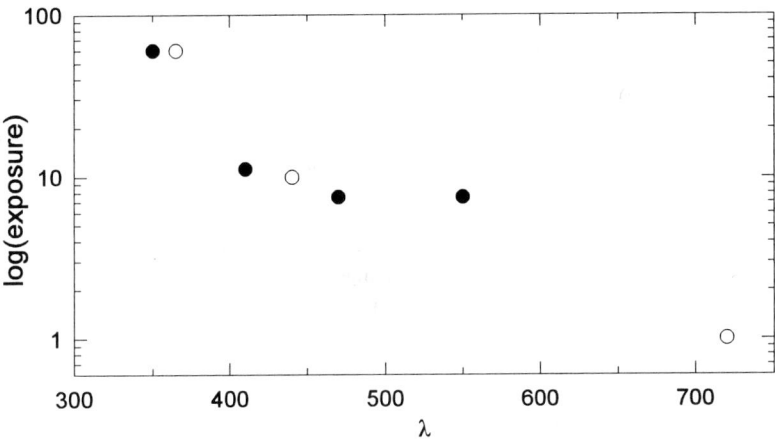

Fig. 6. Normalised exposure times (in seconds) for CCD frames: filled circles from Jønch-Sørensen (1994), open circles from Surma (1993)

Part of the problem of standardization is, of course, caused by the sometimes very short lengths of the observing blocks at large telescopes. As Hainaut et al. (1994) put it: "it is essential to have complete nights instead of half-nights, because sharing the nights makes it very difficult to obtain a sufficient number of calibration frames".

An aspect directly related to standardization is a realistic determination of the flux threshold. The high QE of CCDs has enticed users to press to the faintest possible limits, where systematic errors can lie hidden in the sky noise (Young 1994). The faintest flux level achievable in a given integration time depends on the number of exposures, the CCD and background noise and systematics, and the details of the detection and filters. At the faint limit, there is a tradeoff between angular resolution, confidence of detection, and limiting surface brightness. The limiting magnitude in any given band is also different for the various degrees of filtering of the data: detection, evaluation, splitting, classification, or photometry (Tyson 1988). A very interesting pilot program has been undertaken by Hainaut et al. (1994), who tried to push the performance of the ESO NTT to its limit by searching for comets beyond heliocentric distance $r \sim 10$ AU (in uncrowded areas, $|b| > 15°$). During three half-nights they used a 1024 x 1024 TK1024M chip, and applied a very tedious procedure of data reduction.

However, no candidate cometary images were found on any of the frames. It should be stressed that in such programs, *"non-detection"* must be considered as a real data point, and that the derived "limiting magnitude", with its intrinsic mean error, is directly used for putting constraints on proposed models.

7. CONCLUSIONS

We have highlighted some procedures and pitfalls of CCD photometry. Many a CCD observer does not achieve a complete photometric standardization of the data, but relies on the support of nearby small telescopes, in a way very similar to relying on smaller telescopes to get the best possible ephemerides for moving objects. These facts show that observers are convinced of the fact that their time on large telescopes is too expensive to spend it on accurate standardization and extinction, an attitude that only gives further support to the idea that it is cost effective to have an additional 0.5-m telescope dedicated to monitoring the extinctione over the sky during the night (see also Penny 1993).

In fact, the above also shows that, when working to millimag accuracy levels in photometry, too much is left to the individual observer, such as scheduling of extinction and standard star measurements, and calibration of the instrument. To achieve good photometry, not only suitable standard sequences must be available, but also standardized operation in the dome and at the workstation afterwards is essential to get data of good quality and homogeneity (Tinbergen 1993).

In this context is the increasing occurrence of closing small telescopes for reasons of economy when building large ones - causing large scientific loss for little gain - resembling more than anything else a situation of being invited to a great dinner where the host is saving on salt and pepper.

REFERENCES

Florentin Nielsen, R., Norregaard, P. and Olsen, E.H. 1987 The Messenger 50, 45
Fosbury, A. 1994, private communication
Glass, I. S. 1993 in Stellar Photometry - Current Techniques and Future Developments,. C. J. Butler and I. Elliott, eds., Cambridge University Press, Cambridge, p.154
Hainaut, O., West, R. M., Smette, A. and Marsden, B. G. 1994 A&A 289, 311
Hodapp, K. -W., Rayner, J. and Irwin, E. 1992 PASP 104, 441
Howell, S. B , 1990 in CCDS in Astronomy. II., A. G. Davis Philip, D. S. Hayes and S. J. Adelman, eds., L. Davis Press, Schenectady, p. 133
Jønch-Sørensen, H. 1994 A&A 292, 363
Menzies, J. W. 1993 in Precision Photometry, D. Kilkenny, E. Lastovica and J. W. Menzies, eds., SAAO, p. 87
Olsen, E. H. 1993 A&AS 102, 89
Penny, A .J., 1993, in Stellar Photometry - Current Techniques and Future Developments, C. J. Butler and I. Elliott, eds., Cambridge University Press, Cambridge, p. 146
Schwartz, H. E. and Abbott, T. M. C. 1993 The Messenger 71, 53
Sinachopoulos, D. 1994, personal communication
Sneden, C. 1990 in CCDS in Astronomy. II., A. G. Davis Philip, D. S. Hayes and S .J. Adelman, eds., L. Davis Press, Schenectady, p. 221

Stetson, P. B. 1990 in CCDS in Astronomy. II., A. G. Davis Philip, D. S. Hayes and S. J. Adelman, eds., L. Davis Press, Schenectady, p. 71
Stetson, P. B. 1993 in Stellar Photometry - Current Techniques and Future Developments, C. J. Butler and I. Elliott, eds., Cambridge University Press,Cambridge, p. 291
Surma, P. 1993 A&A 278, 654
Tinbergen, J. 1993 in Stellar Photometry - Current Techniques and Future Developments, C. J. Butler and I. Elliott, Cambridge University Press, Cambridge, p.130
Tobin, W. 1993 in Stellar Photometry - Current Techniques and Future Developments, C. J. Butler and I. Elliott, eds., Cambridge University Press, Cambridge, p. 304
Tyson, J. A. 1988 AJ 96, 1
Tyson, J. A. 1990 JOSA 7, 1231
Tyson, N. D. and Gal, R. R. 1993 AJ 105, 1206
Van Dessel, E., Sinachopoulos, D. and Prado, P. 1992 in IAU Colloquium No. 135, Complementary approaches to Double and Multiple Star Research, ASP Conf. Series 32, 362
Walker, A. R. 1986 in IAU Symposium No. 118, Instrumentation and Research Programmes for Small Telescopes, J. B. Hearnshaw and P. L. Cottrell, eds., D. Reidel, Dordrecht, p. 33
Walker, A. R. 1993 in IAU Colloquium No. 136, Stellar Photometry - Current Techniques and Future Developments, C. J. Butler and I. Elliott, eds., Cambridge University Press, Cambridge, p. 278
Young, A. T. 1994 Reports on Astronomy, XXIIA, 229

DISCUSSION

CULLUM: At what level do errors due to dome flatfields (for example due to CCD fringing and the different optical paths for measurements and calibrations) occur?

STERKEN: The best paper on this topic is Tobin's Dublin paper. I believe he reaches 0.3% accuracy.

YOUNG: Temperature effects in filters are more important than in detectors, but are rarely either measured or controlled.

STERKEN: Yes, especially in glass filters, as you showed decades ago!

YOUNG: Using a different telescope to measure extinction introduces the problem of transforming between the two instrumental systems.

STERKEN: That is right, the extinction correction is a transformation in itself.

PENNY: A flatfield on the sky has illumination flat to about 1%, due to scattered light in the telescope. This can be seen by comparing dusk and dawn sky flats, where the bright part of the sky is on opposite sides of the telescope.

YOUNG: Twilight flats can be dangerous, because the spectral distribution of twilight is very different from that of stars, particularly because of ozone absorption.

STERKEN: So is the case of a dome flat: the incandescent lamp will rarely match the energy distribution of a star, or that of a galaxy.

VERSTRAELEN: You mentioned the importance of pixel size to avoid under- or oversampling. Can you give an upper and lower limit for the pixel size to obtain a well sampled image?

STERKEN: It depends on the seeing, but also on the quality of optics: if the image field is not flat (as in the case of a fast Schmidt-Casseggrain telescope), image sizes at the edges will be larger than in the center, thus undersampling may happen in some areas, and in others not. Half a second of arc is often a guideline for an upper limit.

BESSELL: KPNO recommends using stacked median filtered sky from all exposures during the night for flatfields. Can you comment on this?

WALKER: This technique in general works well. I have found it useful to smooth the resulting frame and use it as a correction to a twilight flat. If you want to use the frame directly you must have plenty of sky, otherwise the S/N in the flatfield will be poor.

FLORENTIN-NIELSEN: Many telescopes have not got proper baffling of the optics. To what extent does that affect the phometric accuracy?

STERKEN: Improper baffling, especially for open-tube type telescopes may lead to deterioration of accuracy due to stray light from the dome (computer displays and some lighting) or from the Moon. One can calibrate the effect by performing sky measurements while modulating the light source in the dome, though one should not forget that there is also contamination by light leaks in the photometer housing.

TOBIN: I'd just like to reiterate that producing a flat flatfield is not trivial. Initially our dome and sky flats agreed to better than 1%, but moving a star around the chip showed the flats were both warped by several percent. Annular baffling to reduce scattered light is very important.

CHOOSING FILTERS TO MAKE CCD PHOTOMETRY TRANSFORMABLE TO OTHER DETECTORS

Andrew T. Young

San Diego State University

ABSTRACT: Transforming CCD data is difficult. Until recently, the general transformation problem was so poorly understood that there were no established design criteria. Now, the Hilbert-space approach to photometry tells us to match the standard passbands in the least-squares sense.

Unfortunately, the spectral responsivities of CCDs vary much more than do conventional detectors, and must be known to design filters. Each individual chip generally requires its own set of filters, if results are to be better than second-rate. Even so, spatial variations in spectral response limit the accuracy that can be reached, unless color terms are determined for individual pixels. Such CCD calibration problems involve the generally neglected optical properties of CCDs and the systems in which they are used.

Large observatories will have to establish calibration laboratories, and to abandon the common practice of moving "standard" filter sets from one instrument to another. Smaller institutions will need to purchase the services of such calibration laboratories if they are to produce accurate CCD photometry. These are major changes in the way astronomers are accustomed to working; but such changes are necessary to stem the backlash developing against CCD photometry.

1. INTRODUCTION

While CCD photometry can reach a *precision* better than a millimagnitude, provided that Herculean efforts are made to keep star images fixed on the chip, and to cancel a host of systematic errors (Gilliland et al. 1993), the *accuracy* reached is usually inferior to that of ordinary photoelectric photometry. For example, although Gilliland et al. (1991) previously achieved millimagnitude precision, the accuracy reached was an unspectacular 0.018 to 0.038 mag per standard star. In their latest effort, Gilliland et al. (1993) did not even attempt to produce standardized photometry on any existing system, but deliberately sacrificed accuracy to obtain the highest possible precision.

However, many problems require accuracy as well as (or even instead of) precision. Any application in which color indices are used to classify stars requires accurate colors. Distance determination is a prime example: corrections must be made for reddening, so the observations must be referred to a standard system.

Transforming CCD instrumental magnitudes to standard systems is difficult (Bessell 1990,

1993a, b). Indeed, until recently, the general understanding of transformations was so poor that there was neither an established criterion for reproducing a photometric system with a new detector, nor a sound theoretical basis for transforming data taken with any instrument to a standard system. Transformations have been done entirely *empirically*, using one or more convenient color indices (and perhaps other parameters, such as estimates of reddening or metallicity) as independent variables, without regard for any physical model.

As long as they used instrumentation similar to that which defined a system, most observers managed to ignore the complexities of transformations. However, now that red-sensitive detectors are commonplace, transformation errors are larger, and the inadequacy of a purely empirical approach is more evident. Transformation errors due to mismatched passbands – the so-called conformity errors (Manfroid and Sterken 1992) – often exceed a tenth of a magnitude, and can no longer be ignored.

2. TRANSFORMATIONS

The standard mythology holds that transformations "ought" to be linear and single-valued, although King (1952) showed more than 40 years ago that this *cannot* be true; the only linear, single-valued transformation is the identity transformation, when the instrumental passbands exactly match the standard ones. Nevertheless, wishful thinking continues to triumph over scientific analysis. Recent attempts to reproduce older photometric systems with new detectors have re-emphasized the importance of nonlinear terms (cf. Bessell 1990), and the multi-valuedness of transformations – especially for reddened or chemically peculiar stars – has long been known.

The extension of the standard Strömgren-King theory to higher-order terms (Young 1992a) revealed serious weaknesses in the series expansion on which it is based. First, convergence of the series is exceedingly slow, so that terms depending on third- and fourth-order central moments are easily observable. Second, high-order terms are made worse by the nonlinearity of converting from intensities to magnitudes. Third, this nonlinearity mixes together terms of all orders, so that both nonlinear and multivalued effects are inherent in transformations based on color indices.

The linear color term in transformations is proportional to the difference between the centroid wavelengths of the instrumental and standard bands. But higher-order terms depend on not only the difference of higher central moments, measured from a common reference wavelength, but also powers and cross-products of low-order spectral derivatives (Young, 1992a). Consequently, while a small linear color term is usually necessary to obtain small higher-order terms, it is by no means sufficient. The rule that color coefficients must be kept below 0.1 is a useful guide, but is not adequate to guarantee accurate transformations.

To indicate the size of higher-order terms, consider the two passbands shown below (Fig. 1). The passband in the upper half of the figure consists of two rectangular portions (shaded), and the lower, of three. Together, the two passbands add up to a single rectangular function, indicated by the dashed outline.

Both passbands are symmetric about the same wavelength, so they have the same centroid, and all their odd-order central moments vanish. Also, they have been constructed so that their

CHOOSING FILTERS TO MAKE CCD PHOTOMETRY TRANSFORMABLE

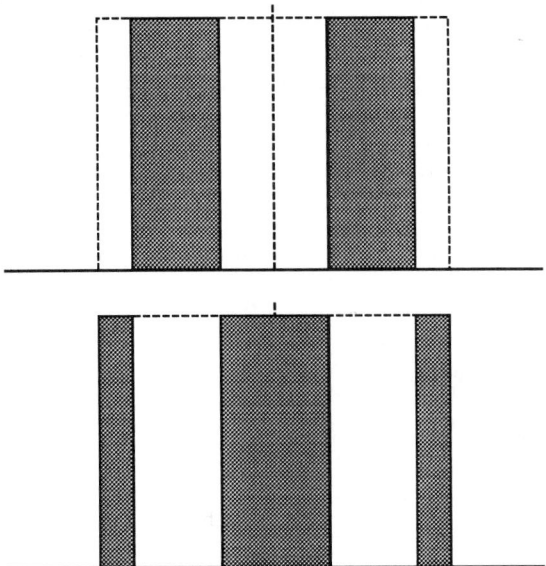

Fig. 1. Two passbands (shaded) having the same centroid and second moment (effective band width). Note that the two passbands do not overlap, and so are orthogonal.

second moments are equal. This means that the "band width" that appears in the classical series expansion (King 1952, Young 1992a) is the same for both bands. Furthermore, their fourth moments differ by less than 20%; so any differences between measurements made with these bands must be mainly due to 6th- and 8th-order terms. Given these similarities, one might not expect that measurements made with the two passbands could differ very much. To see how large such differences can be, I have scaled the passbands of Fig. 1 to a width of 124 nm and a center wavelength of 441 nm, so that each is an approximation to the B band, having the same centroid and mean-square width as B. The bands were then multiplied by each of the stellar spectral irradiance functions tabulated by Gunn and Stryker (1983) and integrated, and the magnitude differences between the two simulated passbands were calculated and plotted as a function of (B-V).

Each band averages over 62 nm of spectrum, so one might expect stellar spectral features to be largely averaged out. If high-order terms were negligible, the two magnitudes should be very similar, and the plot of magnitude difference would be a horizontal straight line.

Nevertheless, as Fig. 2 shows, the differences span a tenth of a magnitude for K stars, and more than 0.2 mag for late M stars. Even for earlier spectral types, there are large nonlinearities, and significant spread (multivaluedness due to conformity error). Among the A stars, we see effects due to luminosity and metallicity that amount to 0.06 mag. If one of these passbands were a standard band, and the other were the corresponding instrumental one, there would be serious difficulties in transforming between them, in spite of their equal effective wavelengths and band widths. These large differences are Manfroid and Sterken's "conformity errors". Evidently, stellar spectra are complicated enough that the terms due to their high-order derivatives are quite substantial.

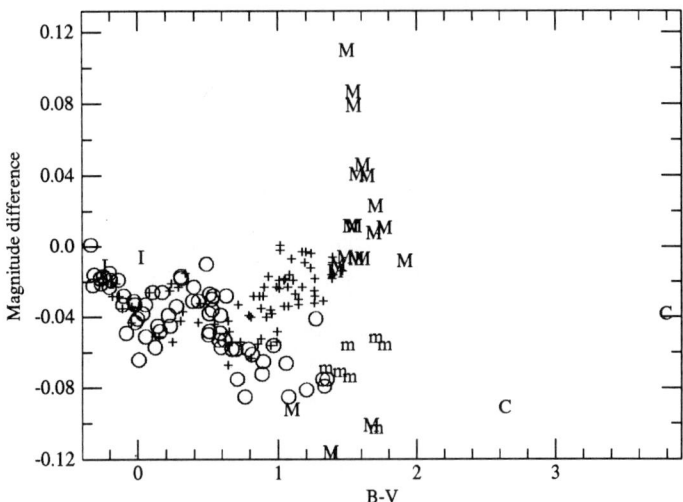

Fig. 2. Difference between magnitudes measured through the two passbands of Fig. 1, scaled to match the centroid and second moment of the B band.

Furthermore, the higher-order derivatives of the (smoothed) stellar spectral energy distribution that appear in the theory are badly estimated by conventional color and curvature indices, which suffer from severe aliasing due to undersampling of the spectrum. As I emphasized at the previous General Assembly (Young 1992b), this means that the information required to do accurate transformations is never acquired in the first place in conventional photometry.

In an effort to avoid the nonlinearities introduced by using magnitudes and colors, I have recently adopted a more fundamental point of view (Young 1994). What we are trying to measure is the spectral energy distribution of a star's light; this is a function, not a scalar quantity. What we actually measure is a function of this function, namely the functional

$$L = \int_0^\infty I(\lambda) R(\lambda) d\lambda, \qquad 1$$

which is a scalar quantity that depends on $R(\lambda)$, our instrumental function. At best, we can measure a few such scalar values through a small set of passbands.

If we regard the functions $I(\lambda)$ and $R(\lambda)$ as vectors in Hilbert space, the measured "intensity" is the projection of I on R. A set of instrumental passbands can be regarded as basis vectors that define a space of low dimensionality. Functional analysis then shows that our conformity errors are proportional to the sine of the angle between one of the standard basis vectors and its projection into our instrumental space. This projection is simply the linear combination of our instrumental vectors that approximate the standard vector in the least-squares sense.

From this point of view, the problem with the passbands shown in Fig. 1 is that they are orthogonal to each other; they measure entirely different spectral regions. Even though these regions are interleaved, they do not overlap. The large conformity errors shown in Fig. 2 are caused by the orthogonality of the two response functions in Fig. 1.

To minimize conformity errors, we must choose a set of instrumental response functions so that a linear combination of them is a good least-squares approximation to each standard response function. As Fig. 2 shows, just matching low-order moments is not enough; the instrumental passbands must accurately match the standard ones in all details. This rule provides a simple and reliable filter-design criterion for reproducing any photometric system with a new detector whose spectral responsivity is known. It also shows that using a detector with unknown spectral response is a recipe for failure.

Furthermore, as the intensity that would have been measured with the standard passband is the same linear combination of the intensities measured with the instrumental passbands, the Hilbert-space view provides a simple, *linear* transformation algorithm for data reduction. (If we want magnitudes and colors, we can take logarithms of the intensities after transforming them to the standard system.)

Good approximations require instrumental passbands with substantial overlap. However, such overlapping passbands are needed in any case to avoid aliasing problems due to undersampling the stellar spectrum. Future photometric systems will provide proper sampling, but at present one must deal with undersampled systems, where conformity errors are inevitable and uncorrectable. The only way to minimize the damage is to minimize the mean-square difference between the instrumental and the standard passbands. At least, we now know what to strive for.

Numerical experiments show that conformity errors for "ordinary" stars are typically about a tenth of the sine of the angle between instrumental and standard response vectors; the angle is typically 10 degrees or so, for conventional detectors, and the transformation errors are typically 0.01 or 0.02 mag (Beckert and Newberry 1989). Of course, for stars with spectral peculiarities, the errors can be an order of magnitude larger (comparable to the sine of the angle). However, for CCDs, the angles for bands like U and I can be on the order of 15 degrees. Thus we expect conformity errors on the order of 0.03 mag for ordinary stars, and tenths of a magnitude for peculiar stars.

These conformity errors are systematic errors for individual stars that depend on the properties of individual stellar spectra. Although they look like "noise" in plots like Fig. 2, they cannot be reduced by making more measurements, using a larger telescope, or measuring more standard stars.

3. MATCHING UNDERSAMPLED SYSTEMS

Some systems were originally defined by the properties of available detectors such as the eye and the unsensitized photographic plate; later, the red and blue cutoffs of photomultipliers such as the 1P21 were used. The spectral response of silicon detectors is quite different from these, so the difference must be made up by proper filtering. In particular, the large infrared response of silicon photodiodes, combined with the low effectiveness of infrared-absorbing glasses, greatly exacerbates red-leak problems.

CCDs differ from photomultipliers in a second, important way. The spectral response of a photocathode depends on only a few factors: the band gap of the material, the electron affinity, and the thickness of the material are the most important. In contrast, the spectral response of a simple silicon photodiode depends on about a dozen different parameters (Geist et al. 1982, Schaefer 1984); the optical effects of the electrodes used in CCDs (Kesler and Lomheim 1986) introduce many more parameters. If the device is coated or treated to enhance the blue response, still more parameters affect the spectral response.

Consequently, even nominally similar CCD devices differ considerably more in spectral response than do photomultipliers. Furthermore, the variety of construction and use configurations (front or back illumination; thick or thin; surface or buried channel; various blue-enhancement techniques) produces a much wider variety of spectral responses than are available in photocathodes. Therefore, as has been repeatedly emphasized by Bessell (1993a, b), "it is best to use different glass filter recipes for different CCDs." Walker (1993) gives some particular examples.

The wide range in CCD spectral responses means that the problem of transforming CCD photometry cannot be solved by replacing existing photometric systems by a new "CCD system", as has sometimes been suggested. Because the actual spectral response of a CCD can be quite different from another nominally identical device (see, for example, Schumann and Lomheim 1989), the design of satisfactory filters for matching any standard photometric system requires that each chip's response be individually measured. Few observatories are able to make such measurements today.

Furthermore, the factory tolerances on even the best colored glasses are so lax that they correspond to more than a factor of two in thickness for most of the sharp-cutoff glasses, which are the ones that cause the largest conformity errors. A specification such as "2 mm of GG 495" may in fact be satisfied by something between 1 and 4 mm of GG 495, depending on the particular sample. Here again, accurate spectral-transmittance measurements, free of scattered light, must be made on the actual piece of glass to be used, before the correct thickness can be calculated.

4. SPATIAL VARIATIONS

An equally serious problem, which seems to have been almost completely neglected, is the marked variation in spectral response among pixels of a single CCD chip. Because the chip is inherently a multilayer optical device, within which standing waves form, the response of a given pixel depends on the strength of the standing wave in its active volume. Because of the high refractive index of silicon, and the high reflectance of any metallic layers in the electrode structure, the contrast of the Edser-Butler fringes formed within the chip can be tens of per cent. Anti-reflection coatings can reduce fringe contrast, but are effective only in a restricted wavelength range; because CCDs are used over at least a factor of two in wavelength, fringe problems cannot be completely eliminated.

Furthermore, because the multilayer structure of a CCD is several wavelengths thick, the fringes formed are of moderately high order, on the order of 10 or 20. That means that small variations in the thickness of any given layer can make large changes in the fringe phase. Kesler and Lomheim (1986) show an example in which a change in thickness of one layer from 300 nm to 350 nm suffices to reverse the phase of fringes that have a period of about 200 nm in the spectrum. Thus, while one pixel in the device could be twice as sensitive at 750 nm as at 650, the reverse could be true for another pixel in the same device. In such a CCD, pixel-to-pixel variations of spectral response within the passband of a typical broadband filter are quite serious.

When a CCD is illuminated by monochromatic light, the Edser-Butler fringes are aliased into the spatial domain by spatial variations in the layer thicknesses. This is the origin of the notorious night-sky fringes often observed at long wavelengths. However, observers who treat the fringes as a flatfielding problem fail to recognize that the night-sky fringes are really only an alias of the real problem, which is strong and rapid variations in spectral response across the chip.

Moreover, the observed contrast of the night-sky fringes is only a lower limit to the inherent contrast of the Edser-Butler channels, for two reasons. First, the filter passband usually contains several significant night-sky lines, whose fringes are more or less out of phase with one another. Wavelength averaging can then greatly reduce fringe contrast in wide-band images, though they may still be visible in narrow-band images. For example, Gullixson (1992) shows dome flats that are fringeless in R and I, but show pronounced fringing at H-alpha.

Second, because the fringes are of moderately high order, they are smoothed out to some extent by the range of angles subtended by the pupil at the detector. While the pupil is completely filled by an extended source like the night sky or a flatfield exposure, it is not filled by a star image that is not perfectly focused. Each pixel in the chip can be regarded as a two-dimensional Foucault tester; if the image is slightly out of focus, only a portion of the pupil will be seen by a particular pixel. Defocusing therefore reduces the solid-angle averaging of the Edser-Butler fringes. Because a star image does not illuminate the detector in the same way as a flatfield, the basic principle that calibration and measurement exposures be made in the same way is violated, and trouble may be expected. Perhaps some flatfielding problems come from variations in focus across the chip.

Third, because stars are generally unresolved, starlight is spatially coherent across the entire telescope pupil, while extended sources like the night sky and calibration targets are incoherent. Thus, the electrode structure of a CCD chip can act as a lamellar grating and produce fringes in starlight that have much less contrast for extended sources. Once again, the flatfielding sources are not comparable to stars, and can fail to show fringes that are significant for star images.

If a CCD shows night-sky fringes in some passband, it is prudent to assume that pixel-to-pixel spectral-response variations are at least as large for stars as the sky-fringe contrast. Then, to obtain accurate transformations, one should determine at least a color term for each pixel individually; an average for the entire chip is not likely to be adequate. It might be possible to determine such color terms by observing both "red" and "blue" flatfields for each passband affected by internal fringes. This problem is briefly alluded to by Massey and Jacoby (1992) in connection with large-scale flatfielding differences between passbands, but they do not actually suggest determining separate transformation terms for each pixel.

An additional complication is that the active volume changes with the exposure level (Percival and Nordsieck 1980, Kesler and Lomheim 1986). Therefore the contrast and positioning of the fringes depends on exposure. This probably explains the difficulties often experienced in trying to subtract night-sky fringes; they are not a purely additive phenomenon, and so do not "subtract" accurately. Neither are they fixed-position sensitivity variations that can be divided out in flatfielding. One can only agree with Gullixson (1992) that, for accurate work, "just say no to CCDs that produce fringes in dome flatfields." I would go farther and just say no to CCDs that produce fringes in night-sky exposures.

The dependence of responsivity on exposure makes CCDs *inherently* non-linear, in a way that depends on the particular filter used and position on the chip. As the time-dependent active volume depends on voltages, clocking frequencies and waveforms, one expects the nonlinearity to depend on the electronic settings, as was actually found by Gosset and Magain (1993).

Indeed, on close examination, about 1/3 of the CCDs investigated by Abbott and Sinclaire (1993) showed nonlinearities exceeding one percent. Similar results were obtained by Gilliland et al. (1993). In addition, the latter authors remark that "Most CCDs have at least isolated regions that respond nonlinearly to light at low intensity." This is analogous to the "inertia" of photographic plates, though the mechanism is different in the two detectors. Other wavelength-dependent nonlinear effects are known, even in simple photodiodes (Schaefer et al. 1983). Unfortunately, because the nonlinearity is wavelength-dependent, one cannot exactly define a spectral-responsivity function for a CCD. Such properties make CCDs look remarkably like those the detectors they have largely replaced: photographic plates.

5. COMPARISONS TO PHOTOGRAPHY

The careful investigations of CCD properties mentioned above have shown that CCDs share many of the properties of photographic plates, such as wavelength-dependent nonlinearity. The problems arising from internal standing waves are inherent in the optical physics of the devices, and are not due to "manufacturing defects" that can be eliminated; major changes in device design and technology would be required to avoid these problems. To be sure, the problems with CCDs are about an order of magnitude smaller than those of photography. Nevertheless, those who require accuracy in their data may come to regard CCD photometry much as photographic photometry is regarded today: a sort of second-rate substitute for the real thing.

Indeed, at the NATO Workshop on Long-Term Photometry of Variable Stars in Ghent last November, I was surprised to hear so many critical remarks about CCDs in the corridor discussions. I have heard similar remarks from laboratory photometrists, to the effect that no one should ever expect CCDs to produce as good results as do PMTs. There is the danger that

CCDs were over-sold at first by their enthusiasts, and that a backlash is developing because of the continued difficulty in demonstrating really accurate results from CCDs.

If CCDs are to earn the full confidence of the photometric community, attention must be paid to their optical properties as well as their electrical ones. Not only antireflection coatings, but even changes in basic device design and construction may be required to eliminate problems due to internal fringes.

In any case, it appears that just as much care is required to extract good photometry from CCDs as is required in photographic photometry. Because many of the problems that arise are similar, it would be useful for CCD observers to study the problems long known to exist with photography, to avoid wasting effort on problems that have previously been well studied.

6. OTHER OPTICAL PROBLEMS

Some flatfielding problems clearly can be traced to stray light (Tobin 1993, Walker 1993). Proper baffling in telescopes, and light-tight camera construction, are required to avoid these problems. Tobin et al. (1993) have shown that filter nonuniformities are significant; filters near the focal plane must be repositioned to better than 100 microns to avoid measurable errors.

One should also remember that the periodic structure of a CCD forms a diffraction grating. There should therefore be considerable polarization effects, particularly at wavelengths where a diffracted order passes off the surface of the grating; because the grating is relatively coarse, several such anomalies are to be expected within a broad photometric passband. Furthermore, the interference effects within each pixel may differ depending on the polarization state. Tobin (1993) reports pixel-to-pixel polarization effects between 0.7 and 3%, but more work on this problem is clearly needed.

7. DISCUSSION

The need to tailor filters to individual CCD chips requires substantial operational changes at most observatories. First of all, accurate laboratory calibration facilities are needed to measure the spectral responses of individual detectors; but such facilities are uncommon today, even at large observatories.

Second, the idea of "standard" filter sets must be abandoned; filters cannot be moved from one instrument to another with impunity. Designing filters for each chip requires measuring individual glass samples, and finishing them to the correct thickness, because catalog tolerances on Schott sharp-cutoff glasses correspond to a factor of more than two on either side of the nominal catalog thickness. Photometric filters should *never* be ordered by glass type and thickness alone.

In narrowband work, the passband can be a fraction of an Edser-Butler channel in width. Then one might improve accuracy by determining the local spectral slope for each pixel. Efforts should be made to develop calibration techniques to do this. However, it would still be necessary to design filters to take account of each CCD's average spectral response.

Even with optimal filters for each chip, spatial variations contribute substantial conformity errors. The obvious way to reduce the effect of pixel-to-pixel differences in spectral response is to average over a large number of pixels, as was done by Gilliland et al. (1993), who essentially resorted to Fabry photometry – a lesson learned from the problems of photographic photometry. In wideband photometry, these variations apparently set a fundamental limit to the accuracy that can be achieved with direct imaging. One should always bear in mind Fred Vrba's remark: "Imaging is not photometry."

This work was partly supported by NSF Grant AST-8913050.

REFERENCES

Abbott, T. M. C. and Sinclaire, P. 1993 The Messenger 73, 17
Beckert, D. C. and Newberry, M. V. 1989 PASP 101, 849
Bessell, M. S. 1990 PASP 102, 1181
Bessell, M. S. 1993a in Stellar Photometry - Current Techniques and Future Developments, C. J. Butler and I. Elliott, eds., Cambridge University Press, Cambridge, p. 22
Bessell, M. S. 1993b in Precision Photometry, D. Kilkenny, E. Lastovica, and J. W. Menzies, eds., South African Astronomical Observatory, Capetown, p. 46
Geist, J., Gladden, W. K., and Zalewski, E. F. 1982 J. Opt. Soc. Amer. 72, 1068
Gilliland, R. L., Brown, T. M., Duncan, D. K., Suntzeff, N. B., Lockwood, G. W., Thompson, D. T., Schild, R. E., Jeffrey, W. A. and Penprase, B. E. 1991 AJ 101, 541
Gilliland, R. L., Brown, T. M., Kjeldsen, H., McCarthy, J. K., Peri, M. L., Belmonte, J. A., Vidal, I., Cram, L. E., Palmer, J., Frandsen, S., Parthasarathy, M., Petro, L., Schneider, H., Stetson, P. B., and Weiss, W. 1993 AJ 106, 2441
Gosset, E., and Magain, P. 1993 The Messenger 73, 13
Gullixson, C. A. 1992 in Astronomical CCD Observing and Reduction Techniques, S. B. Howell, ed., ASP, San Francisco, p. 130
Gunn, J. E. and Stryker, L. L. 1983 ApJS 52, 121
Kesler, M. P., and Lomheim, T. S. 1986 Appl. Opt. 25, 3653
King, I. 1952 AJ 57, 253
Manfroid, J. and Sterken, C. 1992 A&A 258, 600
Massey, P. and Jacoby, G. H. 1992 in Astronomical CCD Observing and Reduction Techniques,. S. B. Howell, ed., ASP, San Francisco, p. 240
Percival, J. W. and Nordsieck, K. H. 1980 PASP 92, 362
Schaefer, A. R. 1984 in Workshop on Improvements to Photometry, W. J. Borucki and A. T. Young, ed., NASA Ames Research Center, Moffett Field, p. 193
Schaefer, A. R., Zalewski, E. F. and Geist, J. 1983 Appl. Opt. 22, 1232
Schumann, L. W. and Lomheim, T. S. 1989 Appl. Opt. 28, 1701
Tobin, W., Kershaw, G. M., Ritchie, R. A., Ma, L., Graham, G. J. and Hemmingsen, S. B. 1993 in Poster Papers on Stellar Photometry, I. Elliott and C. J. Butler, eds., Dublin Institute for Advanced Studies, Dublin, p. 304
Tobin, W. 1993 in Stellar Photometry - Current Techniques and Future Developments, C. J. Butler and I. Elliott, eds., Cambridge University Press, Cambridge. p. 304
Walker, A. R. 1993 in Stellar Photometry - Current Techniques and Future Developments, C. J. Butler and I. Elliott, eds., Cambridge University Press, Cambridge, p. 278
Young, A. T. 1992a A&A 257, 366
Young, A. T. 1992b in Automated Telescopes for Photometry and Imaging,, S. J. Adelman, R. J. Dukes, and C. J. Adelman, eds., ASP Conf. Series, San Francisco, p. 73

CHOOSING FILTERS TO MAKE CCD PHOTOMETRY TRANSFORMABLE

Young, A. T. 1994 A&A, 288, 683

DISCUSSION

OSWALT: Where in relative importance does the lack of faint and lack of red standard stars fall compared to other sources of error in transforming CCD observations to a standard system? At present stars can be observed which are at least several magnitudes fainter and at least 1 - 2 mag redder [in (V-I)] than published lists of standard stars.

YOUNG: This is a big problem, especially for the M stars that appear in great numbers at faint magnitudes. Arlo Landolt has been working on faint CCD standards.

JORDEN: You discussed "fringes" and showed curves of CCD spectral response for a frontside CCD. I consider that the term "fringing" normally refers to long wavelength internal fringes ($\lambda > 700$ nm). In backside CCDs, the variation of spectral response for a thick, frontside CCD is somewhat different.

YOUNG: The fringes are the spatial alias of Edsen-Butler fringes (standing waves) inside the chip, and occurs for both front and back side illumination. They are usually worse for front side illumination.

JORDEN: Do you consider that when astronomers publish photometric data, they should also include the spectral response of the CCD and filter transmission characteristics?

YOUNG: Absolutely!

TOBIN: The CCD quantum efficiency is a crucial part of the passband. How accurate are the QE curves supplied by manufacturers, especially with respect to shape?

YOUNG: I can't say from my own experience; but my impression from published work is that manufacturers' curves are only a rough guide – perhaps accurate only to some tens of percent.

STERKEN: The shortcomings in photometry are not only the CCD and the filters! Besides some of the problems illustrated there is a huge additional shortcoming in education and training of students as photometrists – that is, as well from the point of view of understanding transformation, as from the point of view of observing experience. Enhanced by the all-present high tech data acquisition and reduction tools, larger errors are introduced of non-detector origin than by the whole CCD camera itself. The CCD is becoming a very reliable detector; it is the human interface that is a problem.

YOUNG: Yes, users need to be well trained in the basics of photography. Experience with photographic photometry is very useful, as many of the same problems occur with CCDs, but at a lower level. Photometry shows you what to watch out for.

TOBIN: I've found large differences in UV Schott glasses.

FLORENTIN-NIELSEN: Concerning the QE response, it is not so difficult to measure absolute QE (at say 10 wavelengths) of the CCD. It is particularly important in case you operate UV flooded CCDs when UV response may vary considerably with time.

AUTOMATED CCD SCANNING FOR NEAR EARTH ASTEROIDS

Robert Jedicke

Lunar and Planetary Laboratory, University of Arizona

ABSTRACT: The Spacewatch group at the University of Arizona's Lunar and Planetary Laboratory was probably the first (1984) to implement CCD-scanning in a major astronomical program. In the past three years, using a Tektronix 2048 x 2048 CCD, the program has discovered $\sim$ 45% of the new Earth approaching asteroids, measured astrometric positions for over 50,000 main belt asteroids, discovered two of the three known Centaurs, and found evidence for an unheralded population of small ($\sim$ 10-m) objects in the inner solar system. This success is due to the automated Moving Object Detection Program (MODP) which searches successive scans over the same region for objects showing consistent motion. While visual examination of photographic plates may have a higher efficiency, an automated routine for detecting moving objects does not tire and is repeatable. Our recent work quantifies the efficiency of MODP as a function of the asteroid's magnitude, rate of motion, and orbital parameters. Other work suggests that the observational detection of faint, trailed, Fast Moving Objects may be improved by incorporating linear darkfield subtraction and flatfielding during scans.

1. INTRODUCTION

In 1981 Spacewatch began the first long term, CCD-based, discovery program for Near Earth Asteroids (NEA) utilizing the Steward Observatory's 0.91-m f/5 Newtonian at Kitt Peak. The RCA SID 53612 512 x 320 pixel CCD used at the outset lacked the areal coverage necessary for detection of the sparsely distributed and faint NEAs. In 1988 a thick, front-illuminated, Tektronix TK2048SP 2048 x 2048 pixel CCD was delivered and the software required for real-time analysis of the images was developed by D. L. Rabinowitz (1993a). The first automated discovery of an NEO (1989 UP) was on October 27th, 1989, and since that time Spacewatch has found over 75 NEAs, two Centaur class asteroids, two comets, and recorded astrometric positions for over 50,000 main belt asteroids. Yet another twofold improvement in the NEA discovery rate was achieved in 1991 with the delivery of a thinned, back-illuminated, Tektronix TK2048EB1 2048 x 2048 pixel CCD with twice the quantum efficiency of the earlier Tektronix device.

Until recently, there existed little quantitative understanding of the absolute efficiency of the detector and software systems even though this information is critical to a meaningful analysis of the data. Like photographic surveys for moving asteroids which are limited by the magnitude cutoff of the system for faint objects and by over-exposure for bright objects, a CCD search is similarly limited at faint magnitudes and by pixel saturation for bright objects. Asteroid detection is also affected by crowded star fields, seeing conditions, etc. An advantage to the use of an automated routine for locating the objects is that it does not tire and its results are repeatable. This study presents a determination of Spacewatch's asteroid detection efficiency using the set of numbered asteroids as a benchmark.

2. AUTOMATED DETECTION TECHNIQUE

Spacewatch operates the telescope and CCD in a *drift scan* mode where the telescope drive is turned off during an exposure. Instead of tracking the telescope with the stars, the charges generated on the CCD representing the star's image are transferred across the face of the CCD at the same rate as that of the star's image due to sidereal motion. When the star's image passes off the CCD the charge is read out and almost instantaneously displayed on a computer screen. An image of the sky is built up during a scan which has a fiducial extent of about 32' in declination and a practicable length of up to forty-five minutes in right ascension.

Each scan is repeated three times over the same section of sky. While the image is being read from the CCD and displayed on the computer screen, MODP searches through the image seeking out point sources and also streaks which may be consistent with an interpretation as nearby Fast Moving Objects (FMO). The streak detection mode is efficient for bright objects moving with angular rates of motion (ω) greater than about 2°/day. Streaks are identified in real-time within each of the three passes of a scan and their locations are highlighted in order to notify the observer who decides on the credibility of the object.

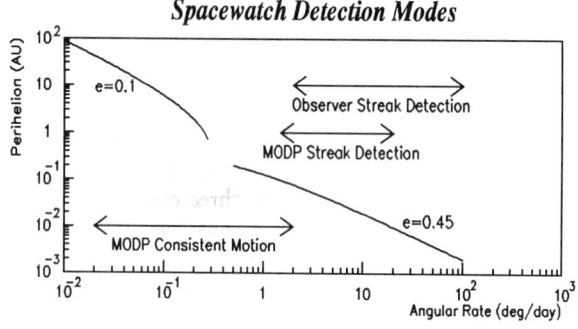

Fig. 1. Spacewatch NEA detection modes as a function of rate of motion and distance. Distances were calculated for the eccentricity specified on the figure and for objects at perihelion and opposition.

In the range 0°.02/day $< \omega <$ 2°/day MODP detects objects through their consistent motion from pass to pass. Point sources identified in the second pass are matched with sources in the first pass. Most of these are stars which align exactly with one another except for a small offset (normally less than about 10") in right ascension and declination. The offset is determined and corrected in software. When there is no direct match in the first pass for a source in the second pass, MODP searches (normally a radius of 120 pixels) for a second unmatched point source. When an unmatched point source is identified in both passes MODP assumes that they are the same object and predicts its location in the third pass assuming that it moves at a consistent rate. During the third pass MODP matches triplets of otherwise unmatched point sources with consistent motion and informs the observer of the location, brightness, and direction of motion of the new object.

A third mode of NEA detection relies entirely on the observer to identify faint and/or long trails missed by the automated streak detection. These Very Fast Moving Objects (VFMO) are typically within 0.2 AU of the Earth, are moving at $\omega > 2°$/day, and have signal to noise (S/N) in a single pixel between one and three. Their proximity to the Earth, small size, and extreme rate of motion make them exciting objects to follow during the few days in which they are visible.

Fig. 1 indicates the modes of NEA detection used by Spacewatch as a function of the range in rate of motion and distance to the object. The distance has been calculated for objects found at opposition and at perihelion. In the range $0.01 < \omega < 0.3$ (in °/day) the distance has been calculated for asteroids with e = 0.1 while for $0.5 < \omega < 100$ the eccentricity is set equal to the mean for the Spacewatch NEAs (e = 0.45). The limit at the lowest rate of motion is determined by the rate required for the object to move two pixels during the time interval between passes. The limit at the highest rate of motion is set by the ability to distinguish between meteors or faint satellites (which traverse the entire field of view in declination or about 2048 lines in right ascension) and VFMOs (which are required to begin and end within a single frame). At the highest rates of motion the system has been limited by the ability of the observer to identify the object and calculate its position at the middle of another scan though new software now automates this process as well.

3. AUTOMATED DETECTION EFFICIENCY

The principle of "blinking" images of the same piece of sky to detect objects which move against the background of stationary stars was established more than sixty years ago. Except for the introduction of the *blink comparator* this tradition continues essentially unaltered at observatories around the world. Two of the three contemporary NEA discovery programs (both of them utilizing the 18" Schmidt at Palomar) use a stereoscopic viewing device which makes moving objects appear to "float" above the background. When carefully implemented, the blinking technique can be very efficient in the range of magnitudes from where the photographic images are not overexposed to close to the limiting magnitude of the system. For example, the Palomar Leiden Survey of Faint Minor Planets (Van Houten et al. 1970) reported efficiencies of 100% in the range $14.0 < m < 19.5$ and 70% in their last bin where $19.5 < m < 20.0$.

The efficiency of the automated Spacewatch CCD system has been determined using the database of 5856 numbered asteroids as a benchmark. Scans are performed preferentially near opposition to take advantage of the opposition effect (where asteroid apparent magnitudes are at their brightest) and along the ecliptic. These two strategies also favor the recovery of known asteroids. Since each of the numbered asteroids has a well determined orbit it is possible to predict their apparent location to within a few arcseconds. Thus, a search was initiated through the Spacewatch archive of astrometric positions from the past two years for serendipitous observations of known objects. Every asteroid which should have appeared in a scan was recorded, as well the measured position, magnitude, rate of motion, etc., for those objects which were actually detected by MODP. Dividing the observed by the expected distribution determines the efficiency of the system.

Fig. 2a shows the detection efficiency across the CCD. The exceedingly low efficiency in the first bin is due to a "forest" of cosmetic defects in the first 220 columns of the CCD. All other efficiency plots in this paper represent only the CCD columns between 221 and 2048. The efficiency is consistent with being constant and equal to about 55% over the rest of the CCD.

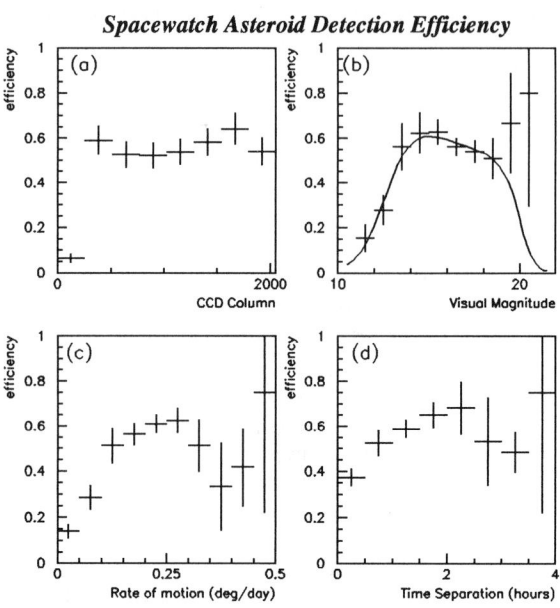

Fig. 2a. Spacewatch asteroid detection efficiency as a function of CCD column. b. efficiency vs. asteroid's rate of motion. c. efficiency vs. apparent magnitude. d. efficiency vs. time separation between observations.

Fig. 2b gives the efficiency as a function of the visual magnitude of the asteroid. A plateau is visible between magnitude 13.5 and 18.5 along with a steep drop for bright objects with the efficiency reaching about 10% of its maximum near magnitude 10.5. The apparently high efficiencies in the last two bins are significantly different from the efficiency determined by the method of Rabinowitz (1993b). His technique extrapolated the observed magnitude distribution of all asteroids detected by Spacewatch, from a range in which the detection efficiency was thought to be nearly constant, to the high magnitude limit. Dividing the actual by the extrapolated distribution gave an indication of the efficiency at faint magnitudes. The problem with his technique is that it assumes the efficiency is flat in some otherwise unknown range of magnitude, that the real distribution of asteroid magnitudes is represented accurately by a power law over the entire range, and that the choice of scanning regions has not biased the apparent magnitude distribution in an unusual manner. Unfortunately, at this time, it represents the only available measure of the efficiency near the magnitude cutoff of the Spacewatch system. The curve superposed on the efficiency vs. magnitude plot is a fit to a function of the form:

$$\epsilon(m) = N(sm+b)\left[\frac{1}{1+\exp(\frac{m-m_H}{w_H})} - \frac{1}{1+\exp(\frac{m-m_L}{w_L})}\right]$$

where N is an overall normalization, $(sm + B)$ provides the sloping "plateau", and the last term provides a "smoothed box car" function spanning the range $m_L < m < m_H$). The efficiency falls off at the lower and upper edges over a range given by w_L and w_H respectively. In practice, m_H and w_H were fixed at values consistent with the shape of the efficiency curve obtained using the method of Rabinowitz. The values of the parameters are: $N = 0.95 \pm 0.51$, $s = -0.032 \pm 0.010$, $b = 1.1 \pm 0.3$, $m_L = 12.7 \pm 0.3$, $w_L = 0.73 \pm 0.22$, $m_H = 20.0$, $w_H = 0.4$.

The ill-defined cutoff at high magnitudes is due to the intrinsic limits of the Spacewatch camera and variability of local conditions. The drop at bright magnitudes is due to saturation of the point sources at visual magnitudes of about 12. The "plateau" is barely consistent with being flat and the maximum efficiency is about 62% at magnitude 15.

The efficiency as a function of the rate of motion of the asteroid is shown in Fig. 2c. Since numbered asteroids were used in this study it is no surprise that this range of rates is a good representation of typical main belt rates observed at opposition. The reduced efficiency for low rates of motion is due to the requirement that an object must move at least a couple pixels during the interval between passes in order to be detected as non-stationary. Since these plots represent an average over all scan intervals (which range between 10 and 90 minutes) the minimum detectable rate varies by about an order of magnitude and this is reflected in the gradual increase in efficiency to about 50% near 0.125°/day. The efficiency for detecting typical MB objects ($\omega \sim 0.2°$/day) is good. The apparent decrease in efficiency for faster rates may be due to the fact that when the time interval between scans is large, faster objects move outside the 120 pixel search radius. It is also possible that the decrease in efficiency is due to a problem in determining the centroid of slightly trailed objects. If the centroid is improperly determined, the link between scans, which depends upon an accurate measure of the rate of consistent motion, may be inefficient.

Finally, the last of the plots in Fig. 2 shows the efficiency as a function of the time separation between the first and third of the observations. The efficiency for MB asteroids increases to a maximum at about two hours because the slower moving asteroids are detected more often when there is sufficient time between the observations. The decrease in efficiency at greater times is due to the fixed search radius within MODP. It is difficult to have a variable search radius due to the fact that the number of false candidates increases as the square of the radius as does the computing time required for the search. Thus, the empirical setting of 120 pixels maximum search radius creates an effective variable limit on the maximum detectable asteroid rate which is inversely proportional to the time separation between scans.

Figs. 3a-c show the efficiency of the Spacewatch camera and software systems as a function of three orbital parameters for the numbered asteroids. If all the asteroids were on purely circular orbits there would be an exact correlation between their distance and rate of motion at opposition. Since the efficiency is rate dependent this would cause an apparent relationship between the detection probability and the orbit of an asteroid. In actuality, the spread in eccentricities of asteroids with a given semi-major axis within the main belt is wide enough to mask the rate dependent effect. In addition, if the software preferentially found objects moving in either right ascension or declination there would be a bias towards or away (respectively)

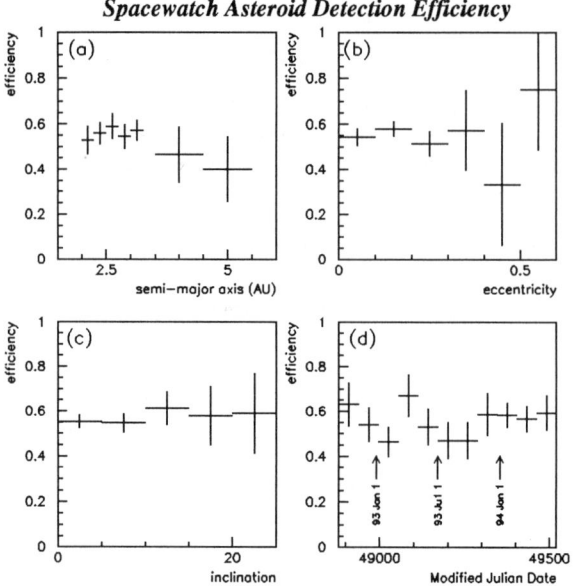

Fig. 3a. Spacewatch asteroid detection efficiency as a function the semi-major axis of an asteroid's orbit, b. efficiency vs. asteroid's orbital eccentricity, c. efficiency vs. asteroid's orbital inclination, d. efficiency vs. Modified Julian Date.

from objects of low orbital inclination. Fortunately, the efficiency is consistent with being constant in all three parameters of semi-major axis, eccentricity and inclination.

The first five data points in Fig. 3a are equally distributed through the main belt while the last two points correspond to the Hilda and Trojan mean semi-major axes respectively. Although the data are consistent with being constant, the fact that Trojan asteroids are almost twice as distant as the average main belt asteroid does result in their rate of motion being about one-half that of the main belt asteroids at opposition. Thus, short intervals between successive observations does bias against detection of the Trojan group of asteroids.

In the interest of successfully debiasing observations for detection efficiency it is desirable to have a system operating consistently over time. Fig. 3d shows the efficiency as a function of time and indicates that the efficiency has been more consistent since this study was performed in late 1993. Unavoidable temporal variations in the detection efficiency may continue to linger due to seasonal changes in weather/seeing and also in the density of stars per unit area. When seeing conditions are bad the stellar point spread function is spread out over more pixels which decreases the probability that an asteroid image will be detected by examining the individual pixel's signal-to-noise (S/N $\equiv \sigma$) ratio. The inverse relation between stellar image density (stars/area) and detection efficiency is due to two factors: (1) the opportunity for increased confusion between faint stars (which may appear in one image and not in the next due to seeing variations) and asteroid candidates and (2) the increased likelihood of at least one of the asteroid's images lying within the saturated area of a star's image.

The plots in Figs. 2 and 3 represent the efficiency for the automated real-time detection of asteroids by MODP using the consistent motion technique. The technique is amenable to quantitative determinations of its absolute efficiency using the numbered main belt asteroids. However, the results are not strictly applicable to the majority of Spacewatch's NEAs since they are distinguished as such from the plethora of main belt objects by their characteristically high rates of motion - in the upper range of rates explored (and beyond) using the technique described here.

Many of the NEAs found by Spacewatch are close enough to the Earth that their images are trailed due to motion during a typical 150 second exposure. The efficiency of MODP's streak detection ability will be difficult to calculate as will the efficiency of the observer's ability to detect faint trailed asteroid images. Not only might there be differences in the ability of each observer to locate and identify these faint trails at the one and two σ level, there may be subtle variations in the mind's perception of faint trails in the vertical, horizontal and intermediate directions. Faint meteors, cosmic rays, strangely "linear" edge on spiral galaxies, subtle variations in the sky background, etc., may inadvertently pique the interest of an observer and, in dense regions of the sky, litter the CCD image with false FMO and VFMO candidates. All of these effects will introduce a bias into the observations which makes it difficult to interpret the data.

Many of the VFMOs detected by the observers have been trailed over hundreds of pixels with peak S/N in individual pixels along the trail of only one or two. Their detection relies on a good background subtraction before being displayed. At the present time, the correction is applied by removing the median background signal in bands 50 pixels high over the full 2048 column width of the CCD. Since the images are formed by drift scanning, every pixel in a CCD column contributes an equal share to the final image. In this way, pixel efficiency variations in the CCD row direction are averaged and, to first order for a good CCD, there is no need for flat-fielding. In the Spacewatch system, a remnant 1 - 2% difference in the CCD response is visible and should be applied as a one-dimensional column correction in order to flatfield the resulting image. Applying this correction will provide greater opportunity for discerning the faint VFMOs within the background noise at the 1 - 2 σ level. The detection of the NEAs through consistent motion is not affected by the column dependent flatfield because MODP does a local background subtraction around each possible source.

4. CONCLUSIONS

Any analysis of the Spacewatch data to determine a magnitude-frequency relationship must compensate for the efficiency of the detector system as a function of apparent magnitude. In addition, studies of the orbital element distributions require knowledge of the efficiency as a function of those observable parameters which depend upon the significant orbital elements. This study has presented a technique which is being used to determine the magnitude-frequency relation of the main belt asteroids and which will be equally applicable to any set of Spacewatch asteroids detected through consistent motion by MODP (NEAs, Trojans, etc.). Other techniques are being developed to better explore the efficiency in the faint magnitude limit.

The mean efficiency of about 57% in the range from visual magnitude 13.5 to 18.5 implies that the real-time detection system is missing about 1/3 of the available asteroids. The most likely culprit for the loss of these asteroids is the routine which searches for sources within the CCD image by comparing the signal within individual pixels to the background noise level. When MODP was written there was insufficient computing resources available for a more sophisticated search for point sources and an impression that the search must be performed in real-time. With the purchase of faster computers and the development of new software it may be possible to achieve a significantly higher detection efficiency.

REFERENCES

Rabinowitz, D. L. 1993a Nature 363, 701
Rabinowitz, D. L. 1993b AJ 407, 412
Van Houten, C. J., Van Houten-Groenevald, I., Herget, P. and Gehrels, T. 1970 A&AS 2, 339

DISCUSSION

PHILIP: Is there any interest in your group in the stellar information contained in your scans? You should have much interesting information concerning some types of variable stars if you scan areas three times.

JEDICKE: There is a lot of interest in our group in all of the information contained in our scans - there is just not enough time to utilize it fully! We are always seeking useful collaborations.

GUSEVA: What is the precision of your positions and magnitudes? May be, the precision in magnitudes is worse due to the drift-scanning mode?

JEDICKE: Our astrometric positions are good to about 1/4" while our magnitudes are probably good to about 0.3 to 0.5 magnitudes. The drift scanning technique does not contribute to the error in the measured magnitudes.

IWERT: What would you call the ideal CCD for this program and of what kind of options in the CCD and the controller you could think to carry out this program even more effectively in the future?

JEDICKE: The ideal CCD and controller system is intricately linked to the telescope. In general, large scale or mosaic CCDs, with fast and flexible readout for both drift scan and stare modes of operation are desirable.

SCHILLING: What would be detection frequency of objects on a collision course? (based on current statistics)

JEDICKE: I believe that the detection frequency of objects which will actually collide with the Earth is very small at current rates. The smallest objects detected by spacewatch are amongst the largest meteors - about five - ten meters in diameter. These objects strike the Earth perhaps once a year. The probability of detecting one prior to impact is very low.

SZECSENYI-NAGY: What are the limitations of this program in sky coverage? I mean what is the declination zone you are able to "space watch" regularly?

JEDICKE: The declination scanning limit is set by the amount of "smearing" we accept in the CCD image due to the effect of differential rotation in declination across the 0.5 width of the CCD in declination. We try to scan only within 20° of the equator.

CRAWFORD: Does the observer see space debris streaks? Are they being logged?

JEDICKE: The space debris is being seen and logged but not followed up.

CCD TIME-SERIES PHOTOMETRY OF ASTRONOMICAL SOURCES

Steve B. Howell

Planetary Science Institute

ABSTRACT: CCDs are essentially the only instrument available today for photometry at most observatories; they are also becoming more readily available to amateurs as well. Thus, obtaining good photometric data with these two-dimensional devices is something we all need to understand. The history of and recent developments in CCD time-series photometry will be reviewed with some comments on future directions.

1. BACKGROUND

The first reference to performing two-D photometry that I believe exists is in a paper by Nather (1972) in which he states, "Multi-channel detectors can be of real benefit in the photometry of faint objects...", and, "the accuracy of the photometric measurement is limited by the amount of "sky" present in the [measurement] diaphragm,...the selection of a "virtual" diaphragm can materially improve the accuracy, if chosen to minimize the statistical error..." Well, that about says it all right there. It took a number of years for CCD technology and the rest of us to catch up with these principles, but we finally made it.

Howell and Jacoby (1986) presented the first detailed treatment of using CCDs for time-series photometry. They show in their examples that marginally cloudy nights, instrumental effects, and color terms are essentially neutralized and mostly avoided if one uses differential photometric techniques. Of course, on a CCD all objects have the same exposure time, so the S/N values varies with object brightness. Therefore, Howell et al. (1988) presented the necessary quantitative basis for working with unequal S/N value and the correct error analysis methods for such two-D differential measures.

There are many details of CCD photometric observations that are different from PMTs. These are not always intuitive but must be understood in order to gather meaningful, well understood data. Howell (1992) discusses the relative merits of two-D detectors over PMTs and Kreidl (1993) compares CCDs to PMTs for photometric observations of astronomical sources.

The history of and recent developments in CCD time-series photometry are reviewed with some comments on future directions. Many astronomical projects have built on these foundations and the current literature is full of uses for CCD time-series photometry. These range from variability of extragalactic objects (e.g., Miller et al. 1992) to rotation periods for comets in our solar system. Gilliland and Brown (1988, 1992) have made extensive use of differential techniques in their work with ensemble photometry in clusters where they have reached RMS precisions of millimags! The field of differential CCD photometry is certainly brightening!

2. PHOTOMETRIC TECHNIQUES

For brighter, high signal-to-noise sources, the data collection and reduction processes are robust and use of even a marginally correct "CCD Eq." and almost any good software package for reduction can provide good photometric results and error assessment. For fainter, lower signal-to-noise sources or undersampled data, however, there are many factors that one must take into account to get correct answers *and* correct error estimates. Methods to deal with these types of sources both in terms of a proper "CCD Eq." for use before (to predict the outcome) and after (to provide proper error estimates) is necessary. One also has to be very careful in terms of the software used for data extraction and reduction. Sources with signal-to-noise values of 20 or lower can provide very accurate data sets through the use of specific techniques such as optimum data extraction techniques and growth curve (aperture) corrections.

Howell (1989) discusses in detail the method of optimum data extraction. Merline and Howell (1995) present a number of examples and reinforce the findings that the optimum radius for point-source extraction is near one FWHM (standard wisdom generally uses a radius of ~three FWHM) and that the optimum radius is in general, different for each point source. Fig. 1 shows the results for stars of various magnitudes: V = 18, 17, 16, 15 and 14 from top to bottom. The stars have a FWHM of 2.4 pixels and we see that the optimum extraction radius (i.e., the least magnitude error) occurs near one FWHM, but not at it exactly. Fig. 2 shows some examples of the resultant S/N ratio for optimum vs. standard aperture extraction for three cases. The three cases are listed on the figure as sequences of numbers corresponding to the following: telescope aperture in meters, f-ratio, detector read noise, detector dark current in electrons/sec, detector gain in electrons/DN, sky noise in electrons/sec, and pixel scale in arc sec/pixel. One can easily see that an increase of > two in S/N can be achieved through the use of the optimum extraction technique.

DaCosta et al. (1982) first discussed the use of growth curves for the reconstruction of stellar profiles in the case of low S/N objects. Howell (1989) talks about growth curve usage for faint and crowded stars and Stetson (1990) discusses an implementation of using growth curves to get better results from faint and crowded stars in clusters. The basic idea is one of obtaining growth curves for brighter point sources in a given CCD frame (essentially another version of 2-D profile fitting but with azimuthal averaging), and then using these curves to guide fits for the fainter, not well detected point sources. See Stetson (1990) for further details.

The CCD equation itself (listed in many places, including typical observatory users manuals) has also been re-examined to look for further improvements that might be made in observational predictions and in error analysis (e.g., Newberry 1991). Merline and Howell (1995) present a detailed look at this equation and the errors involved and they discuss some generally overlooked terms. These include terms for the error contribution due to the detector gain and the effect of the number of background pixels used.

An area that is getting more attention these days, especially as some older telescopes are being outfitted with CCDs, is that of undersampled data. Generally this is only thought about in terms of data from satellites and spacecraft (e.g., HST/WFPC, Holtzman 1990). Buonanno and Iannicola (1989) provided an initial look into the case of undersampling for ground-based data as well. Generally this problem is encountered for large field-of-view telescopes, such as

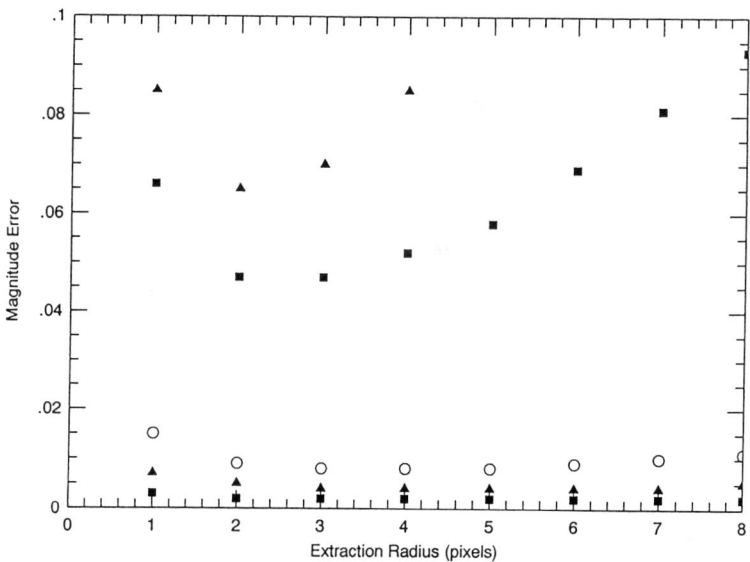

Fig. 1. Magnitude error as a function of Extraction Radius.

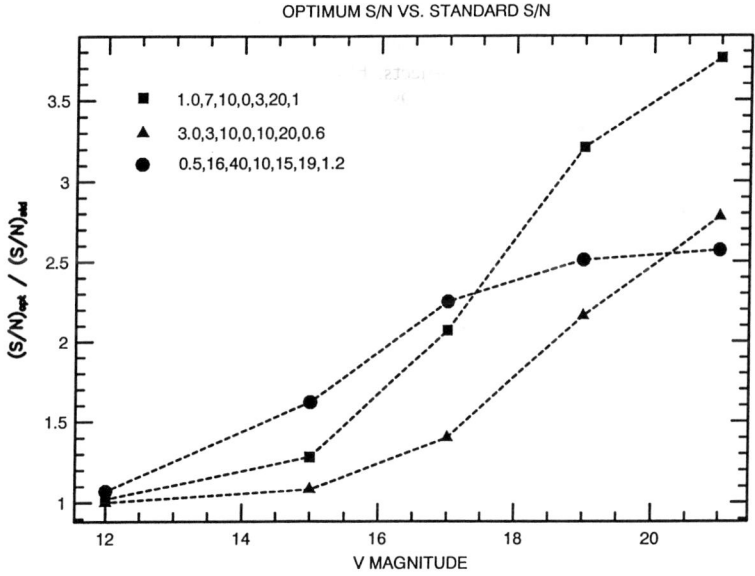

Fig. 2. Optimum S/N vs Standard S/N as a function of magnitude.

CCDs used in Schmidt telescopes. The term undersampled can be defined by the following parameter:

$$r = \frac{FWHM}{p}$$

where *FWHM* is the full-width at half-maximum of the point spread function and p is the pixel size, both in the same units. For the sampling parameter $r \leq 1.5$, errors are highly likely to be expected in the photometry from use of standard analysis techniques. Most popular software packages for CCD reduction are NOT prepared to deal well with undersampled data. Astrometric and photometric errors tend to track each other and, for CCD data with poor sampling, Gaussian fits, for example, are no longer valid, partial pixel handling by the software becomes critical, X,Y moment determinations for centering are suspect, and detection and differentiation of low S/N point sources is difficult. Further work in the area of undersampled data is needed.

3. CONCLUSIONS

The future of CCD photometric observations is only limited by our imagination and how well we can make use of software to perform real-time functions at the telescope and reduction and analysis of the resultant images. The next few years will be a golden time in working with CCDs. For example, Fig. 3 shows a simulation from Howell (1995) of the possibility of relatively easy detection of transits of extra-solar planets from a ground-based telescope. This figure shows a typical M2 star of 15th magnitude being transited by three solar system sized giant planets. To give you an idea of what the x-axis scale might be, a Jupiter-Sun transit viewed from outside our solar system would take about 30 hrs. Large CCD mosaics are already collecting more data in a few nights of observing than many years of single CCD observations combined, with the data collection, reduction, and analysis is becoming routinely automated. Also, the promise of CCDs in space or on the moon allows for a new horizon of photometric possibilities (Granados and Borucki 1994).

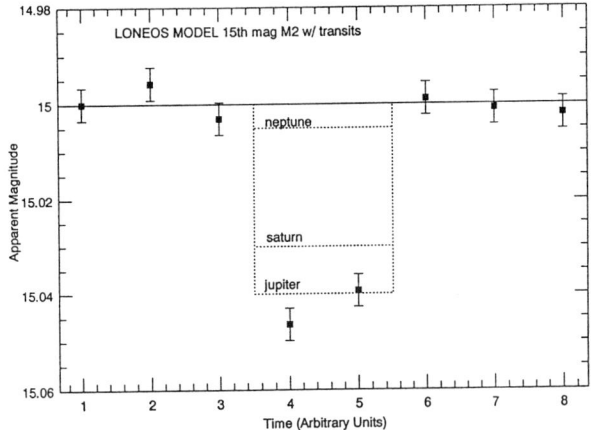

Fig. 3. Apparent Magnitude vs time.

REFERENCES

Buonanno, R. and Iannicola, G. 1989 PASP 101, 294
DaCosta, G. S., Ortolani, S. and Mould, J. 1982 ApJ 257, 633
Gilliland, R. L. and Brown, T. 1988 PASP 100, 754
Gilliland, R. L. and Brown, T. 1992 PASP 104, 582
Granados, A. and Borucki, W. 1994 in Proceedings of the First FRESIP Workshop, A. Granados and W. Borucki, eds., NASA Conference Publication 10148
Holtzman, J. A. 1990 PASP 102, 806
Howell, S. B. 1995, to be published
Howell, S. B. and Jacoby, G. 1986 PASP 98, 802
Howell, S. B., Mitchell, K. J. and Wanock, A.. 1988 AJ 95, 247
Howell, S. B. 1989 PASP 101, 616
Howell, S. B. 1992 in Astronomical CCD Observing and Reduction Techniques, S. B. Howell, ed., ASP Conference Series, Vol. 23.
Kreidl, T. J. 1993 in IAU Colloquium No. 136, Stellar Photometry - Current Techniques and Future Developments, C. J. Butler and I. Elliott, eds., Cambridge University Press, Cambridge, p. 311
Merline, W. J. and Howell, S. B. 1995 Exp. Ast., in press
Miller, H. R., Noble, J. C., Carini, M. T., Urry, C. M., Sokoloski, J. and Howell, S. B. 1992 in Testing the AGN Paradigm, AIP Conf. Proceedings No. 254, S. S. Holt, S. G. Neff and C. M. Urrey, eds., AIP, New York
Nather, R. E. 1972 PASP 84, 149
Newberry, M. V. 1991 PASP 103, 122
Stetson, P. B. 1990 PASP 102, 932

DISCUSSION

JEDICKE: Is the MACHO data suitable for detections of Jupiter-like objects around stars in the Magellanic Clouds?

HOWELL: It may be and I am currently working with that group on such possibilities.

JEDICKE: How will you account for the undersampling of the LONEOS data in photometry intended for discovery of Jupiter-like transits of other stars?

HOWELL: The pixel size will be 1.8 arc sec and the seeing is typically 1.6 arc sec, so the data will be marginally undersampled. Techniques to deal with these types of data have been developed for other projects and include optimum aperture extraction.

TANCREDI: What is the typical transit duration of a Jupiter-sized planet?

HOWELL: For the Sun-Jupiter system viewed from a star, the transit time across the Sun's equator would be 30 hours.

CRAWFORD: In the result of finding one or two Jupiters per year, what one specific assumption of how many telescope hours, field site, etc?

HOWELL: That number is based on using the assumptions (best current guesses) of a Schmidt with a mosaic of four 2048 x 2048 CCDs, covering 1000 sq. degrees/night.

HIGH PRECISION STELLAR PHOTOMETRY WITH CCDs. I.

A. J. Penny

Rutherford Appleton Laboratory

ABSTRACT: This talk is about the limits to the precision of stellar photometry in comparing one star with another in a single CCD frame. This is concerned with bright stars, and concentrates on three problems at the 0.1 percent level of accuracy: how to flatfield; how to deal with varying point-spread-functions that vary across an image; how to deal with the fact that the response inside a pixel is not uniform. The first is the well-known difficulty of getting a uniform illumination across the CCD to use as a flatfield; the use of a rotatable CCD mounting and of drift-scanning is discussed. The second depends on the ability to detect and define small, but significant, changes in the PSF. The third is the fact that the pixels of optical CCDs can have non-uniformities inside them of ten percent, and these when folded with the PSF produce systematic errors significant at the 0.1 percent level; with infra-red arrays these problems can be much worse. The use of software to model these variations and reduce these errors is described.

HIGH PRECISION STELLAR PHOTOMETRY WITH CCDs. II.

Michael S. Bessell

Mount Stromlo and Siding Spring Observatories

ABSTRACT: CCD photometry is capable of high internal precision, however there are several important requirements necessary to attain high precision in standardized photometry. Firstly, the CCD passbands must match as closely as possible the standard passbands; secondly, new faint standards must be set up in several declination zones and thirdly, for convenience a sufficient number of standards covering a good range in color should be obtained on a single CCD frame so that several different frames should suffice for standardization. Landolt has taken the first steps in defining several such fields. The small systematic differences between different UBVRI systems have been examined and transformations can be applied to the photometry of Landolt and Bessell to place it on the Cape - SAAO system.

1. INTRODUCTION

Modern broad-band photoelectric photometry such as the UBVRI system established by Cousins at the Cape and extended by his colleagues at Sutherland (e.g. Cousins 1980, Menzies et al. 1991) and Landolt's equatorial version of the UBV and the Cape RI system (Landolt 1983, 1992) and the Washington system CMT_1T_2 (e.g. Geisler 1990) have been shown to be of high precision and capable of tackling a wide range of astrophysical problems. However, just as widespread adoption of the Cousins system for photoelectric photometry is achieved, photometry is increasingly being done with CCDs and we are now faced with the task of ensuring that the internally precise CCD photometry can be precisely transformed onto the standard system. In this paper I would like to discuss the steps that must be taken to attain this end.

2. SCARCITY OF SUITABLE CCD STANDARDS

The first problem is the scarcity of standards relevant for CCD work. Most of the E-region standards (Menzies et al. 1989) and many of the equatorial standards (Menzies et al. 1991) are too bright for CCD work with 1 - 3-m class telescopes and the faintest stars in these e-regions (48 stars fainter than tenth mag) cover only a small range in color. Just as importantly, we need to make many standards per frame. Single star observations with a CCD are very limiting because few CCD observers are prepared to observe as many individual standard stars as normally observed by photoelectric observers for adequate standardization. It is very wasteful of both observing time (large CCDs can take several minutes to read out a complete frame) and reduction time to have only a single standard in a frame.

Landolt (1992) has recently provided excellent UBVRI standards in several convenient CCD-sized fields. This photometry is based on the equatorial standards of Landolt (1973, 1983) which have been carefully compared with the Cousins system by Menzies et al. (1991). Some systematic differences are evident and probable reasons for the differences in the UBV colors have been discussed by Menzies et al. (1991), Menzies (1993) and Bessell (1990b). That comparison was essentially for stars bluer than (V-I) = 2.0 (earlier than spectral type M1). For the redder stars, larger systematic differences in (V-R) and (V-I) of up to 0.05 mags between the photometry of Landolt, Bessell and Weis exist. Such differences are not unexpected given that there were few stars in Cousins lists as red as (V-I) = 2.0 and none redder than (V-I) = 2.7 (the latest M dwarfs have (V-I) = 4.4). However, careful intercomparisons between the various data sets enables good transformations to be made onto the Cousins (Cape-SAAO) system.

3. TRANSFORMATION OF LANDOLT UBVRI TO THE CAPE SYSTEM

Laing (1989) has published UBVRI photometry for about 400 faint nearby stars. From the 260 stars common to Bessell (1990a) it has been possible to correct Bessell's photometry for stars with (V-I) ≥ 1.50 and place it on the Cape system. Then by combining Bessell's corrected photometry for about 20 M dwarfs common to Landolt with the Menzies et al. (1989, 1991) data for the bluer stars it has been possible to derive polynomial fits to the differences (Cape - Landolt) and so transform Landolt's photometry onto the standard Cape system (Bessell 1995). The coefficients of the polynomials are given in Table 1. These functions permit transformations to within 0.01 mag for most stars [0.02 for (U-B)] and can be used for standard colors in Landolt's CCD-sized fields until direct CCD measurements within the Cape system are available. The transformed Landolt (1992) data is available by anonymous ftp from pub/bessell at MERLIN@ANU.EDU.AU.

TABLE 1

Coefficients of polynomials to correct Landolt colors to Cape colors.

	A0	A1	A2	A3	A4	A5
(B-V)	-0.005155883	-0.000743423	-0.069862929	-0.055528320	-0.004085570	-0.001869369
(U-B)	-0.013029154	-0.010560242	-0.018461003	-0.022517123	-0.0032750981	
(V-R)	-0.000595208	-0.006163583	-0.027564391	-0.038906376	-0.008676337	
(R-I)	-0.000548123	-0.002518544				
(V-I)	-0.001104410	-0.012484664	-0.005182837			

$$X_{Cape} = XL_{Landolt} + A0 + A1*X_L + A2*X_L^2 + A3*X_L^3 + A4*X_L^4 + A5*X_L^5$$

Bessell and Weis (1989) derived precise transformations between the supposed Cousins system colors of Bessell (1990a) and Kron system colors of Weis (1984, 1986, 1987). With the readjustments to Bessell (1990a) colors, transformations between the Weis Kron system and the standard Cape system have also been rederived (Bessell 1995). Landolt's equatorial fields are an excellent start for CCD standards but it will be advantageous to derive more standard stars within these fields and within additional fields containing some of the very red stars. CCD-sized standard fields need also to be set up in the E-regions, the F-regions and near the Magellanic Clouds in the southern hemisphere and in other declination zones in the north.

Graham (1982) has provided some valuable faint standards in E-region fields. However, these stars have a very restricted color range (there are no very blue stars) limiting their usefulness. Comparison between Menzies et al. (1989) and Graham indicate systematic differences for red stars in all colors. Graham and Landolt used the same B filter that appears to have introduced systematic differences in the (B-V) and also the (U-B) colors. Walker (1991) has examined the (U-B) differences in detail and recommends the following transformations be applied to Graham's (U-B) colors:

(U-B) = (U-B)$_G$ - 0.130(B-V) + 0.015, for 0.0 < (B-V) < 0.2
(U-B) = (U-B)$_G$ + 0.046(B-V) - 0.020, for 0.2 < (B-V) < 1.0
(U-B) = (U-B)$_G$ - 0.060(B-V) + 0.086, for 1.0 < (B-V) < 1.7.

Graham's fields are again a good start for E-region standards; however, they also need to be supplemented with more stars per field and with some bluer and redder stars.

It must be emphasized that all the standards we discussed above are from photoelectric photometry and have therefore been observed through large apertures of at least 14 arcsec. As inspection of deep CCD frames of many of the stars indicates that there are often faint companions within such an aperture; the published magnitudes must include the contribution of these companions. Therefore when using the photoelectric standards as CCD standards, the CCD derived magnitude for that standard must be evaluated for a 14 arcsec aperture (or whatever aperture size was used) and not for a small aperture of from profile fitting. For this and other reasons it would be preferable to use CCD photometry instead of photoelectric photometry when setting up CCD standards. But before attempting that, we must ensure that the passbands used for CCD photometry are as close as practicable to those of the standard photoelectric system or at least understand the systematic differences that will result from mismatched passbands.

4. THE PASSBANDS OF CCD PHOTOMETRIC SYSTEMS

Bessell (1990b) has devised passbands that represent the standard UBVRI system. Guided by these passbands one can compute CCD passbands by convolving the transmissions of glass filter combinations and the sensitivity functions of CCDs until they match as closely as possible the standard passbands. Synthetic photometry can then be made by folding the CCD passbands with the data in spectrophotometric atlases, such as the Vilnius spectra (Straižys and Sviderskiene 1972) or the Gunn-Stryker spectra (Gunn and Stryker 1983; corrections given by Rufener and Nicolet 1988). Some CCD passbands were discussed in Bessell (1990b) as were differences in photoelectric systems due to mismatched passbands.

One of the additional complications we now face is that there is much diversity in the wavelength response of CCDs. Originally, most CCDs comprised thick Si photoreceptors which provide high R and I response, lower B and V response and little U response, but through a variety of techniques, such as thinning the Si, electrically treating the backsurface, UV flooding or the application of lumogen or dyelaser coatings, it is now possible to obtain a wide variety of response curves. The U and B responses and to a lesser extent the V response show the most variation. Fig. 1 shows the dramatic differences in CCD sensitivities. All the responses differ from that of the GaAs photocathode with which the standard photoelectric observations were made. This response was basically flat between 300 nm and 850 nm and had a rapid cutoff below 870 nm.

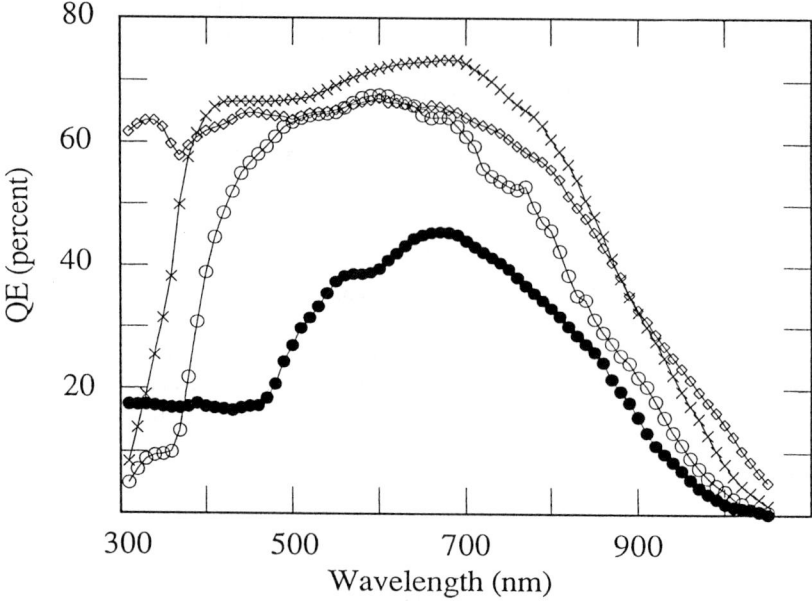

Fig. 1. Sensitivity curves for selected CCDs.

Because of the significantly different responses of some CCDs it has been found for some passbands that it is best to use different glass filter recipes for different CCDs. Walker (1992) shows examples of UBVRI transformations with different CCDs and different filter recipes. Some recipes have been made for total thicknesses of four mm but a thickness of five mm permits more flexibility in recipes and better finetuning of the passbands. In addition, because 1 mm thicknesses of some glasses are difficult to work optically in the large sizes needed for some applications, they need to be replaced by a thicker piece of an alternate glass. Each of the UBVRI CCD passbands will now be discussed in turn.

4.1 The U passband

As pointed out in Bessell (1990b) the photoelectric U band has been quite well defined by most observers and some of the problems in (U-B) systems seem to have arisen from the B band and much from the transformation techniques used; regressions against (U-B) or (B-V) for example. One of the main problems in U photometry with a CCD is the red leak of the UV glasses. The other problem is the low and steeply varying sensitivity of the CCDs across the U band and its variation between CCDs. The coated CCDs and the UV flooded CCDs offer a flat response, so the normal 1 mm UG1 or 3 mm UG5 will define the red cutoff adequately. With the thinned CCDs, the restricted and rapid drop off to shorter wavelengths in some CCDs response necessitates a thicker piece of UV glass to keep the long wavelength side of the band sufficiently blue. The problem with the red leak remains. It was possible with the GaAs photoelectric filter to use liquid or solid copper sulphate to eliminate the red leak. That is still possible, but the larger sizes of some filters and the good imaging quality required of the filters in some optical systems has forced the consideration of alternatives. The Schott BG39 and BG40 glasses with their improved UV transmissions are very useful in this context although they do restrict quite severely the flux shortward of 340 nm; however, the new glass S8612 available from Schott Glass Technologies, Inc., Duryea, Pennsylvania has the same red cutoff as the BG39 glass but has much better UV transmission and seems to offer good transformation possibilities. (I am grateful to Alistair Walker for pointing out this glass to me.) Some interference filter manufacturers offer a coating that appears to eliminate the red leak; however, the reduction in far-red transmission is inadequate for UV observations of red stars with CCDs. The high red sensitivity of some CCDs compared to their UV sensitivity and the low UV flux of some stars (K giants in particular) compared to their red flux can mean that a red leak of a fraction of a percent can still dominate the observation. It is important therefore that the red leak be extremely low for UV work on red objects while for hot objects it is not as important. However, it is best to try for a very low red transmission to avoid later unforseen problems. The 1 mm BG39 recommended in Bessell (1990b) for use with UG1 still has a small red leak for the reddest stars; 3 mm BG40 has negligible leak as does 1.5mm of S8612. The left hand side of Fig. 2 shows the UV transmittance of 1 mm UG, 1 mm BG39 and 1 mm S8612. On the right hand side is shown the far-red transmittances for the same glasses (RH scale); the lowest curve shows the product of 1 mm UG1 and 1 mm S8612, the red-leak. An additional 0.5mm of S8612 or BG39 would apparently reduce the red-leak by a factor of about ten.

Using BG39, 40 or S8612 to stop the red leak causes an unavoidable loss in UV sensitivity and a cutoff in the blue side of the response, but the resultant photometry appears to be transformable, although there is a large non-linearity for the bluest stars. In Fig. 3 are shown the magnitude differences between instrumental u (measured with a UV flashed Tek CCD and glass filter) and standard U. Fig. 4 shows the magnitude differences between the glass and $CuSO_4$ cutoff U filters. Ugl is for the filter comprising 1 mm UG1 + 2 mm S8612, uCu for 1 mm UG1 + 5 mm $CuSO_4$. Individual stars of Menzies et al. (1991) were observed.

Possible combinations for U are:

1 mm UG1 + 5 mm liquid $CuSO_4$ (for thinned, UV flooded or coated CCDs) or
1 mm UG1 + 2 (or 1.5 mm) S8612 (for UV flooded or coated CCDs)

1 mm UG1 + 1 mm BG39 + one mm BG40 (for UV flooded or coated CCDs) or
2 mm UG1 + 3 mm BG40 (for thinned CCDs).

4.2 The B passband

The B passband has continued to cause some difficulties. The variation in the sensitivity of some CCDs across the B band is high and differs a lot between CCDs. This causes quite a large shift in the effective wavelength of BCCD if the same filters are used. As briefly touched on above and discussed in more detail in Bessell (1990b) the B band used by Landolt and Graham in their photoelectric observations introduced systematic differences into their (B-V) and (U-B) colors compared to the standard Cousins system colors for some stars. The glass filter combinations for B of most CCD observers has been closer to those used by the Cape-Sutherland observers than to that of Landolt and Graham, consequently more linear relations are measured after transforming the Landolt and Graham standards. Nevertheless, the reasonable mix of 1 mm GG385 + 1 mm BG12 + 2 mm BG39 (the BG39 is to stop the

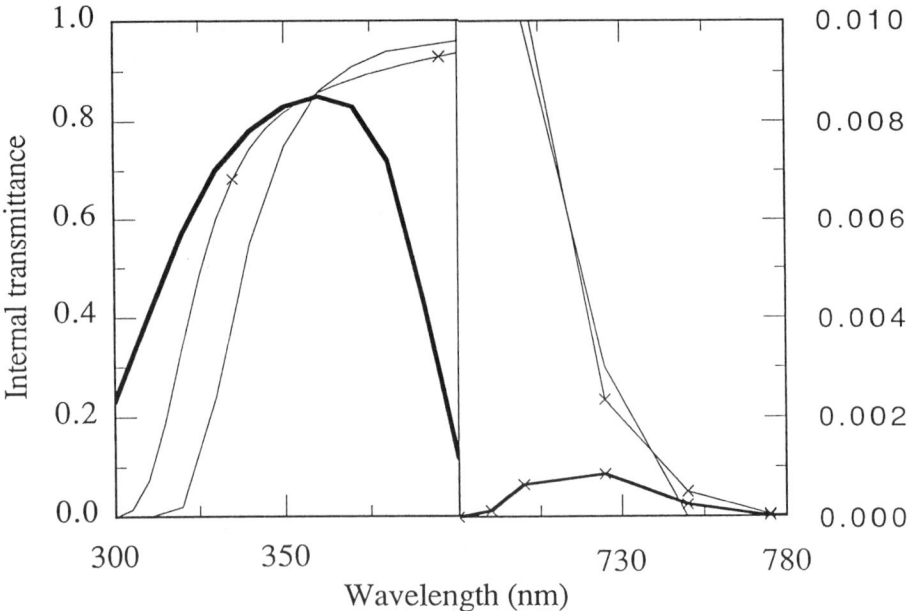

Fig. 2. Transmittance of UG1, BG31 and S8612 glass.

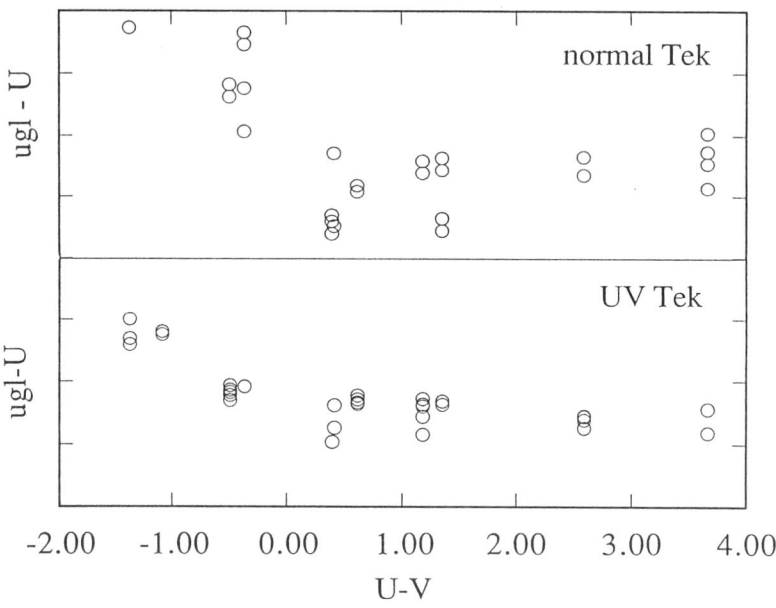

Fig. 3. Differences between natural and standard U magnitudes. Y axis ticks are 0.1 mag.

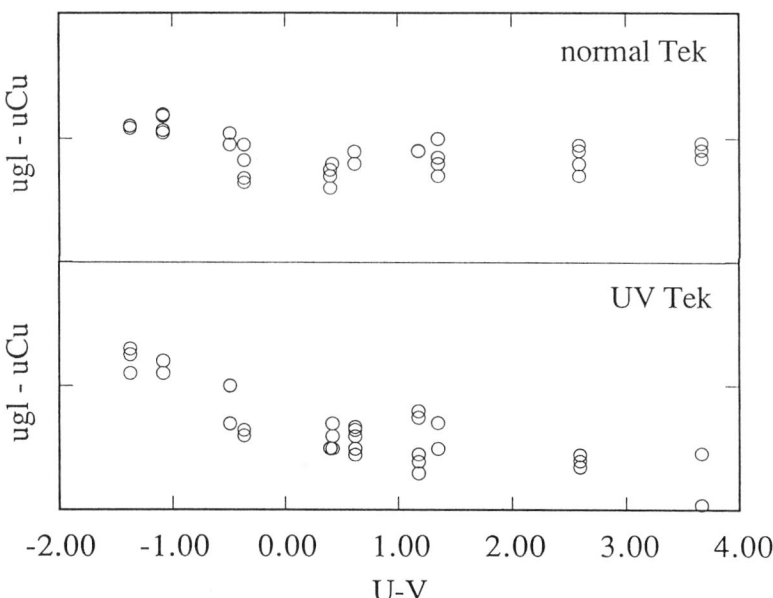

Fig. 4. Differences between glass and CuSO4 blocked responses. Y axis ticks are 0.1 mag.

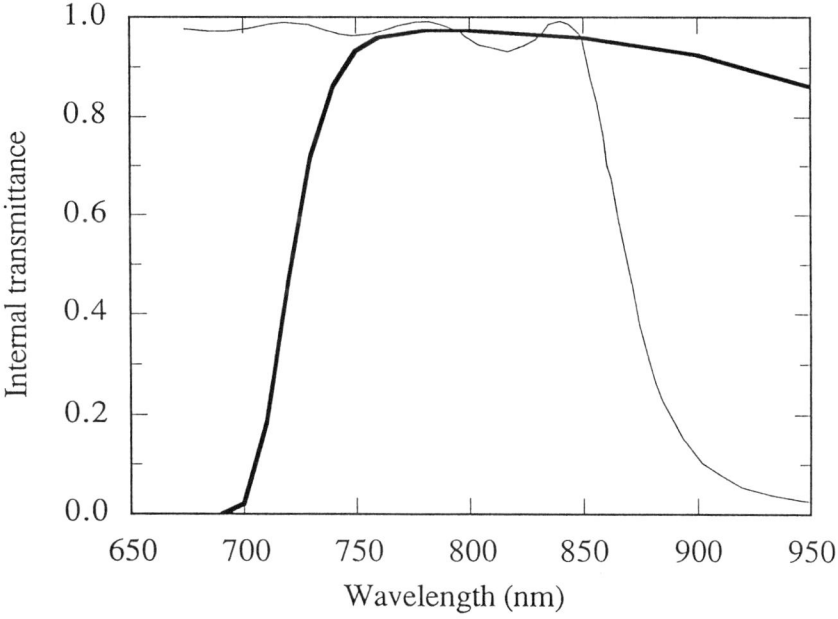

Fig. 5. Transmittance of RG9 glass and interference far-red cutoff filter.

red leak of the BG12) produces a too red response for most CCDs. Walker (1992) notes that with a thinned Tek CCD this filter has a color correction of 0.18(B-V) compared to 0.05(B-V) with a UV flooded TI CCD and 0.07(B-V) with a coated Thomson CCD. Replacing 1 mm BG12 with 2 mm BG1 for the thinned Tek gave 0.07(B-V). We have found that 3 mm BG37 in place of the 1 mm BG12 virtually eliminated the color term in B for our Tek thinned CCD.

When choosing filter combinations one is invariably faced with deciding between a better matched bandpass or higher throughput. Compromises in favor of throughput are usually made and in many cases these do not degrade the standardized photometry. But it is important to examine the residuals of the standard star photometry and the residuals of the synthetic photometry (especially at the moment when there are not enough CCD standard stars of all spectral types) to see in what color range or range of spectral-types, non-linearities will occur. For instance, pushing the blue edge of the B response further to the UV to counter the effective wavelength shifting to the red is not a good idea because it introduces more light near the confluence of the Balmer lines which will introduce systematically different magnitudes for the A and F stars. Possible combinations for B are:

1 mm GG385 + 2 mm BG1 + 2 mm BG39 (thinned CCDs) or
1 mm BG395 + 3 mm BG37 + 1 mm BG39 (thinned CCDs) or
2 mm GG385 + 1 mm BG12 + 2 mm BG39 (coated CCDs).

4.3 The V passband

The different slopes of the responses of different CCDs across the V passband produces small shifts in the effective wavelength of the V band. These shifts can be easily removed by varying the thickness and/or the type of the glass defining the red side of the passband. There are four glasses that can be used, BG18 and BG38 or the newer glasses BG39 and BG40. Their responses are subtly different but quite similar in shape. The older glasses are cheaper.

An important reason for trying hard to get as good a match as possible for the V band is that different colors, (B-V) or (V-I) [V-R] are used when deriving the color term in the transformation equation, and this can lead to systematically different V magnitudes for the reddest stars. For example, if V photometry of standard stars between spectral-type A0 [(B-V), (V-I) = 0] and K0 [(B-V) = 0.98, (V-I) = 1.0) fits a color term of 0.05(B-V) or 0.05(V-I), which may appear acceptably small, then using those relations for an M7 dwarf [(B-V) = 2.0, (V-I) = 4.0) yields a difference of 0.1 mag in the transformed V magnitude. It is clear that differences amounting to more than 0.1 mag do exist in the published V magnitudes of late M stars and the y probably arise in such a way. Calculations using synthetic spectra indicate that the correction based on the (B-V) color and not the red colors is correct. As much CCD photometry of red stars ignores B and observes only VRI, making the color term as small as possible ensures that large systematic errors are avoided. Possible combinations for V are:

2 mm GG495 + 3 mm BG40 (thinned CCD) or
2 mm GG495 + 3 mm BG39 (coated CCD).

4.4 The R Passband

The two glasses OG570 and KG3 of the photoelectric filter can be used but the thickness of KG3 can be altered to match the standard band better. Because of the asymmetric red tail of the Stan2optdard R response and the variation between batches of KG3 glass it was virtually impossible to match the standard R over the full color range using a GaAs tube; that is, all natural photoelectric R systems needed two straight lines to transform to the standard system. It is even more difficult with CCDs, nevertheless, a good match can be made for most stars and the nonlinear effect restricted to the reddest stars as with the photoelectric systems. Possible combinations for R are:

3 mm OG570 + 2 mm KG3
2 mm OG570 + 3 mm KG3.

4.5 The I Passband

Most CCDs appear to have very similar responses from 700 nm to their cutoff near 1000 nm although in practice, because the red tail is sensitive to temperature and the operating temperatures of CCDs differ appreciably, they may not be so similar. However, as the red cutoff in the standard I band was defined by the rapid cutoff in the sensitivity of the GaAs photocathode near 870 nm and the CCD sensitivity extends at least 100 nm further to the red, most I CCD bands have this extended red response.

Synthetic photometry indicates that the main difference caused by the extended sensitivity in the I band is the inclusion of light above the Paschen discontinuity in early-type stars and the inclusion of additional emission lines in emission lines objects. Additional problems arise due to the night-sky emission that is brighter at longer wavelengths and which lowers the S/N for faint stars by increasing the background relative to the star and by increasing the fringing in thinned CCDs. The fringing in I from the night sky will also be different from that of the twilight sky often used for flat fielding further degrading the S/N. Special multilayer coatings can be made to approximate the cutoff of the GaAs tube and eliminate the reddest sky emission. One such commonly available filter is sold by the Rolyn Optics Company as a hot mirror (#60.5050) and we have successfully used this filter in conjunction with 2 mm RG9 to produce an I CCD passband quite similar to the standard I band. More recently we have taken delivery of several I filters with a 13 layer coating of non-quarter-wave multilayer-optimized stack centered near 870 nm. The coatings were done by the National Measurement Laboratory in Sydney. In Fig. 5 the transmittance of two mm RG9 and the multilayer coating is shown.

However, with an I passband whose red side is defined by the CCD cutoff, excellent transformations for sources with stellar-like fluxes are still possible provided the non-linearities of the transformation are followed by taking care to observe standards covering the spectral types of interest. High precision is also maintained provided the fringing is not a problem. Possible combinations for I are:

2 mm RG9 + 3 mm fill glass (GG385) or
2 mm RG9 + 1 mm Rolyn 60.5050 (or equivalent) + 2 mm fill.

4.6 The Z Band

The extended red response of CCDs compared with GaAs and S20R photocathodes raises the possibility of a passband beyond 1000 nm which would provide some wavelength over lap with the In Sb and Mercatel IR arrays. We have used 2 mm RG1000 to define a far-red band that was quite useful in isolating extremely red stars, such as miras, near the Galactic center but it had a much lower sensitivity compared to the I band.

4.7 The C Band

The C band of the Washington system has proved very useful for measuring line blanketing in faint red giants. It is more sensitive than B for such work. The red leak of the three glass was still a problem with two mm of BG40 (Bessell 1990b) as a blocker, consequently it is better to use 2 mm BG39, 2 mm S8612 or 3 mm BG40.

5. SUMMARY

Faint standards, many to a CCD frame and with a good range in color are required in several declination zones to enable good standardization of CCD photometry. The equatorial CCD-sized fields of Landolt represent a good start to this endeavor. The Landolt standards have systematic differences from the Cousins system but good transformations are possible for most stars. Graham's E-region standards are also useful but again transformation to the Cousins system is necessary and stars with a wider color range need to be included.

Glass filter combinations can be quite readily devised to closely match the standard UBVRI system but some of these filters should be made for specific CCDs, especially for U and B. Systematically different V magnitudes can result for late-M stars if the color correction to V is large; it is therefore worth matching V as closely as possible. There are also advantages in faint object photometry from rejecting the light redward of 870 nm from the I band using a multilayer stack. The red leak of glass U filters remains a problem; it can be eliminated by using thick enough pieces of BG39/40 or much better, S8610 glass. This lowers the throughput somewhat and leads to non-linearities for the hottest stars but should not cause too severe transformation problems.

General comments on making the glass filters. We use a spectrally transparent two-component epoxy "EPO-TEK 301" (Epoxy Technology Inc) to glue the glasses of each filter mix together. This epoxy, which bonds to most glasses, metals, ceramics and plastics is transparent between 300 nm and 2.6 microns, has a refractive index of 1.54 (the same as crown glasses) and cures overnight at room temperature. Trial mixes of filter glasses can be oiled or greased together temporarily. When used in a converging or telecentric beam, the optical quality of the standard rough-polished filter glasses from Schott are adequate without further polishing, but in the collimated beam of a reimaging camera we have found it necessary to polish the glasses to a few fringes. We have not antireflection coated the filters but tilt them if reflections are a problem.

REFERENCES

Bessell, M. S. 1990a A&AS 83, 357

Bessell, M. S. 1990b PASP 102, 1181
Bessell, M. S. 1995 PASP to be submitted
Bessell, M. S. and Weis, E. W. 1987 PASP 99, 642
Cousins, A. W. J. 1980 MNASSA 39, 22
Geisler, D. 1990 PASP 102, 344
Graham, J. A. 1982 PASP 94, 244
Gunn, J. E. and Stryker L. L. 1983 ApJS 52, 121
Laing, J. D. 1989 SAAO Circ. 13, 29
Landolt, A. U. 1973 AJ 78, 959
Landolt, A. U. 1983 AJ 88, 439
Landolt, A. U. 1992 AJ 104, 340
Menzies, J. W., Banfield, R. M., Cousins, A. W. J. and Laing, J. D. 1989 SAAO Circ. 13, 1
Menzies, J. W., Maran G., Laing, J. D., Coulson, I. M. and Engelbrecht, C. A. 1991 MNRAS 248, 642
Menzies, J. W. 1993 in Precision Photometry, D. Kilkenny, E. Lastovica and J. W. Menzies, eds., South African Observatory, Capetown, p. 35
Rufener F. and Nicolet B. 1988 A&A 206, 357
Straižys, V. and Sviderskiene, Z. 1972 Bull. Vilnius. Astron. Obs. No. 35
Walker, A. R. 1991, private communication
Walker, A. R. 1992 IAU Colloquium No. 136, Stellar Photometry - Current Techniques and Future Developments, C. J. Butler and I. Elliott, eds., Cambridge University Press, p. 278
Weis, E. W. 1984 ApJS 55, 289
Weis, E. W. 1986 AJ 91, 626
Weis, E. W. 1987 AJ 93, 451

DISCUSSION

YOUNG: You remarked on the use of either (U-B) or (B-V) as regression variables in the color transformation. Peter Harmanec suggested the use of two color indices in the regression, and I commend his suggestion to you. This option is available in the PEPSYS package distributed with MIDAS.

BESSELL: This is important. Most data reduction packages have not included such a capability.

FLORENTIN-NIELSEN: The newest thinned TK 1024 have a more flat response in U, with about 30% QE at 3000 Å.

BESSELL: That is good news. The flatter the UV repsonse the better the transformation to the standard system.

PRECISION DIFFERENTIAL CCD PHOTOMETRY

L. A. Balona

South African Astronomical Observatory

ABSTRACT: In this paper we describe the technique and reduction procedure in use at the SAAO for differential CCD photometry. Much of this work involves the detection of microvariables in galactic and Magellanic Cloud open clusters. The large number of frames requires automated cleaning and reduction software (principally the DoPhot package) which is briefly described. The process of selecting comparison stars and the technique of detecting low-amplitude variables is discussed. The software is well suited to be used for on-line reductions at the telescope and a start has been made in this direction. Finally, some astrophysical results involving precision differential CCD photometry are described.

1. INTRODUCTION

Atmospheric extinction and scintillation is one of the largest sources of error in photometric observations. If one is interested in absolute photometry, one needs the best atmospheric conditions to obtain accurate results. However, the dependency on the atmosphere is greatly reduced for differential photometry, particularly with CCD's. A typical extinction coefficient in the B or V band is 0.2. For two stars separated by three arc min (the size of a typical CCD frame), the difference in extinction is only 0.6 mmag at an airmass sec z = 2 which is comparable to the scintillation noise for a 1.0-m telescope, but is typically much smaller than photon noise. It is clear that differential CCD photometry is capable of high precision provided suitably bright comparison stars are visible on the same frame.

For the last few years my colleagues and I have been using an RCA 360 x 512 CCD attached to the 1.0-m telescope at Sutherland to obtain differential photometry of young clusters in the galaxy and the Magellanic Clouds. The program consists of two parts: (a) to obtain accurate uvbyβ photometry of cluster members and (b) to search for microvariables in the clusters. The purpose of obtaining Strömgren photometry of young open clusters is to refine the absolute magnitudes of early-type stars. The survey for microvariables is important in defining the instability strip for β Cep, λ Eri and 53 Per stars. The field of a galactic open cluster is normally too large to be covered by a single CCD frame: a mosaic of overlapping frames is required. The field of young LMC and SMC clusters is covered by a single frame, but is very crowded. In this paper we describe our techniques for differential CCD photometry, illustrating it with some of our results.

2. INSTRUMENTATION AND OBSERVING TECHNIQUE

The CCD camera is attached to a 1.0-m reflector. The filter box is situated some distance above the CCD chip in the converging beam. Ideally, the temperature of the filters should be kept within narrow limits since the characteristics of the narrow band filters vary with temperature, but no provision is made for this. As a result, some systematic errors may be introduced for the β index, but the effect should be negligible for the other Strömgren filters. The CCD chip itself is kept at a temperature of about 158° K.

A CCD can be modeled statistically by two parameters: λ - the number of electrons per ADU and σ_r - the readout noise in electrons. The pixel-to-pixel variance, σ^2, versus the mean ADU count, z, can be used to determine the two constants λ and σ_r. A number of exposures on the darkening sky soon after sunset were obtained through the y filter. As expected, the relationship between σ^2 and z is linear, but there is a discontinuity in slope for z > 17000. This discontinuity remains unexplained. If we restrict the analysis to z < 17000 ADU we obtain:

λ = 10.91 $\pm$ 0.08 electrons/ADU,

σ_r = 87.7 $\pm$ 0.4 electrons.

Flatfield calibrations are made by observing a cloudless twilight sky. Experiments show quite conclusively that the sensitivity variation across the CCD is independent of time. This means that it is permissible to combine several flatfield calibrations for a given filter to improve the signal-to-noise ratio. Generally, from five to ten calibrations per filter, each with a mean ADU count of 15000, were combined.

The RCA chip has a problem with charge transfer at low light levels which results in a nonlinearity for counts less than about 50 ADU. It is therefore necessary to expose the CCD to a uniform source of illumination to bring the ADU count above this level prior to each exposure. This is accomplished by a momentary illumination of the CCD by four LED's. These "preflashing" exposures need to be calibrated at regular intervals as the illumination level is somewhat dependent on time.

The calibration of the electronic bias is made using an overscan strip on the CCD. We assume that the bias is constant along the readout direction but varies perpendicular to this direction. The variation is modeled from the overscan using a suitable running mean.

Before a frame is subjected to astrophysical analysis, it undergoes the following "cleaning" processes:

a) electronic bias is subtracted,
b) bad pixels are patched up,
c) the frame is divided by a normalized flatfield and trimmed.

The software for cleaning the frames was designed for processing very large numbers and runs automatically in batch mode on a UNIX workstation.

3. REDUCTION SOFTWARE

The search for short-period microvariables in open clusters demands continuous exposures over several hours and generates large numbers of frames (1000 per week is typical). Software which demands user interaction is not suitable for such a task. The process of identifying stars and performing profile fitting and aperture photometry should run automatically in batch mode. We found DoPhot particularly suitable for this purpose. This software was written by Mateo and Schechter and we are greatly indebted to them for allowing us to use it. A brief description of the software is given in Mateo and Schechter (1989); a more detailed account can be found in Schechter, Mateo and Saha (1993).

The program looks for objects above a certain threshold level. It models particular objects such as a star, double star, cosmic ray, galaxy, etc. For example, the model for a star might be an elliptical Gaussian. The model for a galaxy might also be an elliptical Gaussian, but one which is significantly bigger than a star. Having found the stars, it updates the relevant model parameters and subtracts the modeled stars from the field. The threshold is lowered and the process repeated. In each pass, stars found during previous passes are put back on the frame and improved parameters are re-calculated. Since neighboring fainter objects have now been subtracted, the shape parameters are expected to be better than on the previous pass. Throughout the above process, DoPhot constructs and updates a noise image which provides weights for each pixel used in the non-linear least squares fitting of the stellar profiles. In the final pass, aperture magnitudes are also calculated by replacing each star on the star-subtracted frame and then removing it again. The sky values are therefore uncontaminated by neighboring stars.

In our version of DoPhot, a star is modeled by the function:

$$I(x,y) = a_1 + a_4/(1 + t + \{t^2\}/2 + \{t^{3n}\}/12)$$

where $t = 0.5(\{x^2\}/a_5 + 2a_6xy + \{y^2\}/a_7)$. Here x and y are measured relative to the center of the stellar profile (a_2, a_3). The parameters, a_5, a_6 and a_7, define the PSF. It is found that these parameters are independent of position on the chip. The best estimate of the PSF is obtained by weighting the values of the three parameters for each star according to appropriate statistical criteria and calculating a weighted mean. Initial starting values are automatically provided by fitting a simple Gaussian profile to the two or three brightest stars in the field.

In conventional aperture photometry, the aperture size is normally fixed for the duration of the night even when there is a change of seeing. A change in aperture size introduces a discontinuity in the zero point which is more difficult to model than the continuous change associated with seeing variations. This restriction no longer applies in CCD photometry. In our version of DoPhot, aperture magnitudes are determined by calculating the total count inside a square area of size b_1. The sky is calculated from the weighted mean of the count in the annulus formed by b_1 and an outer square of size b_2. The values of b_1 and b_2 vary with seeing according to the following algorithm:

$$b_1 = 4.5 \, (\{a_5\}\{a_7\})^{0.25}$$

$$b_2 = 4\{b_1\}/3$$

These values were obtained empirically by minimizing the error in a sequence of about a dozen frames of a moderately crowded field. The algorithm works well: typical rms values for all-sky photometry of E-region standards are three to five mmags even in conditions of variable seeing.

Although aperture photometry gives good results for the brightest stars, we found that profile-fitting magnitudes are just as accurate for bright stars and far more accurate for the faintest stars. Generally speaking, the rms scatter for profile-fitting magnitudes is a factor of four to five times smaller for fainter stars. We use aperture photometry only for the purpose of standardization. For differential photometry we only use the magnitudes given by profile fitting.

4. MATCHING STARS IN DIFFERENT FRAMES

The output from DoPhot consists of a file listing the positions and relative fit and aperture magnitudes of all stars in the frame. After a typical observing run there will be a large number of frames of the same field, but DoPhot would have given different numbers to the same stars in different frames. The next step is to obtain a uniform numbering scheme. This is accomplished by selecting one frame as a master copy to which the other frames are matched.

The algorithm to match stars in different frames involves the use of similar triangles. All possible triangles formed by the twenty brightest stars in the fiducial frame are used to construct a graph of smallest angle versus largest angle. The same is done for the frame under study. A cross-correlation between the two graphs is sufficient to identify the stars irrespective of the relative orientation and scale of the two frames. If at least five stars are identified in this way, the relative shift and orientation of the frames can be calculated. The remaining stars are easily matched using a suitable distance criterion. For example, if the positions agree within one or two pixels it can be assumed that it is the same star. The algorithm is very robust and works in all but a very small number of peculiar cases.

5. CHOICE OF COMPARISON STARS

There are two criteria for the choice of comparison stars: they must be visible in practically all frames and they should be relatively bright. The number of comparison stars depends on the field. The initial selection consists usually of the ten brightest stars. The zero point of each frame is adjusted so that the mean magnitude for these stars is zero. The rms scatter for each star is examined and those with the largest scatter removed from the list. The process is repeated until a list of at least five or six good comparison stars is obtained.

As an illustration, we will take some recent work on the galactic open cluster NGC 6134. This cluster was examined for variables by Kjeldsen and Frandsen (1989) who detected three δ Scuti variables in one CCD frame. Subsequently, it was put on the STACC network project which is an attempt to join observers with small telescopes equipped with CCD cameras into teams that can produce light curves for variables (particularly δ Scuti stars). As part of this collaborative effort, S. Frandsen and M. Viskum (Aahrus University) observed the cluster with the Dutch 0.9-m telescope at La Silla and L. Balona and C. Koen (SAAO) with the 1.0-m telescope at Sutherland over a two-week period in May/June 1993. Observations were made through the V filter only; a brief description of the project is given in Frandsen et al. (1994). The field includes three δ Scuti stars. For this field, seven comparison stars were chosen, details of which are given in Table 1. Exposure times were 90 - 120 s for each frame.

TABLE 1

Comparison Stars for a Field in NGC 6134
Results from SAAO Observations

Star	σ	N	V	α	$\sigma\alpha$	σ_c
141	2.5	1236	11.87	-150	009	2.3
129	2.2	1240	12.54	-130	008	2.0
162	2.6	1240	12.65	-162	009	2.3
46	3.4	1240	12.94	-100	013	3.3
87	3.5	1239	13.19	-11	014	3.5
67	3.8	1240	13.98	-257	149	3.8

σ is the standard error of one observation in mmag;
N is the number of observations; V is the magnitude;
α is the correlation with seeing and $\sigma\alpha$ its standard error;
σ_c is the same as σ after decorrelation.

As shown in Table 1, the rms scatter for the comparison stars is only two to four mmag on the average. However, this is contaminated by one night when observations were made through a thin cloud layer. On this night the rms error is close to six mmag. If this night is removed, the mean rms scatter is about 2.5 mmag. This indicates that for exposures as short as two min, the variation of thin cloud cover across the CCD field can significantly decrease the photometric accuracy.

6. SEARCH FOR LOW AMPLITUDE VARIABLES

Once the frames have been standardized in the manner described above, the next step is to search for variability among the stars. An obvious way to do this is to select those stars with the largest rms error taking the magnitude into account. While this works well for high-amplitude variables, it is not a satisfactory way to detect microvariables. The periodic variations of a star with an amplitude of a few mmags is completely swamped by photon noise and will not show any significant increase in rms scatter.

A far better method, and the one used for these projects, is to calculate and display the periodogram of each star. In our software package, twenty periodograms are computed simultaneously and displayed on the screen. Since the time of observation is the same for all stars in the field, the cpu time used for calculating twenty periodograms is not much larger than used for a single periodogram. Visual inspection of the periodograms is used to pick out stars which may be variable. Not all the stars that appear to have significant peaks in the periodogram are variable. Spurious peaks can be a result of contamination by a close companion, if the star is close to the edge of the chip, and many other reasons. Candidate stars selected in this way are examined in more detail to produce a list of certain or probable short-period variables.

7. DECORRELATION

When there are two stars very close to each other, the stars will be treated independently on frames of good seeing, but could be treated as a single star on frames of poor seeing. One way of avoiding this problem is to re-run DoPhot with fixed coordinates. While this can be done, generally speaking the results of the multi-fit solution are not as good, so we have simply omitted these frames for those particular stars. Quite often this situation gives rise to spurious peaks in the peridogram and the star is mistakenly put on the list of possible variables. To avoid this problem we have adopted the technique of testing each variable for a possible correlation between magnitude and seeing. If such a correlation exists, the frames are examined for a close companion. If one is found, nights of poor seeing or omitted: often the star is discarded entirely.

This is a simple illustration of a more general procedure - that of decorrelation. The concept behind decorrelation is that the magnitude should not depend on parameters peculiar to the CCD or the atmosphere. If a correlation is found, it illuminates a potential problem and one must consider a corrective remedy as illustrated above. The question may be asked as to whether the observed scatter is consistent with that expected from photon noise and the properties of the CCD or whether there is a non-random component which could be calibrated and used to reduce the noise level. One may think of, for example, a change of sensitivity across the CCD due to incomplete flatfield calibration. In this case one might expect a correlation between the calculated magnitude and the position of the star on the CCD chip. A plot of magnitude versus X or Y position will then show a correlation. This is a purely hypothetical example which is probably not likely to occur, but it is useful to examine possible correlations between magnitude and other parameters for the comparison stars and the δ Scuti variables in the field of NGC 6134.

We examined possible correlations between the calculated magnitude and the following quantities: seeing, sky level, X position and Y position. Analysis of the comparison stars and δ Scuti variables showed that when a correlation exists, it is mostly with the seeing. Very little, if any, of the variation correlates with the other parameters. However, the correlation varies in strength and sign from star to star as shown in Table 1. The last column of this table shows the quantity α in the linear relationship $V = \alpha x + \beta$, where x is the seeing in arbitrary units. The three brightest stars show a highly significant correlation with seeing, but the sign of the correlation changes from star to star. The correlation has, however, only a minor effect in reducing the overall scatter because the seeing was good for the most of the run.

As another example, we show in Fig. 1 part of the light curve of one of the δ Scuti stars in NGC 6134 using the Sutherland and La Silla data. The bottom curve in the figure is the raw data. The standard deviation of the residuals from a sinusoid fit with P = 0.1324 d is 6.3 mmag. When these residuals are decorrelated for seeing, sky level and position, the scatter is reduced to only 3.7 mmag (top curve). Most of the effect comes from the La Silla data. This shows that decorrelation can offer a significant improvement in the signal-to-noise ratio, but it should be used with caution as there is a possibility of introducing spurious signals.

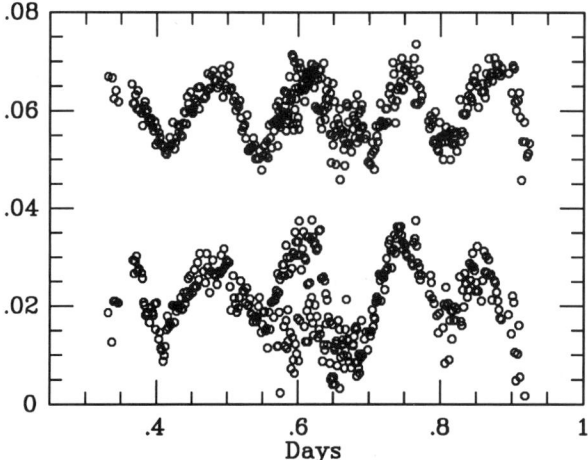

Fig. 1. The relative magnitude variation of a δ Scuti star in NGC 6134 (star 161) during one night from SAAO and ESO observations. Bottom curve - raw data; top curve - decorrelated data.

8. THE NEED FOR ON-LINE REDUCTIONS

Because DoPhot can run without any user intervention, it is well suited to be used for on-line reductions of stellar fields at the telescope. The usefulness of on-line reductions can only be appreciated once one has been exposed to it. For many years we have had on-line reductions on our photoelectric photometers which has resulted in greatly improved observing efficiency. For example, one can tell at a glance the quality if the observations and change observing plans as required. Stellar identification is much easier if one is presented with the magnitude and colors of the star just observed.

There is no doubt that implementation of on-line reductions for the CCD is a more complex problem. It needs to be tackled in several stages. The most important requirement is to provide an indication of the photometric quality of the night using all-sky aperture photometry. This is a relatively simple problem as it involves aperture photometry of the brightest star in the field. The next stage could involve obtaining profile-fitting photometry for a number of stars in the same field. It would not be practicable or necessary to provide the observer with a large list of all the stars identified by DoPhot - only marked stars or perhaps the ten brightest stars need be displayed. With this restriction the reduction time should be only a few seconds using a modern workstation, even in the most crowded fields. The final stage involves the running of DoPhot in the background to provide fully reduced data which may be inspected at any time during the night.

In all this, one needs adequate flatfield calibrations. These would not necessarily be available prior to observing, but for the first two stages an approximate flatfield calibration should provide the required information with sufficient accuracy. Such a project is being

194 L. A. Balona

implemented at the SAAO by Dr. D. O'Donoguhe of the University of Cape Town.

9. SOME ASTROPHYSICAL RESULTS

In Fig. 2 we show the periodograms of four δ Scuti variables in NGC 6134 from the work of Frandsen, Viskum, Balona and Koen described above. These results are from a two-week run involving 1.0-m class telescopes in Chile and South Africa and generated over 2000 frames of the same field in the cluster. These periodograms were calculated using the raw data; somewhat lower noise is obtained using decorrelation. The high-frequency noise level is about 0.5 mmag, allowing detection of periodicities with amplitudes in excess of one mmag for these 12 - 13 mag stars during bright time.

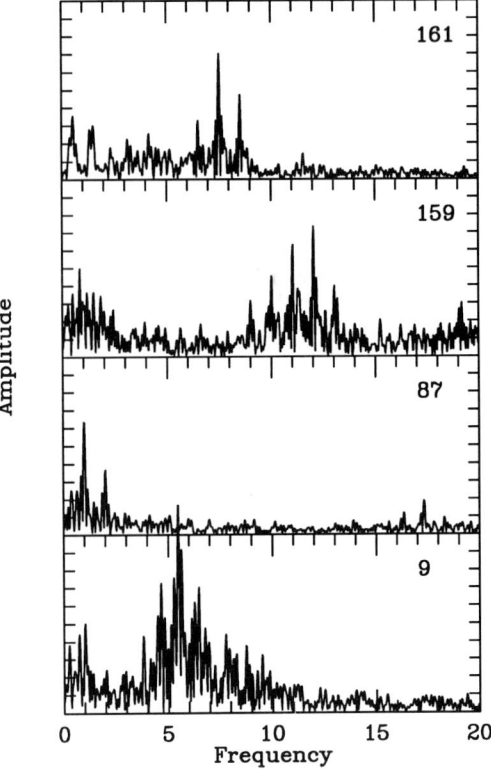

Fig. 2. Periodograms of four δ Scuti stars in NGC 6134 from SAAO and ESO observations. The frequency scale is in cycles/d; the tick marks on the amplitude axis are spaced by one mmag.

A breakthrough in our understanding of pulsation in early-type stars occurred in 1990 when it was realized that the metal opacities in use till then required a significant increase. This provided the driving mechanism for β Cep variables which until this time had been a puzzle. The prediction is that early B-stars will pulsate provided the metal abundance is sufficiently high. To test this idea, Balona (1992, 1993) observed the cluster NGC 330 in the SMC and the clusters NGC 2004 and NGC 2100 in the LMC in the hope of detecting β Cep stars. Even with a 1.0-m telescope and the high readout noise of the RCA CCD, it was possible to detect amplitudes as small as 0.01 for 16th magnitude stars using a careful selection of comparison stars in these rather crowded fields. Although many variable stars, some with small amplitudes, were discovered, none of them satisfied the criteria defining the β Cep class. Statistical analysis shows that the probability of this occurring by chance is less than 0.1 per cent, supporting the theoretical predictions.

A project of obtaining uvbyβ photometry for young open clusters in collaboration with Drs. Koen and Laney (SAAO) is under way. Results from two clusters have already been published: NGC 3293 (Balona 1994) and NGC 4755 (Balona and Koen 1994). These clusters cover a large area of the sky: a mosaic of CCD frames with large overlap was used to transfer zero points from one frame to another. Within each frame comparison stars were used to obtain relative accurate photometry. Each cluster was also observed extensively to search for microvariables. This led to the discovery of one new β Cep variable in NGC 3293 and six new β Cep variables in NGC 4755. No 53 Per variables were found, indicating that moderate and rapid rotation is likely to inhibit pulsations in these mid-B variables.

There is little doubt that the use of differential CCD photometry on small telescopes in bright time can make significant contributions to astrophysics, particularly stellar pulsation. Recent advances in computer technology and intelligent software has removed the main obstacle to efficient use of this technique - the reduction of large numbers of frames. In the near future it will be possible to obtain near-complete reductions at the telescope, further increasing the capabilities of the CCD.

REFERENCES

Balona, L. A. 1992 MNRAS 256, 425
Balona, L. A. 1993 MNRAS 260, 795
Balona, L. A. 1994 MNRAS 267. 1060
Balona, L. A. and Koen, C. 1994 MNRAS 267, 1071
Frandsen, S., Viskum, M., Balona, L. and Koen, C. 1994 Delta Scuti Star Newsletter 7, 5
Mateo, M. and Schechter, P. L. 1989 in 1st ESO/ST-EFC Data Analysis Workshop. P. J. Grosbol, F. Murtagh and R. H. Warmels, eds, ESO, Garching, p. 69
Kjeldsen, H. and Frandsen, S. 1989 The Messenger 57, 48
Schechter, P. L., Mateo, M. and Saha, A. 1993 PASP 105, 1342

DISCUSSION

WALKER: For quick-look establishment of the zero-point while observing, you could just use aperture photometry and read the airmass from the header, to calculate a magnitude.

BALONA: Yes, but one would want this done automatically (i.e. without clicking on the star) at the end of the integration. When following a variable one would probably need the magnitudes of the comparison stars too.

ABBOTT: At which telescope have the ESO data been collected?

BALONA: The 0.9-m Dutch telescope.

ABBOTT: Then that would explain why the decorrelation worked so well - the PSF is highly dependent on position in the image at that telescope.

BALONA: That may be the case.

PHILIP: I have used, some years ago, RCA chips at CTIO and KPNO. I found the u magnitude to have a high rms error, even though the images on the frame looked fine. Do you have similar problems with your RCA chip? The newer TEK chips, however, give good u magnitudes.

BALONA: I have not found any problems peculiar to the u filter.

HOWELL: I agree that while aperture photometry can be much worse than PSF fitting, that is generally only true for "standard" aperture photometry. Use of optimum aperture extraction techniques yields factors of two to three better than "standard", and is computationally faster than detailed PSF fitting.

BALONA: Yes. I was referring to measurements using a fixed aperture.

ARRAY POLARIMETRY AND OPTICAL-DIFFERENCING PHOTOMETRY

J. Tinbergen

Kapteyn Observatory Roden and Sterrewacht Leiden

ABSTRACT: Array detectors have improved the efficiency of optical polarimetry sufficiently for this technique to become part of the standard arsenal of observational facilities. However, we could gain even more: spatially-differentiating photometry can be implemented as an option of array polarimeters and low-noise, high-frame-rate array detectors will allow extremely high precision both in polarimetry and in such differentiating photometry. The latter would be valuable for analyzing many kinds of optical or infrared images of very low contrast; the essence of the technique is to use optical (and extremely stable) means to produce the spatial derivative of the flux image, in the form of a polarization image which is then presented to a "standard" array polarimeter. The polarimeter should incorporate a polarization modulator of sufficient quality for the photometric application in mind. If developed properly, using a state-of-the-art array detector and the most sensitive type of polarization modulator (stress-birefringence), optical differencing will allow levels of relative photometric precision not otherwise obtainable. With the optical differencing option taken out of the beam, the same instrument can be used for high-quality polarimetry.

1. INTRODUCTION

Large telescopes and their instruments (both imagers and spectrographs) are generally built in the first place for "detection of the faintest objects" rather than for "accurate (spectro)photometry of somewhat brighter objects", though the latter often is a scientifically equally valid use of a large telescope. This situation has meant that in many instruments array detectors operate close to undersampling and optics are kept as simple as possible, both features having disastrous consequences for high-precision photometry. Since some observers nevertheless try to use these basically unsuitable instruments for photometry, array detectors are in danger of acquiring a reputation of being unsuitable for photometric applications of the highest accuracy or precision.

There is, however, no proof for this assumption and it is in fact unlikely to be true. Like photomultipliers (PMTs) and unlike the grains of a photographic plate, the pixels of the array are not destroyed by being used as a detector, so we have the option of accurately calibrating our observation by auxiliary measurements. The massive parallellism and the lack of cascaded secondary-emission stages should prove an advantage over PMTs in well-designed instruments. The defects such as within-pixel gain variations (Jorden et al. 1994) appear to be no worse than for PMTs. With sufficient oversampling and properly designed electronics, therefore, arrays are likely to have better short-term gain stability than PMTs, and photometric instruments incorporating them should perform at least as well as the PMT (spectro)-photometers. The essence of good design is to optimize the entire instrument for stable use of the array detectors and to employ calibration in one form or another to extend short-term stability of each pixel to long-term stability of the entire calibrated array.

I have previously carried out, as an intellectual exercise, what one might call a "zero-order system design study" for the case of accurate all-sky spectro-photometry of relatively bright stars in a relatively faint sky (Tinbergen 1993). The main concern in that regime is to calibrate out the system gain variations, designing the entire system to take advantage of the short-term gain stability of the individual detector pixels. In the present paper I explore the regime of high-signal but low-contrast images; with ten-meter-class telescopes and high-QE detectors, this regime occurs more often than one might have supposed. The concrete question I shall examine is: "Can we eliminate a constant background signal to one part in 10^5 or even 10^6, to measure *accurately* the light from point sources as faint as one part in 10^3 or even 10^4 of the background?" Accurate polarimetry is one such situation and another would be a generalization of the chopping-secondary (infrared) and dual-feed (radio) photometric techniques; I shall show that such differential photometry and array polarimetry are intimately related.

The discussion applies to all kinds of images, both in the optical and in the near infrared: direct sky images, spectra, interferograms or hybrids such as long-slit spectra. The only basic requirement is the availability of array detectors and polarization components (polarizers, retarders). The precision one can actually attain will be a function of the properties of available components; examples of such properties are: readout noise as a function of frame-rate for the detectors, optical quality and achromaticity for the polarization components.

2. OPTICAL DIFFERENCING VIA POLARIMETRY

Suppose, for the moment, that we possess an array polarimeter which allows us to determine the *polarized flux* striking any pixel of our array. By inserting a properly-cut calcite plate into the optical beam just in front of an image plane, we can generate from any single flux image two separate half-intensity images of opposite polarizations (Fig. 1); these images will combine in the focal plane, with a slight shear (relative displacement). If, instead of measuring the total flux image, we analyze the *polarized flux image* (by using our array polarimeter), any constant background signal will have disappeared entirely, while a point source has been split into a positive and a negative source (Fig. 2). A constant gradient in the total flux background will show up as a constant signal in the polarized flux image. The separation of the positive and negative peaks, and the sensitivity to gradients, may be controlled by adjusting the thickness of the calcite slab. For infinitesimal thickness, the action of the slab is to spatially differentiate the image when going from total flux to polarized flux; since the slab thickness is always finite, the term "differencing" will be used.

The most important aspect of this differencing action is that it is entirely optical; it is therefore extremely stable and independent of detector properties. *Analyzing* the resultant polarized flux image will involve a detector within the polarimeter, but that is a different matter: "optical differencing" itself is a purely *optical* conversion of a total flux image into a polarized flux image with a different and possibly more useful structure (e.g. when faint point sources are the object of our scientific curiosity and the high background is merely a source of measurement error).

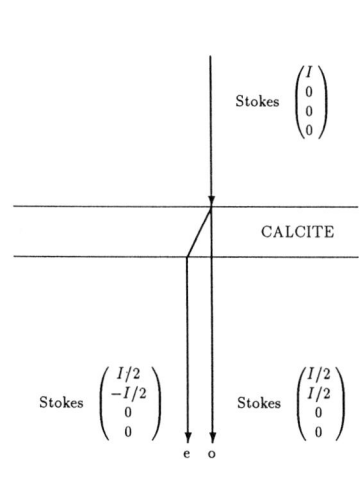

Fig. 1. Double refraction by a calcite plate (Schematic).

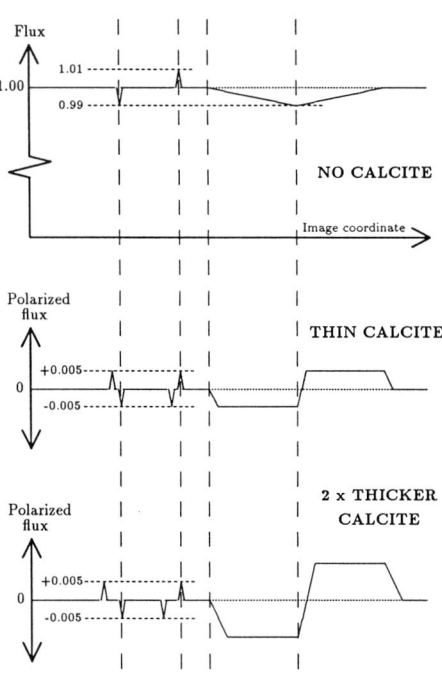

Fig. 2. Optical differencing: the flux and the polarized-flux images. The positive-polarization ray is the ordinary one in the calcite and is not displaced from the total-flux object: the extraordinary ray is assumed to be displaced leftwards.

3. PRECISION OF ARRAY POLARIMETRY

In this section I discuss available precision in array polarimetry. Give or take a factor of two, the same precision will apply to optical differencing.

"Classical" array polarimeters, whether of the direct-imaging type (e.g., Scarrott et al. 1983) or as part of spectrographs (e.g., Tinbergen and Rutten 1992), make use of a two-beam analyzer to produce two separate images of opposite polarization. These two images are both recorded and the polarization information resides in changes of the *flux ratio* of the images as the state of a polarization-modifying component is changed and a second pair of images is recorded. The most common polarization-modifier is a halfwave plate; this is rotated by 45 degrees to switch the linearly polarized part of the flux from one spectrum to the other, thus altering the flux ratio by an amount related to the degree of polarization of the input signal. When computing the degree of polarization from the two pairs of images, the assumption is made that the system gain for any pixel for any given exposure can be factored into a time-dependent but polarization-independent part (including scintillation and extinction noise) and a polarization-dependent but time-independent part (mainly the response of the instrument optics and detector pixels to the two separate beams of polarized light); this is explained in detail in Tinbergen and Rutten (1992). This basic assumption of "two-beam DC polarimetry" is very good, but at some level of precision it will break down, e.g. by sub-pixel-scale image motion as the halfwave plate is rotated or as a result of differential flexure of the instrument between the two exposures, or again by a difference in transmission for the two states of the polarization modifier (e.g. the Fresnel reflection coefficient depends on polarization and rotation of a halfwave plate will also rotate the weak polarization induced by this plate). Experience (partly with the PMT polarimeters) indicates that the assumption is robust to well below 0.1%, but that we shall be lucky if we routinely reach 0.01% in that way; this is two orders of magnitude removed from the goal I set myself in the Introduction.

If we wish to improve on this precision, we must do at least two things: we must ensure that the polarization modifier affects only the polarized signal and does not influence the total flux to any measurable extent, and we should complete our basic polarimetric measurement within a time short compared to all time-constants of unintentional changes in the opto-mechanical system (finally integrating over many such basic measurements to obtain an acceptable S/N ratio).

For PMT polarimeters, there is an almost ideal solution: the stress-birefringence modulator (e.g., Kemp 1969). This consists of a slab of isotropic material (often fused silica), which is periodically stressed mechanically to make it birefringent. The periodic birefringence is so minute (of order one part in 50,000 at maximum) that any influence on the total flux is below all practical limits of detection; also the angular acceptance of the component is very large, for the same reason. In the usual PMT application the device is mechanically resonant at a frequency of tens of kHz, so that the basic measurement (of the pair of fluxes from which the polarized flux is derived) is performed within a fraction of a millisecond and noise due to scintillation and varying extinction is eliminated entirely. PMT polarimeters of this kind have yielded a precision of about one part in 10^6, even without using the information in the second beam of the analyzer (the second beam is then used only to reduce photon noise by the square root of two).

Frame rates of tens of kHz are incompatible with astronomical array detectors, except when charge-switching is practical (e.g., Soucail et al. 1995). However, variants of the stress-birefringence modulator can be constructed to run at a few to a few tens of Hz; these will not be mechanically resonant, but that is not a necessary condition. New CCDs of the frame-transfer design exist which combine a readout noise of order 15 electrons with a frame rate of some tens of Hz; in the near infrared, the PtSi arrays are cited as combining three electrons readout noise with 3 MHz pixel rate (Ueno 1995). Such detectors will allow construction of stress-birefringence polarimeters with hundreds of thousands of parallel channels, one pixel for each (or, more probably, thousands of parallel channels with, say, 100 pixels each, to allow adequate oversampling in search of photometric stability). Such polarimeters will operate at a modulation frequency around 10 Hz, sufficiently high to eliminate almost all of the scintillation noise for large telescopes, even with a single-beam analyzer. A readout noise of about 15 electrons is acceptable: the corresponding noise-equivalent signal of about 200 electrons will allow the user an instantaneous dynamic range within the image from about 1000 electrons to saturation at a few times 100,000 electrons; with control of elementary exposure time (and therefore of modulation frequency) to suit the brightness of the image, such a within-image dynamic range will be enough for the regime we are considering.

A complete system including both optical differencing and an array polarimeter is shown in Fig 3. The only important component shown there but not discussed previously is the input depolarizer. This is needed to ensure exact 50/50 splitting by the calcite plate; it need not be a perfect depolarizer, a "time-averaging linear depolarizer" such as a continuously rotating halfwave plate (Tinbergen 1995) will suffice. The analyzer will generally be of the double-beam type, to use all the available photons and for elimination of "common-mode" noise (such as any remaining scintillation or extinction noise). The total system includes four optical components in addition to the basic spectrograph/photometer, so one must expect a light loss of some 10% even with good anti-reflection coatings on all surfaces; that is the price one has to pay for obtaining high photometric precision in very-low-contrast images. Design problems such as constructing a broadband system have not been mentioned here, but will determine the compromises one will have to accept in any practical system. The "calcite plate" is likely to take one of the forms illustrated in Fig 4, probably in the "Savart plate" modification.

Note: The reader may well wonder what is the reason to go to all the complications of depolarizing a signal, then polarizing it, then analyzing the polarization by what is in essence a differential photometer, in order to obtain differential photometry at the end. The only valid reason is robustness of the instrumental method and therefore of the astronomical results. I have mentioned before that optical differencing is extremely stable and robust. Modulation polarimetry, using a stress-birefringence modulator and a single detector (pixel) for the entire measurement, likewise has been shown (in the PMT case) to be very robust and capable of very high and repeatable precision. So the proposed rather roundabout method is a cascade of robust operational units, each doing that which it does best. Systematic and instability errors are eliminated by doing the basic polarimetric measurement with one detector pixel and within as short a time span as is compatible with the readout noise and the photon stream; the two-beam analyzer is only needed for further reduction of the photon noise and of any remaining common-mode noise.

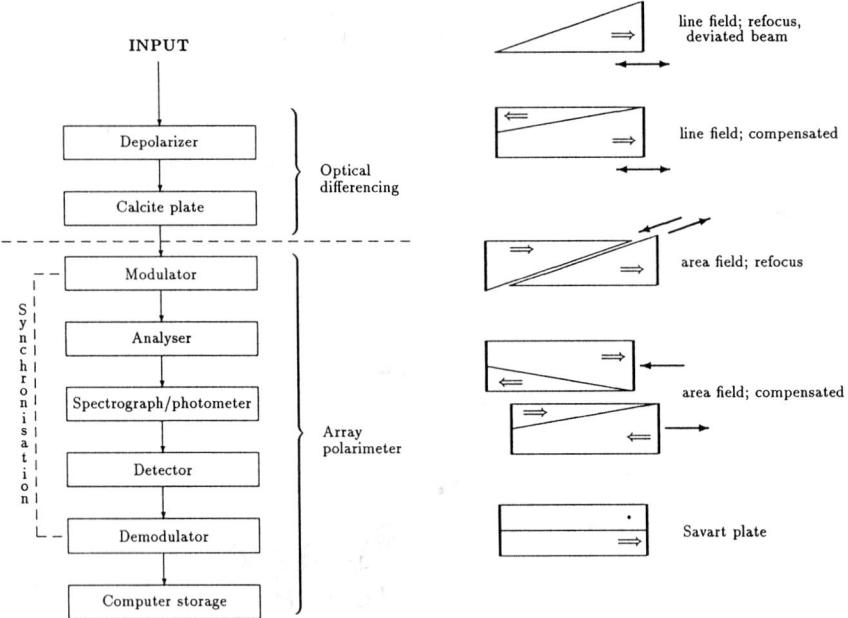

Fig. 4. Compound calcite plates (schematic only)

Fig. 3. System Layout.

4. APPLICATIONS

In this section I explore the potential use of optical differencing; polarimetry is sufficiently well-established to fend for itself. The context is provided by the ten-meter-class of new telescopes, which in many cases provide photon streams sufficient for very high precision indeed on relatively bright objects. The main question will be: what in the present context is meant by "relatively bright"? To me, at any rate, the apparent answer was a surprise; my hope is that appearances do not deceive.

Supposing we can eliminate systematic and instability errors by the techniques discussed, photon noise will be the determining factor of the precision obtained. We are used to 1% precision being called "high" in astronomy, but that is because of our preoccupation with "detection of the faintest objects". A precision (or at least S/N) of 0.1% is common in (spectro)photometry of brighter objects and 0.01% precision is the norm in PMT polarimetry. Large telescopes used on bright objects collect enough photons for all these applications; we may ask ourselves how much further we can go. To put this into perspective, consider the largest practicable signal in stellar astronomy: a zero-magnitude star being observed for six hours by a ten-meter telescope, using a 6000 Å bandwidth and an instrument of 0.5 overall quantum efficiency. The number of photons detected would be of order 3×10^{16} and a precision of better than one part in 10^8 would be possible. We conclude that, even if we set our sights on a precision (very high for astronomy) of one part in 10^6, we still have a reserve factor of 3×10^4 or 11 magnitudes for fainter objects, narrower bandwidth or shorter integration times. This shows that sufficient photons can be collected for a number of applications, but that the largest telescopes will be essential in many cases.

To start with, let us estimate the limiting magnitude for detection of point sources in broad daylight, assuming blue sky to be absolutely uniform and assuming one arcsec seeing. Blue sky is of order of one fourth-magnitude star per square arcsec, which leads to a signal of order 6×10^{14} detected photons and a photon-limited precision of one part in 2.5×10^7. Such precision in measuring the signal from blue sky corresponds to a limiting (one sigma) point-source magnitude of 22. If we reduce the precision to one part in 10^6, we still reach a one-sigma limiting magnitude of 15, now in about 40 seconds if we retain the very wide bandwidth; obviously, other choices can be made.

A similar rough calculation leads to a one-sigma limiting magnitude for point-source detection during full moon of 28.5, using a precision of one part in 10^5. Using the same figure for precision for the analysis of a spectrum with relatively few and shallow lines (absorption or emission), one estimates that a 0.001% band of 30 Å width will still be one-sigma detectable for a tenth magnitude star. Time variations of period 100 seconds of a 0.001% absorption band could be investigated in stars of magnitude four.

A precision of one part in 10^4 used in direct imaging during a really dark night leads to a one-sigma limiting magnitude of 31 in one arcsec seeing and much deeper if the optical differencing technique can be combined with wavefront reconstruction. Other examples can easily be worked out for spectroscopic applications, such as faint interstellar absorption bands, Doppler or Zeeman profiles of stellar or interstellar lines of very low optical depth, etc.

A precision worse than one part in 10^4 does not seem to me to need optical differencing.

Drift-scan techniques should be able to fill the range between one part in 10^4 and one part in 10^3, while conventional flatfielding seems able to cope with precisions worse than one part in 10^3. Since drift-scanning and optical differencing are both one-dimensional techniques, a combination may be both feasible and useful.

Natural applications of optical differencing would be detection of one particular very faint fringe within the grey mist of an interferogram, and sky-line subtraction to very high precision in spectroscopy. Using double-beam analysers to allow much slower modulation and readout, the mid-infrared range with its high thermal background could be a natural application, too; the polarimetric components, to the extent that they are non-lossy, could stay outside the dewar for convenience of operation (cf., Hough et al. 1994 for near-infrared). Speckle methods could conceivably be combined with optical differencing (see discussion at the end of this paper, on confusion).

Having considered photon noise in the final image accumulated in the computer, we need to consider it for the individual images read out from the detector. The minimum signal is taken to be 1000 detected photons per pixel (which may be a superpixel, using binning) per readout. To take the blue sky example: with the 6000 Å bandwidth, the signal is of order 3×10^{10} photons/sec, so narrower bandwidths are no problem whatsoever (essential, in fact). The darkest night-time sky, however, is marginal even for a bandwidth of 6000 Å: 6000 detected photons/sec, allowing at most a 6-Hz frame rate and 3-Hz modulation rate; dual-beam analyzers will often be essential in that application, unless readout noise at about 10-Hz frame rate improves even more. For a typical spectral application of 30 Å bandwidth, a 15th magnitude star still yields 10000 photons/sec, allowing a ten-Hz frame rate.

Scintillation noise is unlikely to be a serious issue. Young (1967) states that for (at that time) "large" telescopes the crossover between scintillation-dominated and photon-dominated noise occurs at 8th to 11th magnitude; since we are concerned only with the residual high-frequency tail of the scintillation power spectrum, the crossover magnitude in our case will be brighter. In addition, as stated above, scintillation noise is "common-mode" for the two beams of a dual-beam analyzer and can be eliminated; actual experience with a ten-meter-class telescope (and its central stop, which determines the high-frequency tail of the scintillation power spectrum) will be needed before such matters of detailed design can be decided.

5. CONCLUSION

I have discussed how frame-transfer CCDs and PtSi arrays may be used to extend the realm of photon-limited precision to situations of very low image contrast. The essential ingredient is adequate very-short-term stability of the gain (detected electrons/photon) of the individual pixels of the array; by sufficiently fast modulation, such short-term stability can be exploited to obtain polarimetric and differential photometric precision. We do not know what the operational very-short-term photometric stability of array detectors is; we suspect it is very good indeed and measuring it seems feasible, to say the least. If we can design instruments that exploit the short-term stability, it seems likely that both polarimetry and optical-differencing photometry can be implemented to levels of precision beyond the reach of any other method. Certain high-precision areas of optical/infrared astronomy would thus become accessible, particularly with the large signals provided by ten-meter-class telescopes. With large telescopes, many such programs could be carried out near full moon or in the daytime, so they would compete only marginally with other programs and the scientific significance of large telescopes would benefit.

ACKNOWLEDGEMENT

An early discussion with Peter Katgert helped to clarify the ideas presented in this paper.

REFERENCES

Hough, J. H., Chrysostomou, A. and Bailey, J. A. 1994 in Infrared Astronomy with Arrays: The Next Generation, I. S. McLean, ed., Astrophys. and Space Science Library 190, Kluwer Academic Pub., Dordrecht, p. 287
Jorden, P. R., Deltorn, J.-M. and Oates, A. P. 1994 in Instrumentation in Astronomy VIII, SPIE. 2198 (in press), paper 57
Kemp, J. C. 1969 Jour. Opt. Soc. Am. 59, 950
Scarrott, S. M., Warren-Smith, R. F., Pallister, W. S., Axon, D. J. and Bingham, R. G. 1983 MNRAS 204, 1163
Soucail, G., Cuillandre, J. C., Fort, B. and Picat, J. P. 1995 in IAU Symposium No. 167, New Developments in Array Technology and Applications, A. G. D. Philip, K. A. Janes and A. R. Upgren, eds, Kluwer Academic Pub., p. 263
Tinbergen, J. 1993 in IAU Colloquium No. 136, Stellar Photometry - Current Techniques and Future Developments, Cambridge University Press, Cambridge. p. 130 and 264
Tinbergen, J. 1995 Astronomical Polarimetry, Cambridge University Press, in preparation
Tinbergen, J. and Rutten, R. G. M. 1992 A User Guide to WHT Spectropolarimetry, La Palma User Manual no 21, Royal Greenwich Observatory, Cambridge
Ueno, M. 1995 in IAU Symposium No. 167, New Developments in Array Technology and Applications, A. G. D. Philip, K. A. Janes and A. R. Upgren, eds, Kluwer Academic Pub., p. 117
Young, A. T. 1967 AJ 72, 747

DISCUSSION

PHILIP: This technique could be used to detect faint members of open clusters, but I imagine that in a globular cluster the images would be too crowded.

TINBERGEN: That is correct. The proposed technique eliminates a high constant background but can do nothing for confusion (it could be operated with wavefront correction, however). If in confusion limited situations one uses CLEAN, the "$\pm$" PSF is no worse than a single peak, I suspect; one does give up the constraint that the image must be positive everywhere.

COSMIC RAY EVENTS AND NATURAL RADIOACTIVITY IN CCD CRYOSTATS

Ralph Florentin-Nielsen, Michael I. Andersen and

Niels Bohr Institute for Astronomy, Physics and Geophysics, Copenhagen University Observatory.

Sven P. Nielsen

Risø National Laboratory, Denmark.

ABSTRACT: We have found that many materials that are most commonly used in CCD cryostats are weakly radioactive and therefore contribute to what is rather liberally labelled as Cosmic Ray Events. Some standard optical glasses that are extensively used for lenses and optically flat windows such as UBK-7 contain large amounts of potassium 40, which renders them useless as windows in CCD dewars. Cobalt 60 is sometimes found in excessive amounts in some steel alloys, even in Covar used as a thermal match to silicon. The choice of materials that emit a minimum of ionizing radiation is discussed.

1. INTRODUCTION

In long CCD exposures one always finds a number of isolated and sharply defined locations on the CCD frame with high exposures - typically a couple of thousand electrons within a few pixels. Such dense spots or tracks are generally labelled Cosmic Ray Events.

In direct imaging these events normally do not pose serious problems, as they can easily be distinguished from stellar images, as they are not circularly symmetric and have considerably sharper profiles than the point spread function, provided that the image scale in the telescope focus is reasonably well sampled by the CCD pixels (2 pixels/FWHM of the PSF or better). If need be Cosmic Ray Events can be removed in the image processing. However, in cases where the tracks coincides spatially with the objects to be studied, they cannot be removed without some loss in signal-to-noise ratio for that object. In spectrographic exposures cosmic ray events are generally a much more serious problem. Exposure times are often quite long, and the spectra and the comparison lines cover a large number of pixels. Hence the risk of having dense tracks right on top of an interesting spectral feature is much higher.

It is therefore important to aim at keeping the rate of Cosmic Ray Events as low as possible. As the majority of Cosmic Rays consist of hard radiation, mainly muons, it is not possible to screen effectively against these at the telescope.

2. LOCALLY GENERATED RADIATION

However, not everything that looks like cosmic ray events actually arrive from space. All kinds of ionizing radiation can create tracks of similar energy and geometric extent as can cosmic rays. In particular, radioactive materials in the CCD cryostat can be a significant - and sometimes dominant - source of the tracks that are commonly named cosmic ray events.

We have undertaken measurements to identify materials used in dewar construction that contain radioactive isotopes, and also carried out experiments to verify to what extent the remaining tracks are truly of cosmic origin.

Cylindrical samples of 25 mm diameter of the most commonly used dewar materials were measured for beta-activity using a calibrated Low-level Geiger-Müller counter at the Risø National Laboratory, Denmark. The results are shown in Table 1.

TABLE 1

Beta ray activity for materials commonly used in CCD cryostats.

Material:	Concentration (Bq/kg)	Detection limit (1.65 sigma) (Bq/kg)
Copper	0.00	0.11
Covar	0.03	0.07
Brass	0.06	0.07
Bronze	0.23	0.09
Titanium	0.28	0.15
AlMgSi alloy	1.3	0.30
Al alloy used by InfraRed Lab.	1.6	0.33

The beta-activity for copper, covar and brass was below the detection limit, whereas the results for the remaining materials lies well above the detection limit. These measurements do not include the identification of the radionuclides in the various samples.

Larger samples were measured in a gamma ray spectrometer, which by analysis of the gamma ray energy spectrum does identify the radionuclides. The results are given in TABLE 2 to one significant digit only since the geometry of the samples did not match exactly that from the calibration.

We have found potassium 40 corresponding to about 4 g potassium and 0.1 Bq Radium 226 in the UBK-7 lens (field flattener, originally intended to be the window of the CCD cryostat for the Danish Faint Object Spectrograph and Camera, DFOSC). The Loral 2k x 2k two side edge buttable CCD, which is mounted in a covar package soldered onto a covar block was found to have a cobalt-60 contamination of 0.7 Bq. The rest of the results are below the detection limit.

3.. CHALK MINE EXPERIMENT

To verify that the "cosmic ray events" of our CCD camera were indeed dominated by true

COSMIC RAY EVENTS AND RADIOACTIVITY IN CCD CRYOSTATS

TABLE 2

Radio isotopes identified by gamma ray spectrometer and their activity in Becquerel (number of disintegrations/sec).

Sample:	K (g)	Ra-226 (Bq)	Th-232 (Bq)	Co-60 (Bq)
Steel 37	< 0.02	< 0.05	< 0.1	< 0.04
AlCuPb alloy	< 0.02	< 0.05	< 0.1	< 0.04
Loral 2k CCD	< 0.02	< 0.07	< 0.1	> 0.7
UBK-7 lens	4.2	> 0.1	< 0.2	< 0.07
Fused silica	< 0.02	< 0.06	< 0.1	< 0.04

cosmic radiation and not by any local source we have operated the camera at 37 meters below ground level in the Mönsted chalk mines. This is the deepest mine in Denmark that can be used for the experiment, and although it is not a very impressive depth the attenuation of 37 meters of soil is sufficient to allow conclusive measurements to be carried out. Fig. 1 shows the total rate of cosmic ray events as a function of their energy in ADUs. The conversion factor to electrons is 1.4 electrons/ADU. The solid line shows the energy spectrum for the counts in the laboratory, and the dashed line the counts in the Mönsted mines.

The CCD used was a thick Loral 2k X 2k with 15 micron pixels. The count rates are shown in Fig. 1.

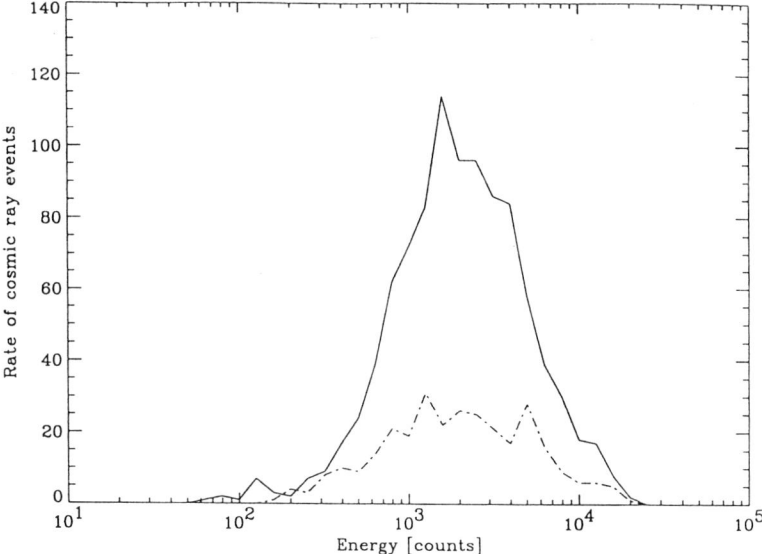

Fig. 1. Energy spectra of Cosmic Ray Events observed at ground level, (sold curve) and 37 m below ground (dashed curve)

The count rates were found to be: in the laboratory in Brorfelde: 1.7 events·cm^{-2}·min^{-1}, At 37 meters below ground: 0.56 events·cm^{-2}·min^{-1}.

The spectral distribution is similar in the two sets of measurements with charges ranging from < 100 - 30,000 electrons, corresponding to energies of the ionizing radiation of: approximately 300 eV - 100 keV, under the assumption that the creation of one electron - hole pair on the average requires 3.5 eV of energy.

Thus, we find a reduction of the count rate by a factor of 3.0. We can therefore safely state that counts due to radioactive contamination is less than one third of what we observe at ground level, and the fact that the energy spectrum is apparently unaltered led us to assume the radioactive counts are well below one third of the count rate found at ground level. For comparison a UBK-7 lens was mounted instead of the fused silica window yielding a count rate of 7.0 events·cm^{-2}·min^{-1} which is more than 12 times higher than the upper limit to the radioactive count rate with the fused silica window in place.

4. CONCLUSIONS

Aluminum, which is the most commonly used material in dewar design has been found to be radioactive to a rather varying degree. First of all, aluminum does contain small traces of uranium 238, but the amount of uranium depends strongly upon where it was mined. A large proportion of all aluminum sold today contains recycled aluminum that has been exposed to radiation from post World War 2 atomic bomb detonations and contain radium. One precaution when using aluminium would be to specify virgin aluminum and to test the material before use.

Electrolytic copper, magnesium and fiber glass composits are ideal materials for use close to the CCD detector from a radiation point of view. Magnesium is highly flammable and chemically very active and is therefore not suited for cryostats. Fiber glass lacks the electrical conductivity which is needed to provide a Faraday cage around the CCD. This leaves copper as the preferred material for the parts of the cryostat that are close to the detector, i.e. the camera head, (ref. Fig. 2). The thermal radiation shield is made of gold plated copper. We shall try to electrolytically deposit a layer of copper on the inside of the camera head made of aluminum.

Covar is a steel alloy with a good match to the thermal expansion coefficient of silicon. We have found large differences in the samples with respect to contents of radioactive cobalt. The new packages for our three side buttable Loral 2k x 2k CCDs will be made of "Invar36", which is similar to covar, but has been found to have an even better match to the thermal expansion of silicon at temperatures around -100° C. To verify whether the invar36 at hand is significantly radioactive we intend to machine a sample of the material to the same dimensions as the cryostat window and insert it instead of the fused silica window and compare the count rates in much the same way as we did with the UBK-7 lens. This way, by bringing up the sample close in front of the CCD itself, one can always effectively test the material for any contamination that would affect the observed rate of cosmic ray events without having to utilize beta counters or gamma spectrometers.

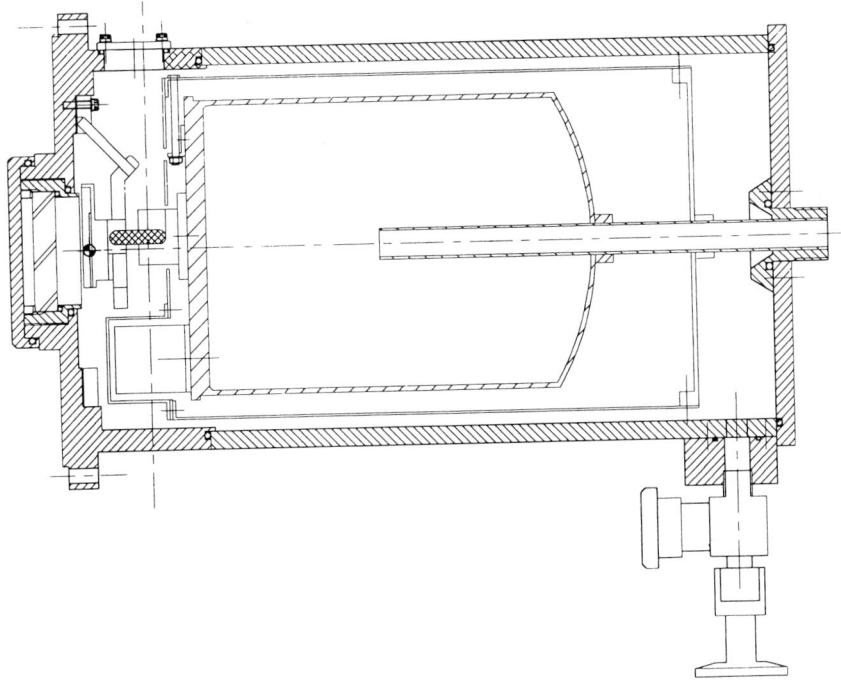

Fig. 2. Liquid Nitrogen cooled Cryostat for Loral and Tektronix CCDs.

DISCUSSION

YOUNG: It seems you are seeing problems we already had 30 years ago with magnetic shields for PMTs. The steel manufacturers put Co60 in their furnace linings; when the radioactivity goes away, it is time to replace the lining. Of course the Co60 ends up in the steel.

FLORENTIN-NIELSEN: Yes, this is the same problem. However, the CCD detector is more sensitive to a variety of radio isotopes.

MOCAM: A 4k x 4k CCD MOSAIC FOR THE CANADA-FRANCE-HAWAII TELESCOPE PRIME FOCUS

J. C. Cuillandre, Y. Melliers,

Observatoire Midi-Pyrénées

R. Murowinski, D. Crampton,

Dominion Astrophysical Observatory

G. Luppino and

Institute for Astronomy, University of Hawaii

R. Arsenault

Canada-France-Hawaii Telescope Corporation

ABSTRACT: MOCAM is a wide field CCD camera, currently nearing completion, which will be offered to the Canada-France-Hawaii Telescope (CFHT) user community in 1995. The project is a collaboration between the CFHT, the Dominion Astronomical Observatory (DAO, Canada), the Institut des Sciences de l'Univers (INSU, France), Laboratoire d'Astrophysique de Toulouse (LAT, France) and the University of Hawaii (UH). In the interests of producing a reliable and effective camera in the shortest time, it was decided to use existing technologies rather than innovative ones. Two-edge buttable 2048 x 2048 15 μm pixel CCDs were obtained from the LORAL aerospace foundry, based on a mask designed by J. Geary at Smithsonian Astrophysical Observatory (SAO). They are mounted in a dewar designed by G. Luppino (UH); the focal plane mounting keeps the mosaic flat to within two pixels and the CCDs are aligned to within two pixels. A mechanical interface designed and fabricated by the DAO holds a 150 mm shutter and a filter wheel which has a positioning repeatability better than five μm.

The four CCDs are operated in parallel by a San Diego GenIII controller adapted by LAT. The mosaic is read out in seven minutes and a single 33 Mb FITS file is generated to enable convenient on-line preprocessing. The user will control the system through a single CFHT Pegasus environment session. The camera field is 14' x 14' with a 0".2 pixel sampling and the readout noise is less than seven electrons. The scientific goals of the initiators of the project are studies of distant clusters, deep galaxy counts and quasars surveys.

1. INTRODUCTION

After the CFHT users meeting in May 1992, the Scientific Advisory Council (SAC) of CFHT recommended building a large mosaic CCD for the prime focus of the telescope, as it had been proposed by a Canadian-French collaboration. The mosaic design consisted of a two

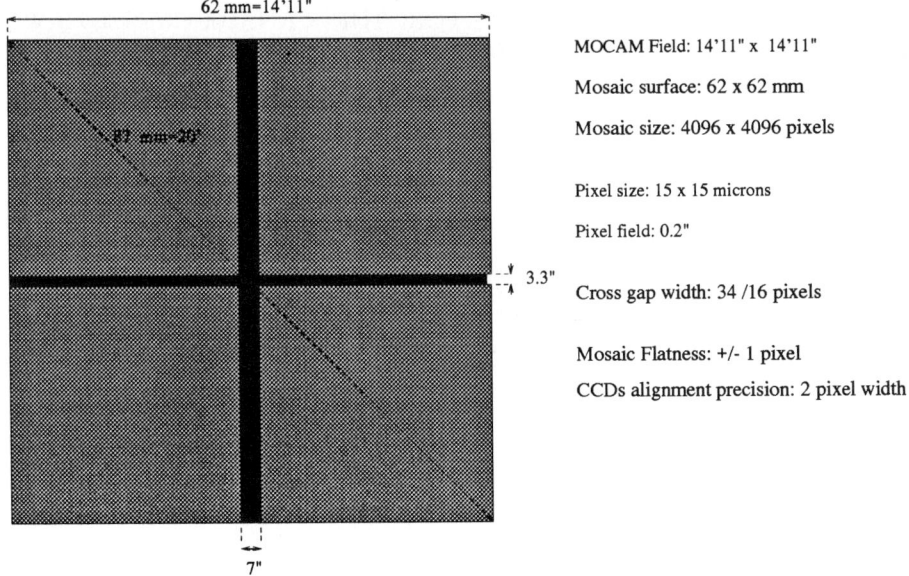

Fig. 1. MOCAM field at the CFHT prime focus.

by two 2048 x 2048 array with 15 micron pixels, giving a total field of view of 14' x 14'. It was found that the scientific impact of such a wide field CCD camera on the world's best image quality telescope would benefit a large number of scientific projects, from asteroids and comets, to quasars or faint galaxies (Crampton 1992). It was also emphasised that the Canadian-French mosaic camera (MOCAM), which covers less than 8% of the total field of view of the prime focus, could be a first step toward the development of a larger CCD camera.

The MOCAM project is now near completion and the camera is ready. MOCAM will be with the dual mosaic cameras designed by Stubbs et al. (1993), the second large CCD mosaic camera in operation on a telescope. Its first light is expected by the end of November 1994 when deep scientific observations of lensing clusters should reveal the full capability of MOCAM before the final acceptance tests and a delivery to CFHT for the users.

3. MOCAM PROJECT OVERVIEW

The objective of this project is to produce a reliable facility instrument for the whole scientific community in the shortest time. The dewar developed by G. Luppino from the University of Hawaii was chosen because its design had been fully developed and optimized (Luppino and Miller 1992) as a prototype of the 8k x 8k mosaic that is currently near completion (Luppino et al. 1994).

The CFHT hardware and software environment were used by the LAT group to develop the parallel readout mode on the SDSU Generation III CCD controller (Leach 1988) adapted by the CFHT for its own use (Kerr et al. 1994). The software was developed under the CFHT

Pegasus software environment. DAO built the filter wheel, its associated hardware and the software that control the filter wheel and the shutter. DAO was also in charge of the selection

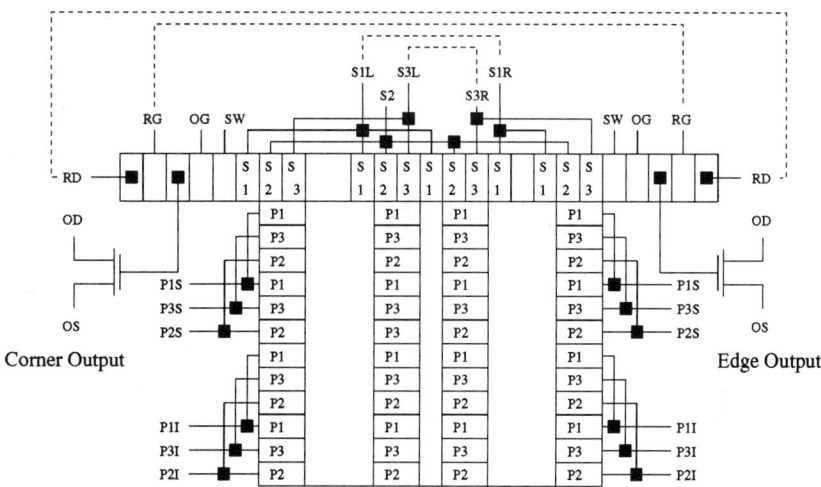

Main characteristics: 2048x2048 pixels 15x15 microns
                        Serial register split in two parts
                        Possibility of frame transfer operations

MOCAM use: Full frame operations (PS=PI)
                        One output used (SL=SR)

Fig. 2. 2k x 2k two-edge buttable LORAL CCD schematic diagram

and cryogenic tests of the individuals CCDs. The whole system was integrated, tested and optimized at Toulouse. Because of staffing priorities at CFHT, their participation in MOCAM was limited in software support to LAT.

4. MOCAM DEWAR AND CCDs

The dewar has been fully described by Luppino and Miller (1992). It was designed to handle a CCD mosaic of four 2k x 2k two-edge buttable LORAL 15 μm pixel CCDs. The charact-eristics are excellent as the hold time with a 3.0 liters nitrogen fill exceeds 24 hours with a fully wired mosaic operating at -100° C. A single high density connector simplifies the connection to the controller and one of the two vacuum ports enables the pressure to be monitored.

Light enters the dewar through a five mm thick 14 cm diameter quartz window. Spot diagrams analysis shows that the effect of the field curvature on the image quality is negligible, but the image quality slightly degrades when moving out to the edge of the CCD field. For an f/4 beam, homogeneous image quality can be obtained on the overall field by focusing on stars close to the mosaic edges. Fig. 1 gives the focal plane characteristics at the CFHT prime focus. The larger gap between the CCDs is 500 μm and thanks to the recent development that uses new techniques to cut the packages (Luppino and Miller 1992) onto which the CCDs are mounted, the alignment between the different CCDs is kept within two pixels and the mosaic

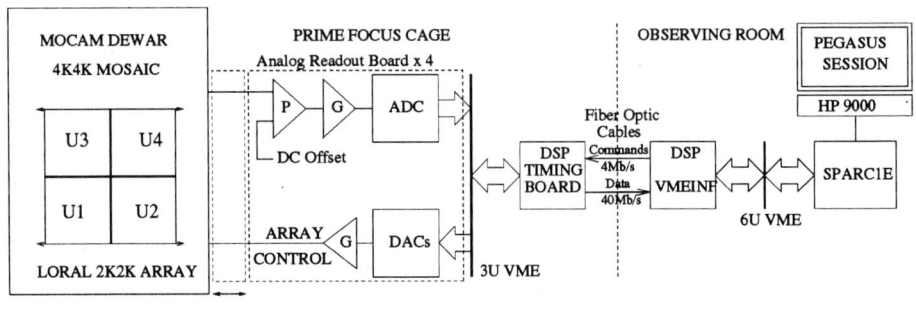

Fig. 3. MOCAM GenIII controller system.

is kept flat to about two pixels, introducing no defocusing effect at the the f/4 prime focus.

The 2k x 2k two edge buttable CCDs were designed by J. Geary (SAO) and produced by LORAL (Geary et al. 1990). These CCDs have two low-noise outputs and are capable of frame storage operations. MOCAM is wired to allow full frame readout through one output per CCD (see Fig. 2). Switching from one CCD output to the other one is achieved simply by downloading a new DSP code to the controller and inverting the two coaxial video cables inside the controller. Frame transfer operations are possible by simply downloading a new DSP code and installing an appropriate mask over the storage regions, but MOCAM is initially planned to be used only in full frame mode with four CCDs parallel readout.

5. CCD CONTROLLER HARDWARE AND SOFTWARE

Fig. 3 shows the MOCAM CFHT's Generation III controller based on the same design of the Redeye controller (Kerr et al. 1994). This multi-amplifier readout system consists of a data acquisition computer running the Pegasus Software system on a Unix platform, communicating through a client-server protocol with the CCD interface control computer that manages the data acquisition and the DSP controller based on the SDSU design.

The four CCDs are read out in parallel but the data are sent sequentially through the single fiber optic connection to the VMEINF SDSU board, which sends them onto the VME bus in DMA mode. At the end of the transfer, the data are descrambled in the CCD computer by a Unix routine. The four CCDs are read out by the four corner outputs with a fixed number of overscan pixels in each column and row. As the CCDs mirror each other about the central gaps, the overscans simulate the mosaic gap cross and then enable to keep the field astrometry in the final single FITS file.

When adapting the CFHT software at LAT, there was a big concern over keeping the existing single CCD readout mode as a possible multi-amplifier readout option. With this aim, the concept of topologies was introduced to address all the physical possibilities available with the mosaic. Each individual CCD has a unique 2^n weight with n = $\{0,..,3\}$ (see Fig. 4, left) so that each mosaic configuration has a unique topology code (Fig. 4, right). In this software operating mode, FOCAM the actual CFHT single CCD camera, would be managed by the

MOCAM: A 4k x 4k MOSAIC FOR THE CFHT

mosaic software as the CCD having the weight one. This concept is used from the higher Unix layer (MOCAM Pegasus session) to the lower one in the DSP code. Because of the hardware design, the four CCDs are always clocked together even in the case of a single raster acquisition. Consequently the data transmission depends on the current topology so that the CCD computer receives only the required data to perform the descrambling process.

Unix is not suited for real time process and some unexpected loss of data occurred when sending large amounts of data through the fiber optic. We solved this problem by slowing down the DSP controller so that the CCD computer became the communication master. The

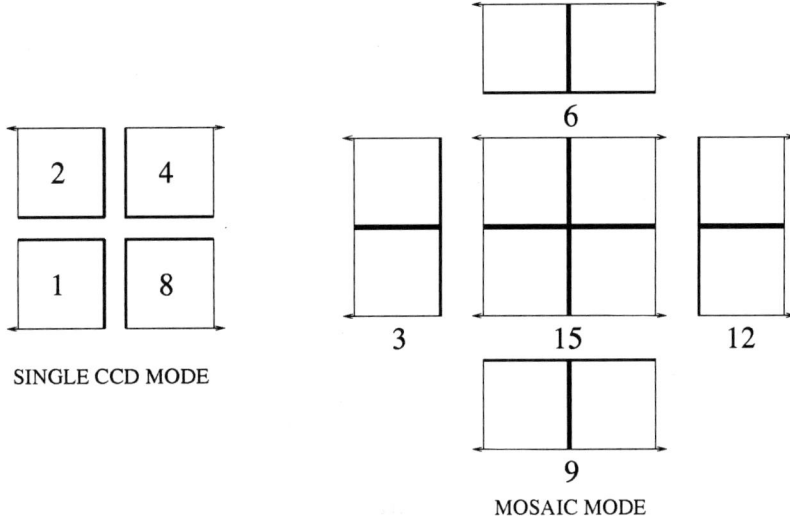

Fig. 4. MOCAM virtual detectors: the topologies concept.

drawback is a higher four pixels mosaic readout time of 48 μs as a delay of at least seven μs is required between each optic data. This problem should be fixed with the new CFHT software release being currently implemented. Fig. 5 shows the mosaic pixel readout cycle, where the overlap of functions are used as much as possible to save time. The full 4k x 4k mosaic can be read out in 3 min 30 sec and the descrambling process adds one mnute more.

Developing MOCAM as a CFHT instrument required use of a two meter long cable from the CCD controller to the dewar. Because there are no electronics inside the dewar, the cable had to be very highly shielded. The four CCD harnesses are completely isolated from each others to avoid interchannel crosstalking through capacitive coupling. High level crosstalking is avoided by a short delay after the video dump to take into account the limited slew rate of the CCDs outputs (Fig. 5). There are a lot of interconnections between the MOCAM devices (see Fig. 6) and to keep the readout noise as low as possible, the best ground configuration found was to connect the dewar body to the digital ground whereas the analog board is the ground node for each CCD harness. The readout noise is 6.7 electrons with small variations between the four CCDs.

6. FILTER WHEEL AND SHUTTER

The filter wheel which also has the function of mechanical interface with the prime focus, incorporates a 15 cm iris shutter. Its opening time of 50 ms and its closing time of 200 ms, which are long, make the standard dome flatfielding operation rather inaccurate for ultra-deep imaging. We recommend that observers use superflats as much as possible rather than twilight flats or dome flats even if some simple computational methods can be applied to correct the shutter response (Surma 1993).

The filter wheel can support five filters but MOCAM will initially use only V, R and I band photometric filters since MOCAM is equipped with four thick CCDs. In a second stage, we plan to mount thin CCDs in the camera head and can make use of U and B filter positions. The filter wheel position is controlled by two stepping motors giving a positioning repeatability of five μm, which is enough to get very reliable flatfields.

The filter wheel and the shutter are run by a Pro-Log Controller communicating over the serial port with a server running on the CFHT serial server host which communicates with a client which can be anywhere over the network (Fig. 6). The Pro-Log shutter control signal comes from the GenIII controller, which manages all exposure tasks.

7. CONCLUSION

Less than three years after the MOCAM proposal, the camera is ready to be used and become a CFHT instrument available for the whole scientific community. Some Canadian and French astronomers have already applied to observe with MOCAM, but we must wait for first observations at the end of November before having a fully operational camera. We hope that in 1995 the first 14' x 14' CCD imaging capability will be available at CFHT and will be a unique instrument on four-meter class telescopes.

For the future, we are now considering a very large CCD camera with a one square degree field of view. We already have compelling evidence that a wide field CCD camera on a high

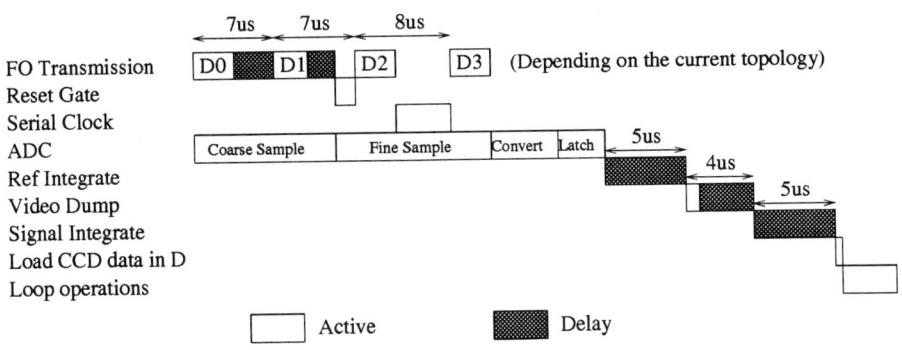

Fig. 5. MOCAM pixel readout diagram.

MOCAM: A 4k x 4k MOSAIC FOR THE CFHT

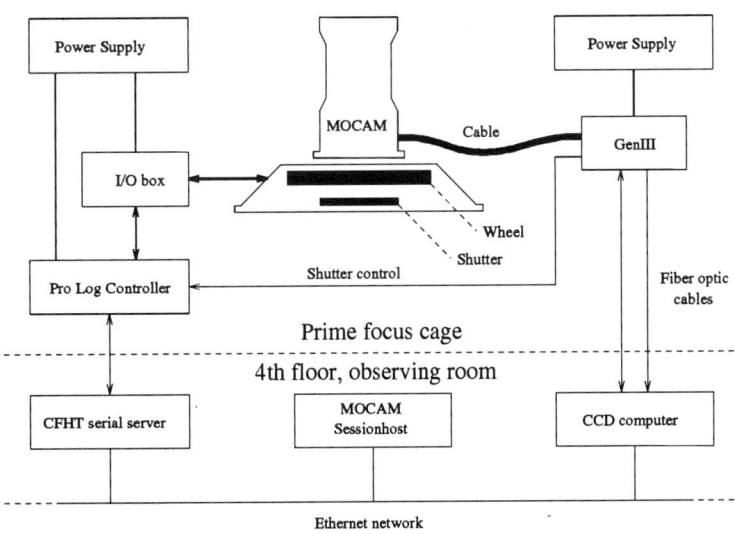

Fig. 6. MOCAM hardware.

image quality telescope such as CFHT can make a major breakthrough in cosmology, because it combines subarcsecond images, a large field of view and a large collecting area. It is almost certain that planetology, galactic and extragalactic astronomy need such instrument. This second generation of wide field CCD camera must be considered seriously for the future.

REFERENCES

Crampton, D. 1992 Proceeding of the Third CFHT User's Meeting, 52
Geary, J. C., Robinson, L. B., Sims, G. R. and Bredthauer, R. A. 1990 SPIE 1242, 38
Kerr, J., Clark, C. C. and Smith, S. S. 1994 SPIE 2198, 980
Leach, R. W. 1988 PASP 100, 1287
Luppino, G. A., Bredthauer, R. A. and Geary, J. C. 1994 SPIE 2198, 810
Luppino, G. A. and Miller, K. R. 1992 PASP 104, 215
Stubbs, C., Marshall, S., Cook, K., Hills, R., Noonan, J., Akerlof, C., Alcock,C., Axelrod, T., Bennet, D., Dagley, K., Freeman, K., Griest, K., Park, H.-S., Perlmutter, S., Peterson, B., Quinn, P., Rodgers, A., Sosin, C. and Sutherland, W. 1993 SPIE 1900, 192
Surma, P. 1993 A&A 278, 654

DISCUSSION

D'ODORICO: In the operation of mosaics of CCDs, interference between the different channels has been sometimes detected. Have you carried out any tests on these objects (bright source in one of the channels, low level pickup noise)?

CUILLANDRE: The four CCDs are driven by four fully independent analog cards; the only common point at the focal plane array is the digital ground. To test high level cross talking, we use a mark with a hole, the illumination level is about three times the saturation level: a human eye check on a SAO image enables one to detect ghost images easily, then simple aperture photometry gives the cross talking ratios. We did not quantify the part of the pickup noise in the total readout noise as the SDSU timing board always address the four analog cards at the same time, even if we read only one CCD. We can easily check it by simply unplugging the analog cards.

ECHELLE SPECTROSCOPY WITH A CCD AT LOW SIGNAL-TO-NOISE RATIO

Didier Queloz

Observatoire de Genève

ABSTRACT: The measurement of some physical parameters of astronomical objects can only be carried out with high resolution spectra. Unfortunately the high dispersion of the light on the detector restricts such observations to relatively bright sources. However, some spectral information can be concentrated into a single spectral "line" by a cross-correlation algorithm, allowing the observation of fainter objects. Such a technique, taken from the CORAVEL optical correlation, is presented. A complete description of the errors of the correlation function parameters is given and the minimum signal-to-noise ratio is also discussed. Finally, a short investigation of the best resolution needed to observe efficiently radial velocities and velocity broadenings is made.

1. INTRODUCTION

The measurement of some physical parameters from stellar spectra, such as accurate radial velocities, broadenings of slow rotators or intrinsic velocity dispersions of unresolved low mass stellar systems (e.g. the core of globular clusters), line asymmetry and detailed chemical composition of stars, can only be carried out with high resolution spectroscopic observations. Unfortunately, these studies are restricted to relatively bright objects due to the high spread of the light over the detector. However, the information about the Doppler shift, the velocity broadening and the metallicity of spectra, is in fact distributed among all the spectral lines and can be concentrated into a few parameters by a cross-correlation algorithm. If the number of lines is large enough, it is possible to extract information from low signal-to-noise (S/N) ratio spectra and to observe faint stars.

Griffin (1967) demonstrated that the optical cross-correlation technique with a mask, first described by Fellget (1953), was able to measure quickly and easily, accurate radial velocities with higher sensitivity than the conventional photographic technique. Ten years later, Baranne et al. (1979) extended this technique to echelle spectrographs by building the CORAVEL and showed that the metallicity could also be extracted from the cross-correlation functions (CCF).

Simkin (1974), Tonry and Davis (1979) and Sargent et al. (1977) gave the first detailed descriptions of correlation techniques to measure the Doppler shift and velocity broadening of digitized stellar spectra. With the advent of CCDs, these techniques have been intensively applied in several modifications and improvements (e.g. Bender 1990 for a review) to various resolutions and wavelength ranges. Recently, Dubath et al. (1990) applied the CORAVEL technique with success to high resolution CCD echelle spectra to measure small velocity broadening of composite spectra.

222   Didier Queloz

The purpose of this paper is to describe in detail the CORAVEL technique applied to digitized spectra. At present, this technique is routinely available with the automatic reduction program of the ELODIE echelle spectrograph at the Observatoire de Haute-Provence in France

2. THE CORAVEL-TYPE NUMERICAL CROSS-CORRELATION

The main difference between the classical numerical cross-correlation techniques and the CORAVEL-type one, is the structure of the template: The CORAVEL-type one is composed of *box-shaped emission lines*, matching the typical lines of cool stars, selected to give the best CCF (e.g. Baranne et al. (1979)). However, this template is only efficient if the number of lines is statistically significant (about 1000 lines). Therefore, this technique is almost restricted to echelle spectra covering a large wavelength domain band and to spectra having a significant number of lines (e.g. spectra of cool stars).

Formally, the cross-correlation process can be written:

$$C(\epsilon) = \frac{R(\epsilon)}{R(\infty)} \qquad 1$$

$$R(\epsilon) = \int_{-\infty}^{+\infty} S(v) M(v - \epsilon) \, dv \qquad 2$$

where C is the CCF, S the spectra and M the template, both expressed in velocity space v. Thanks to the box-shape of the lines in M, Eq. (2) can be simplified to:

$$R(\epsilon) = \sum_i \int_{-\epsilon + u_i - \Delta u_i/2}^{-\epsilon + u_i + \Delta u_i/2} S(v) \, dv \qquad 3$$

where Δu_i is the width of the ith box of the template centred on u_i.

The spectrum is modeled by a set of absorption lines with Gaussian shapes. If only the n_l lines matching the template at $\epsilon' = \epsilon$ are considered, Eq. (3) can be written:

$$R(\epsilon') = \sum_i^{n_l} \int_{-\epsilon' - \Delta u_i/2}^{-\epsilon' + \Delta u_i/2} F_i^c \left(1 - D_i \exp\left(-.5 \frac{(v\, 2.355)^2}{w^2} \right) \right) dv \qquad 4$$

where w is the FWHM of the lines, D their relative intensities and F^c the Flux in the continuum of the spectra. If all the quantities are expressed in function of w: ($\alpha = \Delta u/w$, $\epsilon = \epsilon'/w$ and considering $\Delta u_i \equiv \Delta u$, Eq. (4) can be written:

$$R(\varepsilon) = \sum_i^{n_l} R_i(\varepsilon) \qquad 5$$

$$R_i(\varepsilon) = \int_{\varepsilon - \alpha_i/2}^{\varepsilon + \alpha_i/2} F_i^c \left(1 - D_i \exp\left(-.5\, x^2 (2.355)^2 \right) \right) dx. \qquad 6$$

Finally, the assumption $\alpha_i \equiv \alpha$ allows us to express $C(\epsilon)$:

$$C(\varepsilon) = 1 - \frac{G_\alpha(\varepsilon)}{\alpha} \langle D \rangle \qquad 7$$

where

$$G_\alpha(\varepsilon) = \int_{\varepsilon-\alpha/2}^{\varepsilon+\alpha/2} \exp(-2.77\, x^2)\, dx \qquad 8$$

$$\langle D \rangle = \frac{\sum_{i=1}^{n_l} F_i^c D_i}{\sum_{i=1}^{n_l} F_i^c}. \qquad 9$$

The shape of the CCF is determined from the G_α function which depends on the intrinsic shape of the spectral lines and on the width of the template lines. The depth of the CCF is related to the average $<D>$ of the depths of the spectral lines matching the template, weighted by the flux in the continuum of the spectrum. Therefore, it is possible to carry out a classic spectroscopic analysis on this "concentrated single line" and extract information concerning the spectrum itself (see Fig. 1a). Furthermore, thanks to the large number of blended lines and some template mismatchings, the shape of the CCFs are extremely well approximated by Gaussian functions. However, if the spectral lines are strongly asymmetric, the CCF can significantly deviate from the Gaussian approximation (see Fig. 1b).

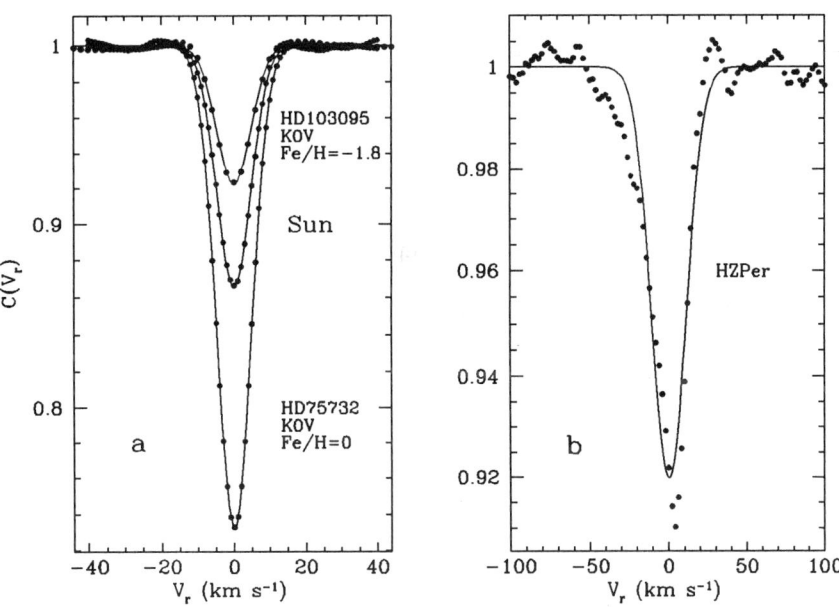

Fig. 1. CCFs for stars with various temperatures and metallicities (dots). The CCFs reproduce well the average behavior of the majority of lines matching the template (neutral lines). The solid lines represent the fitted Gaussians and it is worth noticing that they approximate the CCFs very well. b) CCF of a Cepheid star at a particular pulsating phase where the lines are asymmetric.

2.1 Computing the Error

The estimation of the error on the fitted parameters of the Gaussian, due to photon and CCD read-out noise (ϵ_p), is done with the help of a numerical model. Numerous Monte-Carlo simulations have been carried out using Eq. (7) and the approximation of Gaussian-shaped spectral lines.

The following expression is derived for the rms of the radial velocity ($\epsilon_p (V_r)$) with an uncertainty of 20%:

$$\varepsilon_p(V_r) = .44 \frac{w_i}{D\,(S/N)\,(n_l)^{1/2}} O(s, s_i) \;(\mathrm{km\,s^{-1}}) \qquad 10$$

where w_i is the resolution (FWHM) of the spectrograph in km/s, D the relative depth of the corresponding CCF without noise, (S/N) the average signal-to-noise ratio of the spectrum, n_l the number of spectral lines matching the template, s the number of pixel per resolution element (FWHM) of the spectrum and s_i the instrumental sampling per resolution element (in pixels). The function $O(s,s_i)$ is a correction to take account of oversampled spectra due to the velocity broadening ($s > s_i$) and an intrinsic instrumental sampling greater than two: $O(s,s_i) = 1.6\,s_i^{-1} + 0.2(s/s_i)$ ($2 < s < 10$). It is worth mentioning that for unresolved spectra, observed with an instrumental sampling of two pixels per resolution element, $O(s,s_i) = 1$.

The estimation of the rms on the fitted rms (σ) of the Gaussian is described by the same equation as the radial velocity:

$$\varepsilon_p(\sigma) = \varepsilon_p(V_r). \qquad 11$$

Formally, the value of the error is independent of the α value of the template. The use of larger boxes for the template would reduce the depth of the CCF and the amplitude of the noise in a similar way, only leading to a scaling of the CCF. But the use of a template with $\alpha = 1$ is optimum because it minimizes the blend effect and avoids an undersampling of the CCF. One can notice that the method of the Q-optimization of Baranne et al. (1979) always produces templates with $\alpha \approx 1$.

Equation (10) is only valid for a template with $\alpha = 1$. However, as the depth of the CCF is a slow function of α (see Eqs.(7) and (8)), the equation can also be used with templates ranging from $\alpha = 0.75$ to $\alpha = 1.25$.

It is worth noticing that Eq. (10) reproduces the empirical Eq. (3) of Dubath et al. (1990) and is compatible with the result of Murdoch and Hearnshaw (1991) computed analytically for stellar templates.

2.2 The Signal-to-Noise Limit

The purpose of this paragraph is to estimate the minimum (S/N) of the spectra to clearly distinguish a true CCF from a spurious one due to the noise. A CCF is considered as true when $D > 3.5\,\sigma_c$, where σ_c is the rms in the continuum of the CCF, formally at $C(\infty)$.

The relation between the rms in the continuum of the R function (σ_R) at $R(\infty)$ and the rms of the spectrum due to noise (σ_s) is $\sigma_R = (n_l)^{1/2}(\alpha s)^{1/2}\sigma_s$. The link between σ_R and the rms of $C(\infty)$ is given by $\sigma_c = \sigma_R(\alpha, s, n_l F^c)^{-1}$, where F^c is the flux per pixel in the continuum of the spectra. Finally, with the definition $(S/N) = F^c/\sigma^s$, the following expression describes the rms in the continuum of the CCF:

$$\sigma_c = \frac{1}{(S/N)\,(n_l\,\alpha\,s)^{1/2}}. \qquad 12$$

Using the detection criterion $D > 3.5\sigma_c$, a minimum (S/N) can be defined (see Fig.(2) for illustration):

$$(S/N)_{\min} = \frac{3.5}{D\,(n_l\,\alpha\,s)^{1/2}}. \qquad 13$$

The combination of Eqs.(13) and (10) leads to the definition of the maximal feasible error:

$$\varepsilon_p^{\max}(V_r) \approx .13\, w_i\,(s)^{1/2} O(s, s_i)\,(\mathrm{km\,s}^{-1}). \qquad 14$$

Equation (14) indicates that, for every resolution, there is a typical maximal accuracy beyond which the CCF could be mistaken for a spurious CCF generated from the noise. For the optimization of the telescope time, the concept of the maximal error has strong consequences on the choice of the instruments. For example, an observing program requiring only an velocity accuracy of five km/s should not be carried out with a spectrograph having a resolution greater than 10,000 because a larger integration time is needed to measure reliable velocities with unecessary higher accuracy (see Table 1).

TABLE 1

Maximal error for various resolutions

Resolution R	w_i (km/s)	$\varepsilon_p^{\max}(V_r)$ (km/s)	$\delta(w)$ (kms)
10,000	30	5.5	30.7
20,000	15	2.8	15.5
40,000	07.5	1.4	07.8

Table 1 shows the maximal error for various resolutions beyond which the CCF could be mistaken for a spurious one ($s = 2$ and $s_i = 2$, are used). $\delta(w)$ is the typical maximal error on the velocity broadening measurement, taking a Gaussian broadening function: $\delta(w) = (w_i^2 - (w_i + 2.355\varepsilon_p(\sigma)^{\max})^2)^{1/2}$.

2.3 Observational Strategy

To investigate the optimum instrumentation to measure radial velocities and velocity broadenings, we compute the ratio of the exposure times, for two instrumental resolutions (w_o and w_1) needed to reach the same accuracy.

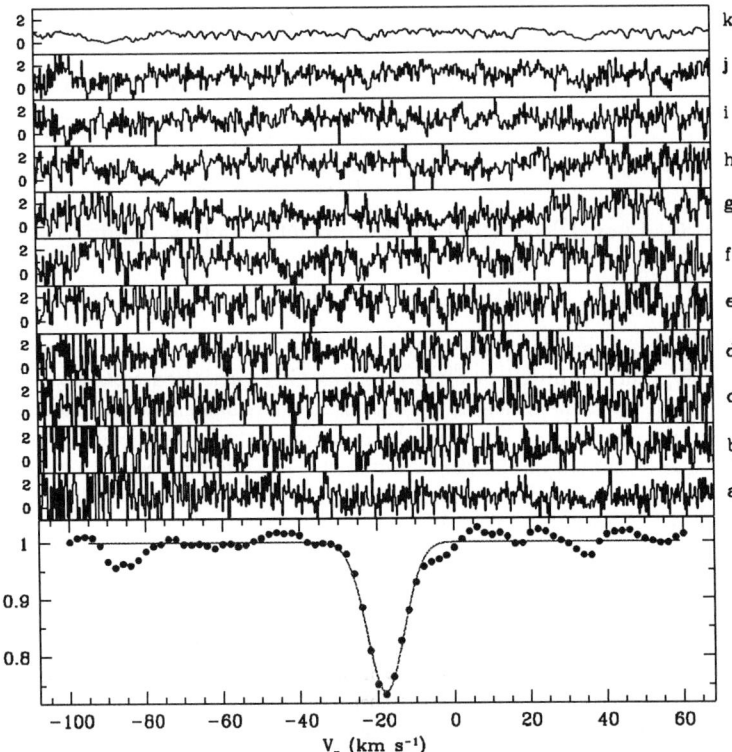

Fig. 2. Illustration of a CFF computed from a K0 III spectrum with (S/N) ≈ 1. The wavelength range used is from 4110 to 4440 Å with a resolution of 40,000. The number of spectrum lines matching the template is about 1000. The ten orders of this spectrum, covering 40 Å each, are displayed from a to j. The last slice (k) displays, for comparison, the same order as j but at a higher signal-to-noise (S/N = 40). In the continuum of the CCF, we observe a typical noise of about 0.02 which corresponds to the expected value from Eq. (12). It is worth mentioning that, for this spectrum, the minimum (S/N) to detect a reliable CCF would be 0.3 (computed from Eq. (13))

For both resolutions we suppose that $(S/N) > (S/N)_{min}$ and thus that the errors are lower that the maximal ones to measure reliable CCFs with the highest resolution.

Two noise regimes are considered: "read-out noise limited" (RONL) and "photon noise limited" (PNL). The first, (RONL), has a $S/N \sim w_i f(w_i/w_s)t)$ and the second, (PNL), has a $(S/N) \sim (w_i f(w_i/w_s)t)^{1/2}$. The function $f(w_i/w_s)$ is a transmission function to take into account the seeing effect, where w_s is the width of the seeing spot projected on the detector in velocity unit and w_i the instrumental resolution. We use the factor given by Allen (1973) to convert the slit width (w_{sl}) to the FWHM of the resolution: $(w_{sl} \approx 1.5w_i)$. We suppose $w_o < w_1$. In the case both spectra are not resolved $(s = s_i)$, we have $O(s, s_i) = 1$ and $D_0 = (w_1/w_0)D_1$, then $w_1^2 S/N_0 = w_0^2 S/N_1$. Therefore we find:

$$\frac{t_0}{t_1} = \frac{f(w_1/w_s)}{f(w_0/w_s)} \frac{w_0}{w_1} \quad \text{(in the RONL regime)}, \qquad 15$$

$$\frac{t_0}{t_1} = \frac{f(w_1/w_s)}{f(w_0/w_s)} \left(\frac{w_0}{w_1}\right)^3 \quad \text{(in the PNL regime)}, \qquad 16$$

If the spectrum is completely resolved by the instrumental resolution w_0 $(s_o > s_i)$ and at the resolution limit with w_1 $(s_1 = s_i)$, we have $D_0 = D_1$ and $O_1(s,s_i) = 1$ and thus $w_1(S/N)_0 = w_0(S/N)_1 O_0(s,s_i)$. Therefore we find:

$$\frac{t_0}{t_1} = \frac{f(w_1/w_s)}{f(w_0/w_s)} O_0(s, s_i) \quad \text{(in the RONL regime)}, \qquad 17$$

$$\frac{t_0}{t_1} = \frac{f(w_1/w_s)}{f(w_0/w_s)} \frac{w_0}{w_1} O_0(s, s_i)^2 \quad \text{(in the PNL regime)}. \qquad 18$$

In Fig. (3), the behavior of t_0/t_1 in these four cases are displayed for two typical seeing values. A general conclusion seems to come into view from these figures: The optimum is rather found towards the high resolutions than the lower ones, but the resolution of the spectrograph must never reach the intrinsic resolution of the spectrum. This behavior of the optimum seems to be independent of the seeing value if the slit width is reasonable compared to the seeing spot.

REFERENCES

Allen, C. W. 1973 Astrophysical Quantities, Athlone Press, London, Ch 4
Baranne, A., Mayor, M. and Poncet, J. L. 1979 Vistas in Astronomy, 23, 279
Bender, R. 1990 A&A 229, 441
Dubath, P., Meylan, G., Mayor, M. and Magain, P. 1990 A&A 239, 142
Fellget, P. 1953 Optica Acta 2, 9
Griffin, R. F. 1967 ApJ 148, 465
Murdoch, K. and Hearnshaw, J. B. 1991 A&AS 186, 137
Sargent, W. L. W., Schechter, P. L., Boksenberg, A. and Shortridge, K. 1977 ApJ 212, 326
Simkin, S. M. 1974 A&A 31, 129
Tonry, J. and Davis, M. 1979 AJ 84, 1511

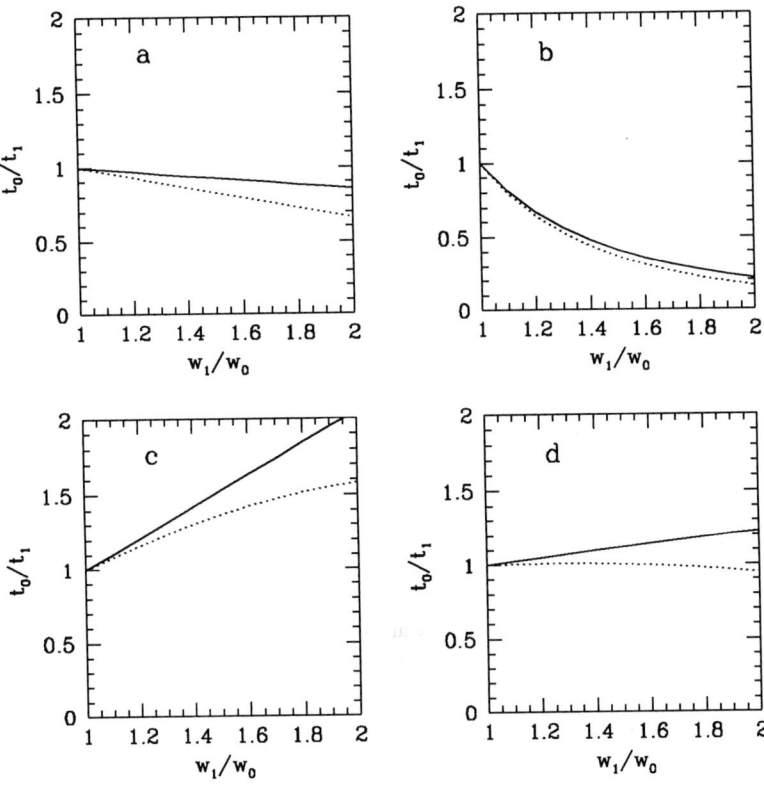

Fig. 3. The results from the Eqs.(15) in a, (16) in b, (17) in c, (18) in d are computed for two different seeing values: two and one times the slit width adjusted to the resolution w_0 (solid and dotted line respectively).

DISCUSSION

UNDERHILL: I suspect that it would be difficult to use this method with early type stars because the lines do not all have the same shape. Is it possible to work in the yellow-green and red regions where the re are numerous telluric lines and lines from diffuse interstellar bands superimposed on the stellar spectrum?

QUELOZ: You can work in any region you like. The only crucial point is having enough lines.

BIKMAEV: Is it possible to obtain simultaneously data on Vr, Vsini and (or) asymmetry and metallicity from one frame?

QUELOZ: Yes, but you may need more than one template.

D'ODORICO: How sensitive is this technique to a changing line profile across the orders or across CCDs (changing focus, etc.)?

QUELOZ: It is sensitive. In fact, if the effect is clearly visible on high signal-to-noise spectra you will also see it by doing cross-correlation on restricted wavelength regions of poor signal-to-noise spectra.

CUBY: Why don't you use a more "realistic" template (synthetic spectrum) instead of boxed shape lines. The solution of the matched filter problem is to correlate with a function of similar shape as the one we want to extract.

QUELOZ: With the large number of lines used and the numerous blends present in cool stars, I do not think there is a big difference. But anyway, following P. Connes (1985, A&AS), it seems that neither the boxes nor the Gaussians are optimized line shapes to extract the radial velocity.

NICMOS3 DETECTOR FOR SPECTROSCOPY

L. Vanzi, A. Marconi and

Universita degli Studi di Firenze, Dipartimento di Astronomia

S. Gennari

Osservatorio Astrofisico di Arcetri

1. INTRODUCTION

We present some results about the use of an engineering grade NICMOS3 array for medium resolution spectroscopy. All the work described here was performed at the Infrared Laboratory of the Osservatorio Astrofisico di Arcetri in Florence and at the Italian Telescope TIRGO (Citterio 1978). The detector has been mounted on the LonGSp (Longslit Gornergrat Spectrometer), an instrument designed and set into operation in Arcetri and described in Gennari and Vanzi (1994).

This paper deals with the characterization of engineering grade arrays and with some problems in the reduction of infrared spectra.

2. CHARACTERIZATION OF THE ARRAY

Our spectrometer uses a subsection of an engineering grade NICMOS3 array as a detector; our purpose is to understand if this kind of array is useful for application to spectroscopy and, meanwhile, to have a good characterization to choose the best available subsection.

2.1 Laboratory Tests

The characterization of the array consisted of evaluating the percentage of bad pixels, the read out noise and the dark current rate.

We acquired dark and flatfield frames, the former without any illumination of the detector and the latter with the uniform illumination of an led diode inside the dewar. The dark frames were taken at three different temperatures (55, 65 and 78° K) with different integration times (stacks of ten frames at 1, 2, 5, 10, 20, 30, 60, 120 and 180 sec.). The flatfield frames were obtained in three different conditions: the first series after a complete reset of the array, the second and the third after a weak and a strong illumination of the detector to test for memory effects. For each series of measurements stacks of ten frames were taken with the same integration times used for dark frames.

Bad pixels were determined from normalized dark and flatfield frames by dividing each quadrant in four by four subareas and defining bad pixels as those exceeding by 4 σ the median value in the subarea. In all of the following analysis bad pixels were not considered and their value was substituted with the median over an 11 x 11 box.

The mean and the standard deviation were computed for each stack (pixel by pixel) after having renormalized the frames to get the same median value in a selected region of the first quadrant. Therefore, the averages over 16 x 16 subareas were taken both for pixel stack mean and standard deviation.

The readout noise was determined as the mean standard deviation of each pixel in the stacks at one and two sec and then averaging over areas of 32 x 32.

The dark current rate and the bias were derived by a linear fit, pixel by pixel, of the stack mean versus integration time, with the error on the mean given by the pixel standard deviation in each stack.

2.2 Results

We characterized two engineering grade arrays and compared the results with the science grade chip presently mounted in the large field camera ARNICA (ARcetri Near Infrared CAmera) at the TIRGO telescope (Lisi et al. 1993). Fig. 1 compares the flatfields obtained with the two engineering grade arrays.

In Table 1 the results of our measurements over the best 100 x 100 pixels subsection of the two engineering grade arrays and over the whole scientific array are presented.

TABLE 1

Measured characteristics of the arrays.

	1 best 100 x 100	2 best 100 x 100	Scientific
Bad Pix (%)	003.0	01.7	01.0
Read Out Noise (e^-)	240	45	45
Dark Current (e^-/s)	040	02	01

In Fig. 2, the dark rate is plotted as a function of temperature. The three different symbols at 78° K indicate frames taken after complete reset (circle), after normal illumination (triangle, like measurements at lower T) and after strong illumination (squares). In the latter case one can notice an increase of the dark current rate by about 20%. Indicative values for the dark rate are 0.1 ADU/s (2 e^-/s).

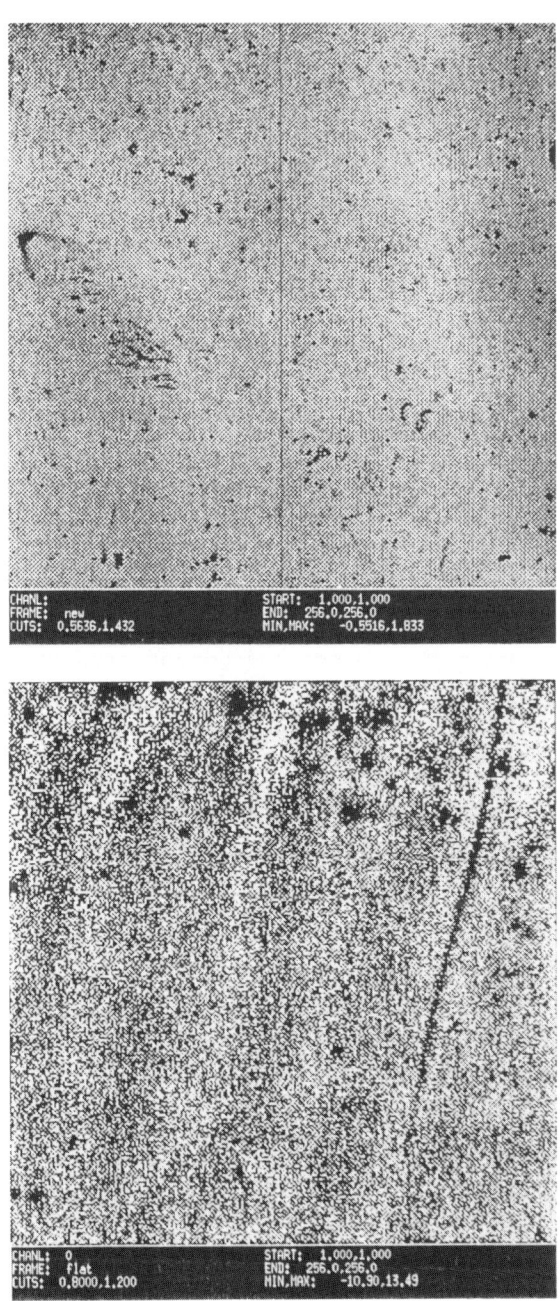

Fig. 1. Flatfield frames of the engineering grade arrays (the one now mounted on LonGSp is in the lower panel).

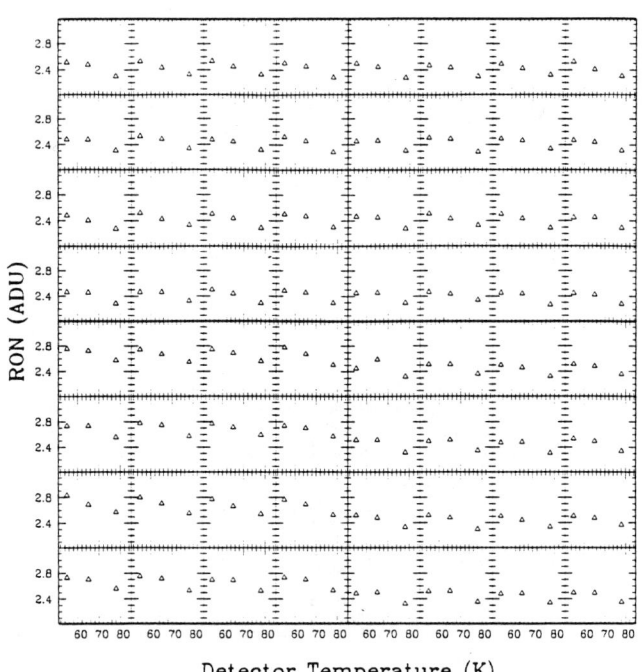

Fig. 2. Dark rate (ADU/sec) of the new engineering grade array as a function of the detector temperature and in three different conditions: after a full reset (circles), after normal (triangles) and strong illumination (squares).

NICMOS3 DETECTOR FOR SPECTROSCOPY

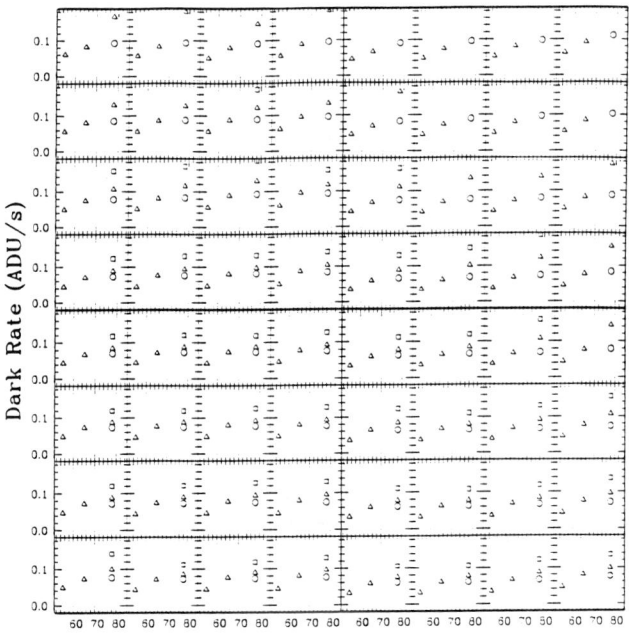

Fig. 3. Dark rate (ADU/sec) of the new engineering grade array as a function of the detector temperature and in three different conditions: after a full reset (circles), after normal (triangles) and strong illumination (squares).

3. DATA REDUCTION

The data reduction is performed with the ESO package MIDAS in the context IRSPEC. The context has been optimized for reduction of IRSPEC data but can be used for LONGSP since the principles of data reduction are quite general and not specific to a particular instrument.

3.1 Data from Observations

Each observation includes object and sky frames with the same integration time taken with the ABBA sequence as well as the spectrum of a standard star and, if necessary, a wavelength calibration lamp. Flatfield and dark frames should be taken, for instance, at the beginning or at the end of the night or every few hours according to the stability of the detector.

3.2 Correction for Bad Pixels

This correction can be performed once a mask of bad pixels is given. Bad pixels are substituted by the average value over the neighboring pixel in the x or y direction. The chosen direction is that with the lower gradient. This method works only with fairly isolated bad pixels and cannot handle cases when the bad pixels are clustered. In such a situation one may use techniques similar to those for cosmic rays removal in CCD frames.

3.3 Flatfield and Dark Current

The original flatframe cleaned by bad pixels is divided by the average of those rows where there will be the object spectrum. After the normalization, all of those pixels that are lower than a threshold (for distance 0.5) are considered vignetted and set to a fixed high value (i.e. 100) so that after flat division the same pixels in the object frame will not be considered. The flat image is usually a measurement of the halogen lamp with counts level as close as possible - and well within a factor of two - to those in the astronomical frames.

3.4 Sky Subtraction

The greatest problem in NIR observations is the subtraction of the OH sky lines which are very bright and variable on time scale of a minute. A simple object-sky subtraction is not always enough to guarantee a complete removal of the emission lines as the sky spectra in the two frames (source and sky) may be different, because of a shift in the wavelength direction due to mechanical flexures in the instrument and of variation of line intensity.

The technique used is to consider the difference between object and sky frames and choose an area of the frame where there are bad subtracted sky lines. The extra noise there is mainly due to residuals of OH lines. The factor to rescale the sky frame and the shift to apply are fixed by minimizing the noise in the previously chosen area. In the spectral region where the background is not too high one also need a dark frame. The procedure can be summarized in the following equation:

$$SKYSUB(x,y) = \frac{OBJ(x+\delta,y) - \alpha SKY(x,y) - (1-\alpha)DARK(x,y)}{FLAT(x,y) - DARK(x,y)} \qquad 1$$

where α is the factor by which the OH line varied (and this correction must not be applied to the dark frame!) and δ is the shift due to grating movements. α and δ are determined by minimizing the noise on the background.

This procedure is not able to correct grating shifts $\geq$ 0.1 pixels during the integration, in that case the problem is still open. This procedure of sky subtraction becomes less and less efficient as the integration time increases, in particular one can notice how the noise gets greater than simple fluctuations of the background when integrating longer than 200 sec (Fig. 3.)

This problem limits the integration time and increases the limiting flux (about 10^{-13} erg s^{-1} cm^{-2} arcsec^{-2} μm^{-1}, preventing us from reaching background limited performances in the J and K bands.

3.5 Rectification of the Frames

The slit images at the various wavelengths ("spectral lines") are tilted as a consequence of the off-axis mount of the grating, and the angle by which they are tilted varies with the position of the grating, i.e. with the wavelength. The tilt angle is computed analytically from the instrumental calibration parameters (on line central wavelength etc.) so the correction does not present particular problems.

3.6 Wavelength Calibration

The wavelength dispersion on the array is linear within a small and totally negligible fraction of the pixel size. Hence, wavelength calibrating simply means modifying the x-start and x-step values (descriptor) of the image. Another advantage is that one can very precisely compute (analytically) the pixel size - in wavelength - once the central wavelength of the frame is known. A quite precise, usually within one pixel, estimate of this quantity is available on-line at the instrument ("mechanical" calibration). One can directly use this information and determine the "mechanical" wavelength calibration. It is possible to determine more precisely the central wavelength of the frame, and hence to obtain a very accurate wavelength calibration, if one has a frame containing lines with known wavelengths; up to 2.3 microns the OH lines in the sky frames are a very convenient calibrator (Oliva and Origlia 1992).

3.7 Calibration with a Standard Star

The calibration procedures up to wavelength calibration followed for the object must be repeated also for a standard star. Having flux calibrated the spectrum of a standard star with known photometric data one divides the object frame by the standard star, pixel by pixel, to achieve the cancellation of atmospheric absorption lines and flux calibrate the spectrum. Possibly there might be small shifts and different widths between absorption features in object and star spectra. In that case one determines by hand the shift and the smoothing to apply to the star to better remove the features.

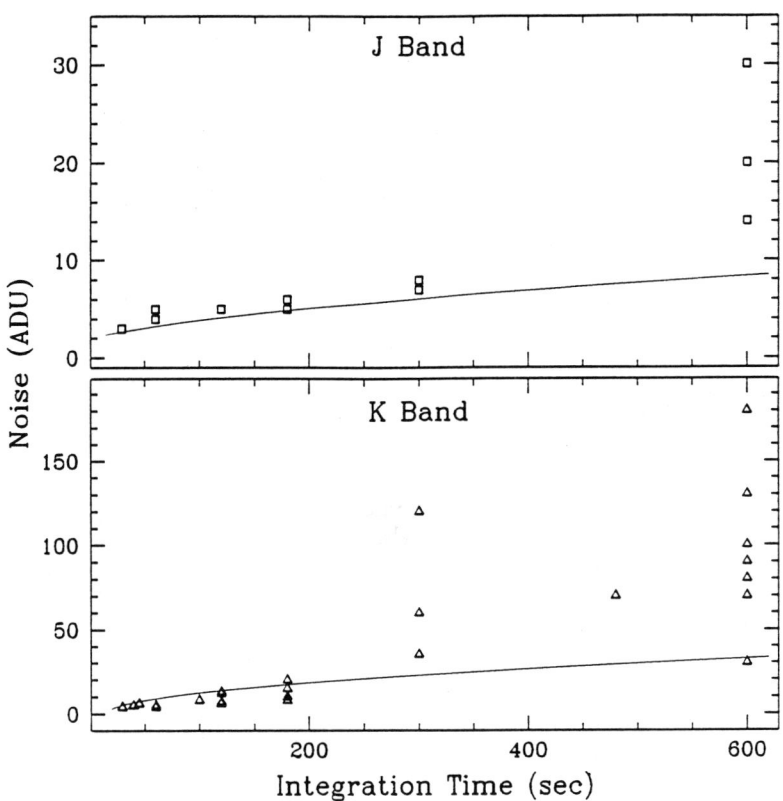

Fig. 4. Noise (ADU) of the sky subtracted images as a function of the integration time. The lines represent the Poissonian component of the background noise: note the high noise due to bad subtraction of OH sky lines after 200 sec of on-chip integration.

4. CONCLUSION

Engineering grade arrays have quite large subsections with performances comparable to a scientific grade array and can be a cheap solution if a large field is not required. Doing infrared spectroscopy at medium resolution with these arrays allows the use of quite long integration times and the limits on the signal to noise ratio are mainly posed by the subtraction of sky lines.

REFERENCES

Citterio, X 1978 Mem. Soc. Astron. It. 49, 57
Gennari S. and Vanzi L. 1994 in Experimental Astronomy 3, 191, InfraRed Astronomy with Arrays, I. McLean, ed.
Lisi, F., Baffa, C. and Hunt, L. K. 1993 Communication 1946.58 at the International Symposium of SPIE, The international Society for Optical Engineering, Orlando (USA) 12-16 April 1993
Oliva, E. and Origlia, L. 1992 A&A 254, 466

DISCUSSION

CORBALLY: Are you aware of any significant differences between your Nicmos-based spectrometer and a similarly based one of George Rieke?

VANZI: I know George Rieke has a similar spectrometer but actually I am not aware if he has similar problems. I know about other spectrometers; it seems they don't have problems of sky subtraction but all use much shorter integration times.

CUBY: Can you comment on persistence effects which may be important after calibrations (flatfields or standard stars) at high flux levels.

VANZI: Yes, they are. What we have observed is that after a strong illumination of the detector you have to reset it many times to remove completely any persistence effect. This depends on how strong the illumination was.

SPECTRAL CLASSIFICATION WITH ARRAY DETECTORS

C. J. Corbally

Vatican Observatory Group
University of Arizona

ABSTRACT. This paper presents the principles, the techniques, and some of the astrophysical potential of spectral classification in the era of array detectors.

1. REMOVING THE MYSTERY OF MK CLASSIFICATION

It is over fifty years since the MKK Atlas (Morgan, Keenan and Kellman 1943) was produced, and for about the first forty, photographic spectrograms dominated the way spectral classification was done. As in other fields of astronomy, the medium has shifted to digital spectra, witness "The MK Process at 50 Years" proceeding's lack of any photographic illustrations, save for the conference photo (Corbally, Gray and Garrison 1994). However, the spectral classification of stars, and in particular that based on the MK System, seemingly wrapped in the gloom of the photographic darkroom, might be thought of as a "black art" (Meyers-Rice and Young 1994) by those unfamiliar with its principles. In fact, the principles of spectral classification are essentially simple, the technique lends itself readily to array detectors, and its potential for astrophysical insight is enormous and enhanced by the digital array medium. I shall try to demonstrate these three points briefly.

1.1 The Principles

Spectral classification takes a morphological approach: that is, it starts with a good look at the specimen, the spectrum of a star. Thus, the classifier can fulfil *the mandate* of the MK System which is "to describe the appearance of the blue-violet spectrum of stars at moderate dispersion by reference to a set of standard stars (Garrison 1985)." The description uses *all* of the information in a stellar spectrum, not just certain ratios of line equivalent widths, and integrates it with a unique perspective. That perspective, sharpened by experience, may decide whether some slightly abnormal feature in a spectrum is significant or perhaps how best to characterize the astrophysics underlying a real abnormality. Thus, the description can complement other, more quantitative techniques, since it has a richness which numbers on their own lack.

The key to describing a spectrum is "by reference to a set of standard stars." So, the description is not arbitrary, but repeatable by different classifiers, whether human or machine (within the limitations of the latter). Garrison (1985, 1994) well describes the MK System's standards and their role in classification. In both papers Garrison makes the necessary point that the calibration of the MK System's classes in terms of temperature and luminosity are independent of the classifications themselves: so, if the calibrations change, the classifications do not.

Setting up and maintaining standards takes research effort. It is worthwhile because fundamental astronomy comes out of careful spectral classifications. For instance, calibrations of the classifications in terms of absolute magnitude and color index lead to spectroscopic parallaxes. Calibrations in terms of T_{eff}, gravity, and composition provide initial input to stellar atmosphere syntheses. Surveys give us stellar populations, the signs of significant events in a galaxy's past. Detailed studies will isolate peculiar stars, and these peculiarities are often signs of significant events in stellar evolution.

1.2 The Method with Digital Spectra

Some corresponding effort, by way of precautions, is also needed by those who use the spectral standards in classifying program stars. In the days of photographic spectra, it was imperative, if the "look" of spectra was to be compared, to observe both standard and program stars in exactly the same way: same telescope, same spectrograph, same emulsion, same developer, etc. The advent of digital spectra has relaxed these requirements into just needing a similar instrumental profile and resolution for the final spectrograms that will be compared. One should be able to deconvolve a spectrum from the instrumental profile, but it is certainly easier to obtain that "similar profile" if spectra from intensified detectors are *not* compared to those from non-intensified detectors. Figure 1 compares an intensified CCD spectrum with that from a naked CCD, and there is a noticeable filling-in of lines for the former. This figure comes from a preliminary study by Garrison et al. (1991), without subsequent quantification of the effect, but it is sufficient to raise a caution over mixing spectra from different detection systems.

Photographic spectra were widened to achieve a high effective signal-to-noise ratio (S/N), and likewise, the S/N for the standards in digital spectra should be high, 200 or greater. It is always a pleasure, and besides aids accuracy (lowering systematic error) and precision (lowering random error), to classify the program-star spectra if they are also of high S/N, but these stars are generally faint and so a compromise in exposure time must be made. With digital spectra, it does not matter that the standard and program spectra have different S/N, since, providing that the standard spectra have high S/N so that the features in at least these are identified unambiguously, the accuracy and precision will depend on the S/N of the program spectra (e.g., noise added to the spectrum of a metal-weak, early A-type star's spectrum can simulate lines and so appear to be a slightly later spectrum). This contrasts with photographic spectra, which need to have similar densities on the plate for a precise comparison and so an accurate classification. (Bracketing exposure times for the standard spectra was a familiar procedure.)

The normal processing of digital spectra should also be done for those to be used in classification, i.e., bias removal, field flattening, wavelength calibration, fluxing (not required if spectra from the same telescope and instrument are compared), pixel binning. Obviously, final resolutions should be the same for standard and program spectra. Rectification (continuum removal) is needed for comparing spectra both by eye and by machine. The more carefully this is done, the easier and more accurate is the comparison, and so effort spent at this stage is usually repaid. Tools for rectification are found in such image-analysis packages as IRAF, but the Fourier division technique (LaSala and Kurtz 1985) can produce more consistent results than "by-hand" methods, though it still tends to "fit" the hydrogen lines, so "dragging" them upwards and altering their profiles. Perhaps the flattest spectra of all are produced via a

SPECTRAL CLASSIFICATION WITH ARRAY DETECTORS

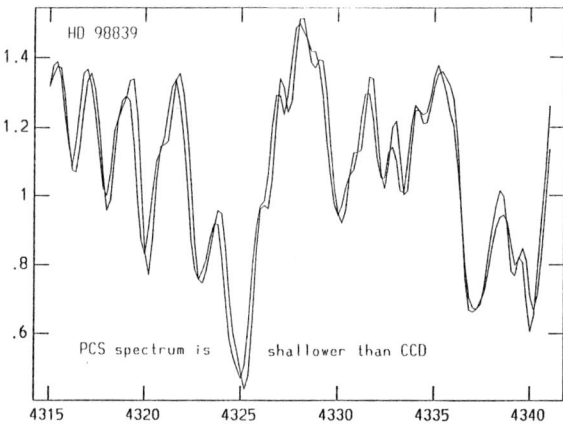

Fig. 1. Comparison of spectra from intensified (PCS) CCD and naked CCD detectors for the same star, HD 98839. The intensified spectrum is shallower than the naked spectrum (Garrison et al. 1991).

program of Richard Gray that uses a grid of synthetic spectra as templates, identifies the continuum points on the synthetic spectra, and then multiplies the observed spectra by factors that will bring those points up to the continuum.

With the digital spectra of both the standard and the program stars to hand, it is now time to see what insights and astrophysics can result from spectral classification.

2. CURRENT DIGITAL SPECTRAL CLASSIFICATION

The following account of advantages and highlights of classification with array detectors is based on the papers given at the MK50 workshop, and those whose interest is aroused are invited to read further in the proceedings (Corbally, Gray and Garrison 1994). The report of the IAU Commission 45 (MacConnell 1994) is a more comprehensive, if compressed, account of current classification work.

2.1 Accessibility of UV and IR

Papers in this Symposium (e.g., Vanzi et al. 1995) have shown how spectroscopy has been given access to wavelength regions beyond the optical. The opportunities have not been lost on classifiers, though now not strictly producing "MK" classes since the spectra are out of the MK System's blue-violet region and 2 Å resolution. Two current "opportunists" are Walborn and Rountree. For instance, Walborn (1994), while first recognizing the O3 stars in the traditional MK region, also finds that they are a class clearly separated from the O4 stars in the UV, based on their wind lines (Fig. 2). Their identification and analysis from the UV as well as the optical gives physical parameters for these, such massive stars (100 - 200 $M_\odot$, T_{eff} = 50 - 60,000° K) and information about their winds. Rountree (Rountree and Sonneborn 1994) has tackled the dwarf and giant B stars, and she has produced a UV spectral atlas for these which does not depend on the wind line strengths and is truly independent of, though happens to run parallel to, the MK System. Thus, Rountree has applied the "MK Process" (see Morgan [1984] for a definition) to a spectral region other than that used for the traditional MK System.

When we move in wavelength redwards of the MK region, we find Torres-Dodgen and Weaver (Torres-Dodgen 1994) making spectral sequences of, of all things, hot stars in the region 5500 - 9000 Å (Fig. 3). However, this makes sense for investigations of hot stars in regions of high extinction.

For M dwarfs it does make sense, more obviously, to move even further into the infrared, and so we find Boeshaar, Kirkpatrick and colleagues (see Boeshaar and Davidge 1994) working down to the two micron region, which is dominated by the molecular bands (e.g. TiO, VO, FeH, CO, and H_2O) though there are some useful atomic lines. Here (Fig. 4), the quest is to identify what features are sensitive to temprature and gravity, and the reasons for their sensitivities - as well as find the Holy Grail of brown dwarfs.

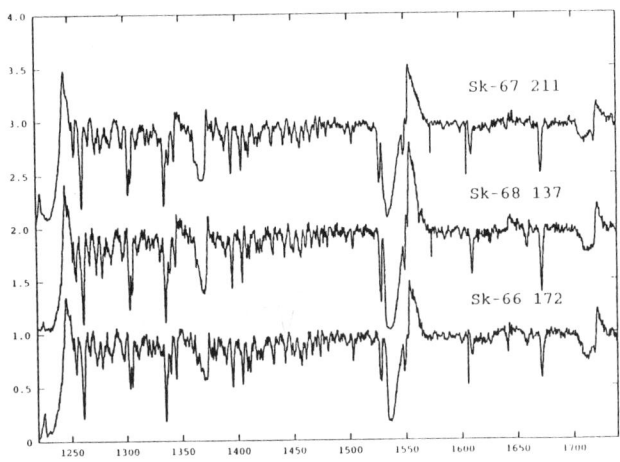

Fig. 2. Rectified HST/FOS spectrograms of three O3 III (f*) stars in the LMC. They show prominent wind lines (from Walborn 1994).

2.2 Precision Classification of Fainter Stars

Gray (1988) noticed that the hydrogen lines of λ Bootis stars, at classification resolution, divide those stars into two types, those with normal or with peculiar profiles. It is reassuring to find that this sometimes subtle distinction is readily seen in a digital spectrum such as Fig. 5. In this figure we find the first identification of a λ Bootis star in an association, that of Orion OB1, lending support to the theory that the mechanism for producing these stars is accretion from a circumstellar envelope with selectively weak metals (Gray and Corbally 1993). Levato et al. (1994) have announced the discovery of two more λ Bootis stars in Orion OB1,

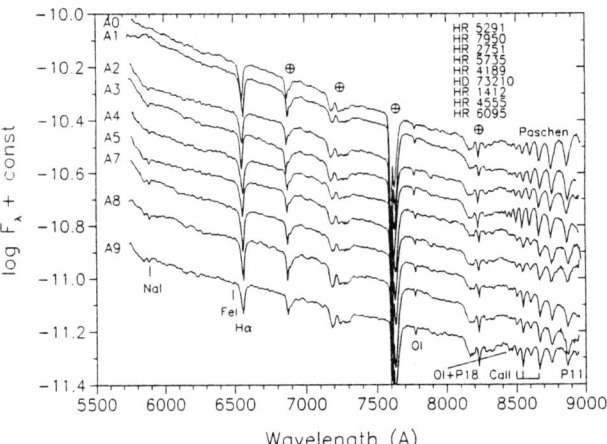

Fig. 3. Temperature sequences for A giant stars in the near infrared (Torres-Dodgen 1994).

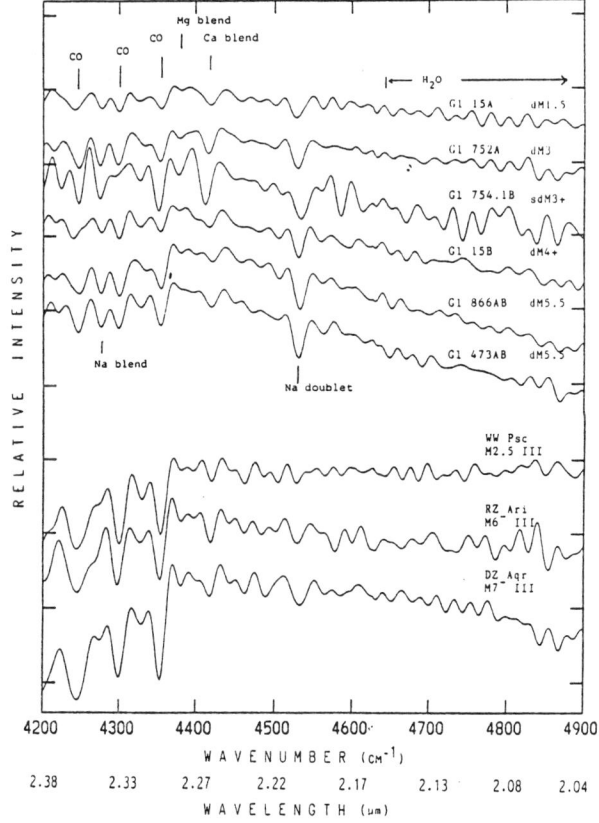

Fig. 4. K-band spectra of M stars at a resolution of ≈ 10 cm^{-1} arranged by luminosity class in order of decreasing temperature (Boeshaar and Davidge 1994).

and currently work is in progress to search in other associations and young clusters to establish the duration of the phenomenon. The sensitivity of CCD detectors a great help in widening searches like this to fainter magnitudes, especially when the detectors are allied with multi-object spectrographs.

Digital spectra lend themselves immediately to comparison with synthetic spectra, as in Fig. 6 where helium-rich and helium-normal models are compared with a Field Horizontal Branch star (Corbally and Gray 1994). While the broad hydrogen-line profiles in the FHB star could be explained by a helium-rich atmosphere, the increased He I strength (not seen in the FHB star) makes the hypothesis fall down. However, this explanation may have more success in understanding the A-type supergiants with anomalously strong hydrogen lines that Humphreys et al. (1991) find in the Magellanic Clouds. The ease of comparing digital spectra with synthetic spectra also makes practicable a new way of calibrating MK spectral types with respect to physical parameters such as T_{eff}, log g, and [M/H] (Gray and Corbally 1994).

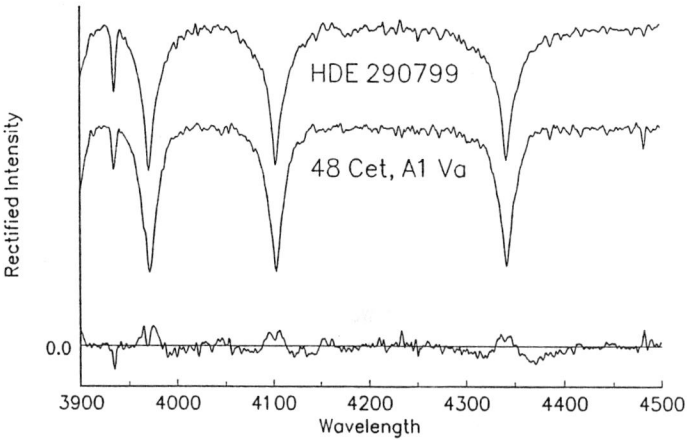

Fig. 5. The spectrum of HDE 290799, a λ Bootis star in Orion OB1, compared with the A1 Va standard. The slightly peculiar nature of the hydrogen-line profiles of HDE 290799 can be seen clearly from the difference spectrum at the bottom of the graph (Gray and Corbally 1993).

2.3 Automatic Classification

Digital spectra are in a form suitable for input into automatic classification, and I look forward to the time when I can drag the filename of a spectrum across a computer screen into a task box labelled "classification." If the computer returns "peculiar" for the spectrum, I shall then enjoy characterizing and investigating the peculiarities by eye, leaving the normal stars for the gathering of statistics. Two main teams are developing the tools to fulfil this dream, tools which are based on the pattern-recognition that is the essence of the MK System's mandate and power. One team employs a technique of weighted metric-distance (LaSala 1994), and the other uses artificial neural nets (von Hippel et al. 1994, Weaver 1994). Both methods are proving fruitful. The prospect for future surveys in and beyond the Milky Way, based on digital spectra, seems good.

2.4 Archiving and Dissemination

Since digital spectra need to be reduced before being archived, this involves more steps than for photographic spectra, which are "reduced" by being developed. While that extra reduction can seem a disadvantage, digital spectra have a clear advantage over photographic ones in their dissemination. Copies can be made and sent easily and are equivalent to the original. Thus, digital spectral archives can let one have one's cake *and* eat it (Griffin 1992). The IAU recognizes the great value of this through its support via Commission 29 of a Working Group on Spectroscopic Data Archives.

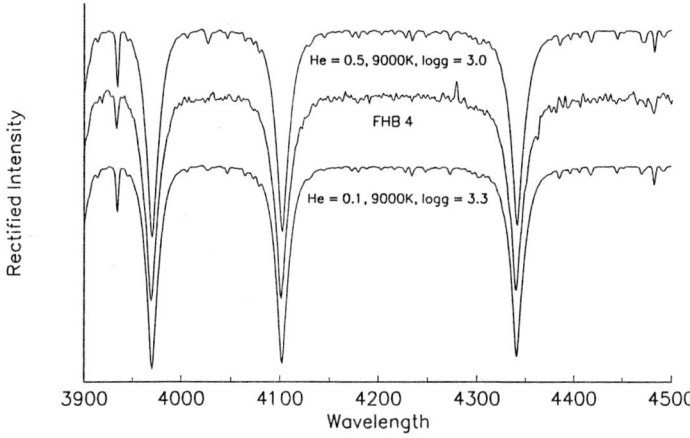

Fig. 6. Comparison of helium-rich and helium-normal synthetic spectra with a FHB star spectrum. Note the broad hydrogen lines (Corbally and Gray 1994).

3. CONCLUSIONS

The MK System lends itself to classification with spectra from array detectors as readily as from its original medium, photographic plates, providing similar care is taken to get consistent, high quality data that are anchored within a grid of the MK standards. We have seen that certain precautions are needed with regard to instrumentation, signal-to-noise ratio, and processing.

Some advantages apparent in current classification with arrays are: a) the ready accessibility of other wavelength regions; b) the opportunity to classify fainter stars than with previous media and still achieve precision spectral classification; c) the ease with which the digital medium lends itself to archiving, dissemination, and automatic classification of spectra.

So, thanks to new detector technology, things bode well for the science that will come from spectral classification in the next fifty years.

ACKNOWLEDGEMENTS

I thank the authors whose Figures are shown here and draw attention to the fuller treatment of the Figures in the references. I also thank Richard Gray whose suggestions improved the final version of this paper.

REFERENCES

Boeshaar, P.C. and Davidge, T. J. 1994 ASP Conf. Ser. 60, 246
Corbally, C. J. and Gray, R. O. 1994 CCP7 Newsletter (St. Andrews), No. 21, 43
Corbally, C. J., Gray, R. O and Garrison, R. F., eds., 1994, The MK Process at 50 Years: A Powerful Tool for Astrophysical Insight, Proceedings of a Vatican Observatory Workshop, ASP Conf. Ser. 60 (MK50)
Garrison, R. F. 1985 in The Calibration of Fundamental Quantities, IAU Symposium 111, D. S. Hayes, L. E. Pasinetti and A. G. D. Philip, Reidel, Dordrecht, p. 17
Garrison, R.F. 1994 in The MK Process at 50 Years: A Powerful Tool for Astrophysical Insight, Proceedings of a Vatican Observatory Workshop, ASP Conf. Ser. 60 (MK50), Corbally, C. J., Gray, R. O and Garrison, R. F., eds., ASP Conf. Ser. 60, 312 ASP Conf. Ser. 60, 3
Garrison, R.F., Beattie, B., Kamper, K., Pedreros, M., and Thomson, J. 1991 Transactions of the IAU, Vol. XXIB, J. Bergeron, ed., Kluwer Academic Pub. Dordrecht, p. 328
Gray, R. O. 1988 AJ 95, 220
Gray, R. O. and Corbally, C. J. 1993 in Peculiar vs Normal Phenomena in A-type and Related Stars, M. Dworetsky, F. Castelli and R. Faraggiana, eds., ASP Conf.Ser. 44, 418
Gray, R. O. and Corbally, C. J. 1994 AJ 107, 742
Griffin, R. E. 1992 Comments Astrophys. 16, No. 3, 167
Humphreys, R. M., Kudritzki, R. P. and Groth, H. G. 1991 A&A 245, 593
LaSala, J. 1994 in The MK Process at 50 Years: A Powerful Tool for Astrophysical Insight, Proceedings of a Vatican Observatory Workshop, (MK50), Corbally, C. J., Gray, R. O and Garrison, R. F., eds., ASP Conf. Ser. 60, 312
LaSala, J. and Kurtz, M. J. 1985 PASP 97, 605
Levato, H., Malaroda, S., Grosso, M. and Morrel, N. 1994 in The MK Process at 50 Years: A Powerful Tool for Astrophysical Insight, Proceedings of a Vatican Observatory Workshop, ASP Conf. Ser. 60 (MK50), Corbally, C. J., Gray, R. O and Garrison, R. F., eds., ASP Conf. Ser. 60, 93
MacConnell, D.J. 1994 Transactions of the IAU, Vol. XXIIA, J. Bergeron, ed., Kluwer Academic Pub, Dordrecht, p. 525
Meyers-Rice, B. A. and Young, E. T. 1994 in The MK Process at 50 Years: A Powerful Tool for Astrophysical Insight, Proceedings of a Vatican Observatory Workshop, (MK50), Corbally, C. J., Gray, R. O and Garrison, R. F., eds., ASP Conf. Ser. 60, 270
Morgan, W. W. 1984 in The MK Process and Stellar Classification, R. F. Garrison, ed., David Dunlap Observatory, Toronto, p. 18
Morgan, W. W., Keenan, P. C. and Kellman, E. 1943 An Atlas of Stellar Spectra, Yerkes Obs., Chicago
Rountree, J. and Sonneborn, G. 1994 in The MK Process at 50 Years: A Powerful Tool for

Astrophysical Insight, Proceedings of a Vatican Observatory Workshop, (MK50), Corbally, C. J., Gray, R. O and Garrison, R. F., eds., ASP Conf. Ser. 60, 277

Torres-Dodgen, A. V. 1994 in The MK Process at 50 Years: A Powerful Tool for Astrophysical Insight, Proceedings of a Vatican Observatory Workshop, (MK50), Corbally, C. J., Gray, R. O and Garrison, R. F., eds., ASP Conf. Ser. 60, 253

Vanzi, L., Gennari, S. and Marconi, A. 1995 in IAU Symposium No. 167, A. G. D. Philip, K. A. Janes and A. R. Upgren, eds., Kluwer Academic Pub., Dordrecht, p. 231

von Hippel, T., Storrie-Lombardi, L. J., Storrie-Lombardi, M. C. and Irwin, M. J. 1994 in The MK Process at 50 Years: A Powerful Tool for Astrophysical Insight, Proceedings of a Vatican Observatory Workshop, (MK50), Corbally, C. J., Gray, R. O and Garrison, R. F., eds., ASP Conf. Ser. 60, 289

Walborn, N. R. 1994 in The MK Process at 50 Years: A Powerful Tool for Astrophysical Insight, Proceedings of a Vatican Observatory Workshop, (MK50), Corbally, C. J., Gray, R. O and Garrison, R. F., eds., ASP Conf. Ser. 60, 84

Weaver, Wm. B. 1994 in The MK Process at 50 Years: A Powerful Tool for Astrophysical Insight, Proceedings of a Vatican Observatory Workshop, (MK50), Corbally, C. J., Gray, R. O and Garrison, R. F., eds., ASP Conf. Ser. 60, 303

DISCUSSION

VERSCHUEREN: When referring to automated classification, is it possible to quantify all MK methodology criteria, e.g., through intensity or equivalent-width ratios of different lines?

CORBALLY: Yes. That first and simpler method of automating classification is called the criterion-evaluation technique. It is useful to a limited extent. I mentioned the newer, pattern-recognition technique since this is closer to visual classification and better.

GARRISON: The criterion-evaluation technique is antithetical to the MK methodology, which is much more powerful. Using a few line ratios quantitatively requires careful fluxing, with all of its problems. Pattern recognition over the entire spectrum includes the experience of the observer and easily isolates peculiarities, which may not be included in the quantitative ratios of the criterion-evaluation method, which is a very simplistic approach to a complex information system. Using quantitative ratios ignores a lot of interesting and important information. Using the entire spectrum in a pattern-recognition methodology, whether visual or automated, is certainly the way to go, leading more surely to discovery of new phenomena. People working in this area include Kurtz, LaSala, von Hippel, and Weaver. The first two use minimum metric-distance techniques and the last two use neural nets. With modern computers, there is no reason to use the overly-simplistic criterion-evaluation approach.

HOWELL: Chris, could you comment on your plot showing that the PCS shows "shallower" profiles than the CCD spectrum? How can you rule out that the cause is in the star itself? What effects for, say, temperature and gravity or MK classification itself, would these changes cause?

CORBALLY: Non-variability in the star itself can be ruled out by the choice of star, e.g., avoiding supergiants. For MK classification, the example of HD 98839 would probably have negligible effect, but I have seen MK-significant differences in spectra from Kitt Peak compared with those from Sutherland.

GARRISON: This was an experiment to see what are the effects of using a photon-counting detector versus a naked CCD. My interpretation of the difference, which is relatively small, is that there is some asymmetric scattering in the intensifier train. The two detectors were used on the same spectrograph, so the intensifier train is the only variable. If the scattering is perfectly symmetric, the centroiding should take care of it, but obviously there is some asymmetry. The effect is small, but if I were doing stellar atmosphere work, I wouldn't go near an intensified photon-counting system, but would stick to a naked CCD.

FLORENTIN-NIELSEN: If you do get a large number of suitable CCD spectra, could you conceivably apply neural networks to do the classification?

CORBALLY: Yes, as in the automatic classification work currently exemplified by von Hippel and Weaver.

WARREN: If you could get your hands on Hipparcos data for a large number of bright stars right now, what would you do with them in terms of calibrating the MK system?

CORBALLY: I should first pick out any supergiants with zero or low reddening, since they occupy the part of the HR diagram for which absolute magnitudes are difficult to calibrate and yet very much needed.

HOUK: Do you envision a survey of fainter spectra being done with CCDs rather than, say, with a four-degree prism? I'm referring to a large, if not all-sky survey.

CORBALLY: When large format CCDs become available on Schmidt telescopes, I should hope for such a fainter spectral survey. Meanwhile, LaSala's and von Hippel's project to classify automatically the non-HD stars on the Michigan spectral survey plates will prove a valuable extension of your own work.

SPECTROSCOPIC OBSERVATIONS OF SOLAR SYSTEM OBJECTS: PUSHING THE LIMITS

Anita L. Cochran

The University of Texas, McDonald Observatory

1. INTRODUCTION

Targets within the solar system generally fall into one of two types: a) major planets (except Pluto) and our Moon; b) minor planets, comets, Pluto and planetary satellites. The first group is noteworthy for being reasonably bright. Most are also spatially extended. The inner planets never achieve large solar elongation. The second group comprises bodies which are generally faint. Comets are spatially extended. The minor planets and comets may be in orbits which are highly inclined or viewed at small solar elongations. Comets may even be in retrograde orbits. Planetary satellites may be bright or faint but suffer from being in the glare of the parent planet.

As with most aspects of astronomy, the study of bodies within our solar system has seen rapid progress with the advent of modern array detectors. Array detectors assist by having good quantum efficiency, facilitating faster observations, increased spectral coverage over plates or photomultipliers, high signal/noise ratio allowing detection of low contrast features, large dynamic range, and spatial coverage which is useful for extended objects or to remove a sloping background.

Much success in solar system observations have come about in the IR. However, that is beyond the scope of this report. In this paper, I will confine myself to the optical region of the spectrum and demonstrate the remarkable advances of the past decade or so which have been made possible through the use of modern array detectors.

2. MAJOR PLANETS: JUPITER

Jupiter is one of the brightest objects in the sky and thus, it would seem to be a simple body to observe. It is spatially extended, reaching a diameter of 45 arcsec at opposition. It is an interesting object to study since it has a dynamically changing atmosphere. However, with a rotation period of 9^h56^m, spectra must be obtained rapidly to avoid spatial smearing.

The spectrum of Jupiter is dominated by molecular features. Many of these molecules, such as CH_4 or NH_3, have dense absorption bands. Thus, it takes reasonably high spectral resolution to resolve the structure of the bands. Some of these molecules are strong absorbers, blocking out views of the underlying planet. At other wavelengths, Jupiter is quite bright. Thus, dynamic range becomes quite important to obtain good spectral data.

The state of the art for observations in the mid-1970s was exemplified by the work of Woodman et al. (1979). In their Fig. 1, they show a spectrum of the central meridian of the equatorial region of Jupiter obtained with a coudé scanner on the 2.7-m telescope at McDonald Observatory. The spectrum covers the spectral range from 6000 - 10,800 Å with a resolving power (R) of 500. The exposure had a duration of seven min.

Inspection of their Fig. 1 shows a high signal/noise ratio spectrum with broad molecular emissions. The 8900 Å CH_4 band reaches to about 10% residual intensity. There is an NH_3 feature at 6450 Å which shows as a double-lobed absorption.

In contrast, Fig. 1 shows a spectrum obtained by W. Cochran and K. Baines using a Cassegrain echelle spectrograph on the McDonald Observatory 2.1-m telescope in April 1993. This spectrum looks, at first glance, to have quite a low signal/noise ratio. However, that is a false impression due to the compressed spectral scale. These data were obtained with R = 60,000 using a cross-dispersed echelle imaged onto a Reticon CCD detector. The spectrum in the upper panel contains the complete 31 orders. Terrestrial and solar lines have been removed. The exposure time was two min.

The bottom panel of Fig. 1 shows the same 6450 Å NH_3 band observed with the Cassegrain echelle which was described above in the discussion of the Woodman et al. (1979) data. In the expanded view, it is obvious that the molecular band is not composed of only two parts but of many individual lines. The noise level of this spectrum can be estimated from looking at the continuum in the bottom panel.

Certainly, this spectrum has a much higher information content than the spectra of Woodman et al. (1979). The higher signal/noise ratio was made possible by the use of a modern array detector. The CCD allows for the inclusion of many orders from the cross-dispersed echelle; it has more sensitivity allowing for much higher spectral resolution in one third the time and has excellent dynamic range to allow for the strong absorbers. However, it does have only one half the spectral coverage of the scanner data, so two spectra will be needed to cover the same spectral range.

More recently, there is a cross-dispersed echelle spectrograph, 2DCoudé, for the 2.7-m telescope which allows 59 echelle orders to be imaged onto a Tektronixs 2048 x 2048 pixel^2 CCD with R = 60,000. This then yields a spectrum of Jupiter in three min with spectral coverage from 0.4 to 1.0 μm.

3. THE MOON

The Moon is perhaps the most familiar of all solar system bodies. Its brightness and size, along with its changing phases have made it an object of notice, study and contemplation since the start of time. Its nearness has also made it a target for exploration.

With the Apollo missions to the Moon, we were able to perform studies of this body which are unrivaled on other extraterrestrial bodies. Samples were returned to the lab and in situ observations were performed. Despite such measurements, at the end of the Apollo missions, we knew that the Moon had an atmosphere (correctly an exosphere since gas-phase collisions are negligible) with a pressure over the subsolar point of 10^6 atoms cm^{-3}. However,

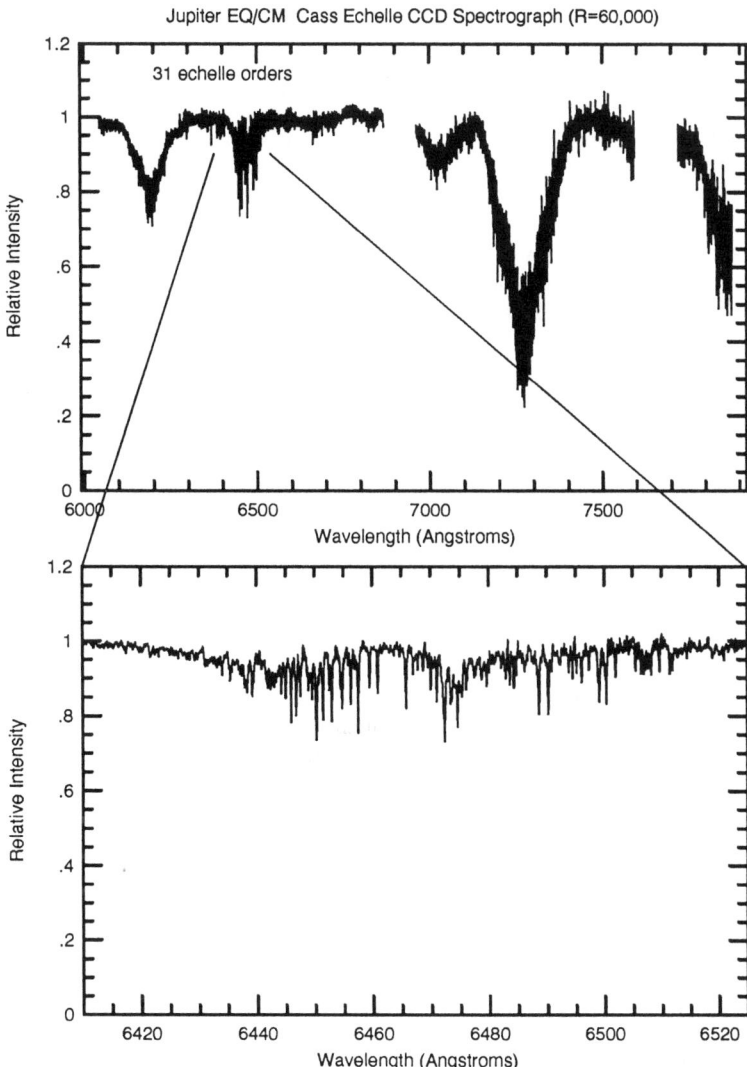

Fig. 1. A spectrum of Jupiter. This spectrum of Jupiter was obtained with a Cassegrain echelle spectrograph on the 2.1-m telescope at McDonald Observatory. The upper panel shows the complete 31 echelle orders obtained with R = 60,000. This exposure time is two min. The lower panel shows a spectral blowup of the region around the 6450 Å NH_3 band. It is obvious from this spectrum that the signal/noise is high and that the NH3 band is composed of many lines. This spectrum is courtesy of W. Cochran and K. Baines.

despite our best efforts, the composition of this atmosphere was unknown at the end of the Apollo missions.

Potter and Morgan (1988) first reported positive evidence for atmospheric constituents on the Moon when they detected sodium and potassium above the limb. The atmosphere is quite tenuous and it required the use of very high spectral resolution (22 - 100 mÅ) observations which also contained spatial information to detect the signal against the bright lunar surface. These observations could only be performed with high signal/noise data since the emissions which they detected are found in the cores of the strong solar absorption lines and the Doppler shifts of the solar and lunar spectra are insufficient to move these features away from the solar absorption. Also, an array detector allowed them to obtain spatial information simultaneously with the spectral information to model the scale height of the gas accurately. These observations are a wonderful example of how, even with a very bright source, some questions can only be answered with modern array detectors.

4. PLANETARY SATELLITES: GALILEAN SATELLITES

The observations of planetary satellites offer some unique problems which are well handled by modern array detectors. Although many planetary satellites are bright, the brightest satellites tend to be in orbits which are quite near the parent planet. Thus, the background of the satellite can show a strong gradient due to scattered planet light. In the past, this gradient was handled by observing at apoapsis, limiting the phase at which a satellite could be observed, or by separate observations of sky background, either at a symmetric distance from the planet or at a later time. Neither of these approaches allows for observations of the highest quality.

With the advent of current high quality array detectors, observations are once more being obtained of objects such as the Galilean satellites, with interesting new results. Spencer et al. (1994) have used a long slit CCD spectrograph at 18 Å resolution to obtain spectra of all four Galilean satellites from 3200 - 7800 Å. Each spectrum is the sum of 20 - 30 spectra of one to two sec each. A comparison solar analogue star was observed with the same instrumental set-up.

Except for Io, not many spectra have been obtained of the Galilean satellites since the work of McFadden et al. (1980). The spectra of McFadden et al. (1980) were comprised of data points spaced about 200 Å apart. In these spectra, Europa, Ganymede and Callisto appeared to have much the same reflectance curves. In contrast, the higher signal/noise spectra of Spencer et al. (1994) show that Ganymede and Callisto have similar spectral reflectances but Europa has a strikingly different spectral reflectance.

The high signal/noise of the Spencer et al. (1984) spectra showed what appeared to be a weak absorption feature in the spectrum of Ganymede. Using spectra at 6 Å resolution, Spencer et al. (1984) confirmed the presence of two absorption features of approximately 1% depth in the spectra of Ganymede which were not present in the spectra of Callisto or Europa. They attribute this absorption feature to solid O_2. This interesting discovery could only have been possible with extremely high signal/noise spectra obtained with modern detectors.

5. MINOR PLANETS

Perhaps the most progress which has been made in the field of solar system research has come with new observations of faint objects. The minor planets constitute a large class of objects which are relatively small bodies with low albedos (3 - 25%). The minor planets are interesting because of their surfaces. Spectra show the signature of mineralogy and lattice structure. These features are generally broad and often shallow. Details on the band centers and depths differentiate between various minerals and phases of the same mineral.

Most of the asteroids are so small that they are irregularly shaped bodies. They may have albedo features on the surface which cause changes in the spectrum. Therefore, it is necessary to obtain spectral information as a function of rotational phase. However, many of the rotation periods are short (five to ten hours) so that it is necessary to obtain the spectra quickly, despite the fact that the objects are faint. In this effort, high quantum efficiency is essential.

In addition to rotational phase effects, there are also orbital phase effects. For main-belt asteroids, phase angle variations are generally small. However, for earth-approaching asteroids, phase angle can change rapidly and the window-of-opportunity of observations is short.

It was not until the late 1970s that observers started to obtain more than UBV photometry on asteroids. Still, UBV photometry played a major role in early classification schemes (Chapman et al. 1975). More recently, several multi-color photometric systems have been developed to determine taxonomic differences of asteroids (Chapman and Gaffey 1979 (26 Color Survey); Tholen 1984 (Eight Color Asteroid Survey = ECAS); Bell et al. 1988 (52 Color Survey)).

Today, several groups have started spectral surveys with CCDs which are taking advantage of the higher spectral resolution and excellent quantum efficiency to enhance our understanding of asteroid spectra. An example, from Sawyer (1991), is shown in Fig. 2. In this figure, I show a comparison of four asteroids which were observed with a CCD spectrograph plotted along with results from the ECAS, 26-color and 52-color surveys, as available. These asteroids are of a variety of types and moderately bright. Inspection of Fig. 2 shows that the agreement between the photometry and spectroscopy for 57 Mnemosyne is quite good and the spectroscopy only yields a little more information. For 41 Daphne, the spectroscopy is able to define better the center and shape of the band at 7000 Å. Clearly, for 65 Cybele and 93 Minerva, the spectroscopy yields information which is not able to be interpreted in the photometry. Vilas and McFadden (1992), have obtained similar mineralogical information from their spectra.

Recently, Xu et al. (1994) have started their Small Main-Belt Asteroid Spectral Survey (SMASS) to extend the classification of asteroids in the belt to smaller sizes. They use a CCD detector on a Cassegrain spectrograph. Whereas the ECAS survey was limited to objects of > 20 km in the inner belt and larger than > 55 km in the outer belt, the SMASS observations have extended to < 5 km in the inner belt and < 25 km in the outer belt. At the same time as SMASS observations have been able probe to smaller objects, they have equaled or bettered the signal/noise of all but the best ECAS observations, and with much higher spectral resolution. With this survey, Binzel and Xu (1993) showed that many asteroids existed which may be collisional fragments of Vesta and the source for basaltic achondritic meteorites. In addition Binzel et al. (1993) have discovered an asteroid having a reflectance spectrum very similar to L6 and LL6 ordinary chondrite meteorites. This discovery is significant since the L6

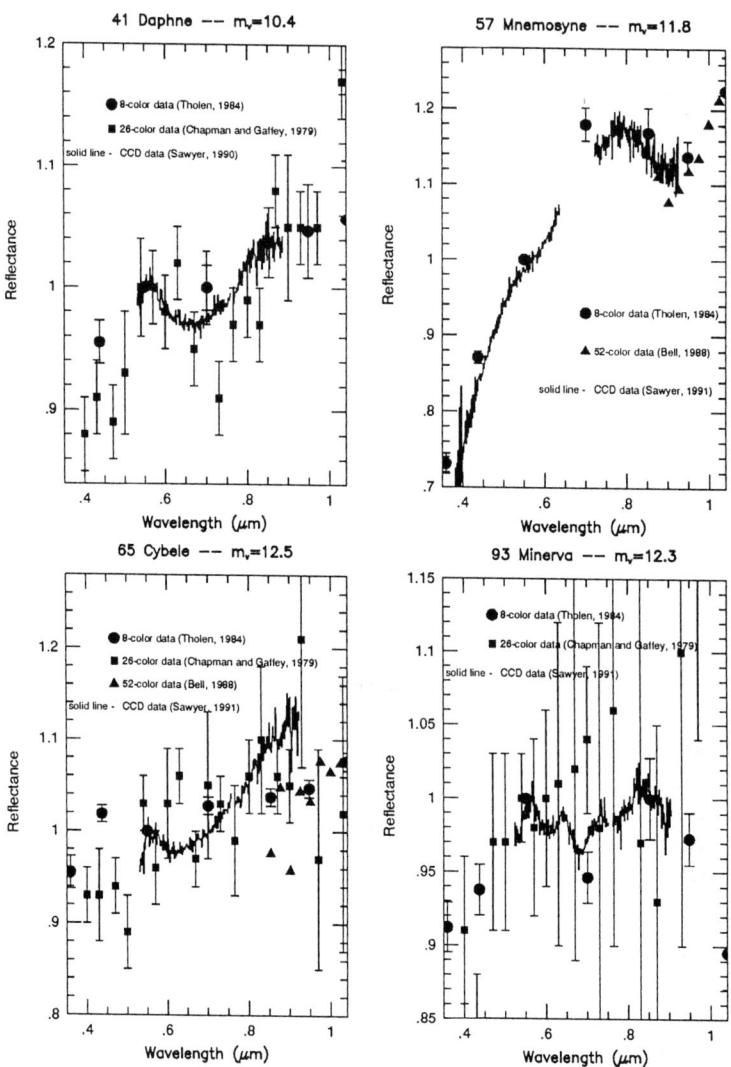

Fig. 2. Comparison of ECAS, 26-color and 52-color survey observations of asteroids and CCD spectra. The CCD observations are of four asteroids from Sawyer (1991). These objects are of a variety of taxonomic types and range in magnitude from 10.4 - 12.5. Clearly, it is possible to identify more accurately features in the CCD spectra than in the multicolor photometry.

and LL6 meteorites are the most abundant type of meteorites and, until the Binzel et al. (1993) observation of this asteroid as part of SMASS, no asteroidal analogue had been identified.

6. COMETS

The problems of observing comets are often quite different than for other solar system bodies. Though some comets are quite bright, most are quite faint. The nuclei are small (of order a few km) and dark (albedo 5%). This nucleus is shrouded by a coma and cannot be viewed from earth except when a comet is far from the Sun, when it is especially faint.

Comets are spatially quite extended objects. Depending on the comet and geocentric distance, cometary comae can subtend several arcmins, while tails can be a degree or longer.

The brightest comets are those which are active near the Sun. Generally, this means that they are brightest at small solar elongation angles and must be observed at large airmass or even during twilight. The orbits can have any inclination, resulting in short windows-of-opportunity for some objects and offering challenges to tracking.

In the absence of chemical reactions, material flows outward from the nucleus as $1/r^2$, where r is the distance to the nucleus. Photochemical reactions destroy species faster. Thus, the outer regions of the coma are substantially fainter than the optocenter, requiring dynamic ranges to be two or more orders of magnitude.

Fig. 3 illustrates the changes in cometary spectroscopy which are achieved in the transition from non-array instruments to array instruments. The left-hand panels in this figure show data obtained with an Intensified Dissector Scanner (IDS) spectrograph. The right-hand panels show data obtained with a long slit spectrograph imaging onto a CCD detector. All observations were obtained with the McDonald Observatory 2.7-m telescope and are of comet Schaumasse.

The IDS was a dual beam spectrograph which obtained two simultaneous spectra 52 arcsec apart. Beam switching was used to alternately image object and sky down each entrance aperture. The telescope was beam switched every 50 seconds and many such observations summed to form the resultant spectra. This cycle yielded two spectra on the optocenter (one through each aperture) and one spectrum each 52 arcsec either side of the optocenter. Of course, there was an overhead to perform the beam switching. The top, left panel shows one of the spectra of the optocenter imaged through a 4 x 4 arcsec^2 slit. The total exposure time was 4800 sec but the elapsed time was 6540 sec. At the same time as the two optocenter spectra were obtained, the two spectra 52 arcsec away were obtained, one of which is shown in the middle panel. The four resultant column densities are shown in the lower panel.

The long slit CCD spectrograph has a slit which is ~ 150 arcsec long with each pixel being 1.3 arcsec in the spatial direction. We used a slit width of two arcsec. Thus, the CCD effective aperture for each pixel is 1.3 x 2 arcsec^2. We exposed for 3600 sec. There is no beam switching overhead so the exposure time and the elapsed time were the same. The top panel shows the spectrum from the optocenter. It looks quite similar to the IDS spectrum in the wavelength overlap region, except that it is clearly slightly higher signal/noise even though the exposure time and elapsed time for the CCD image was much shorter. The middle panel shows the spectrum 52 arcsec from the optocenter, comparable to the IDS "off" position. Clearly, this

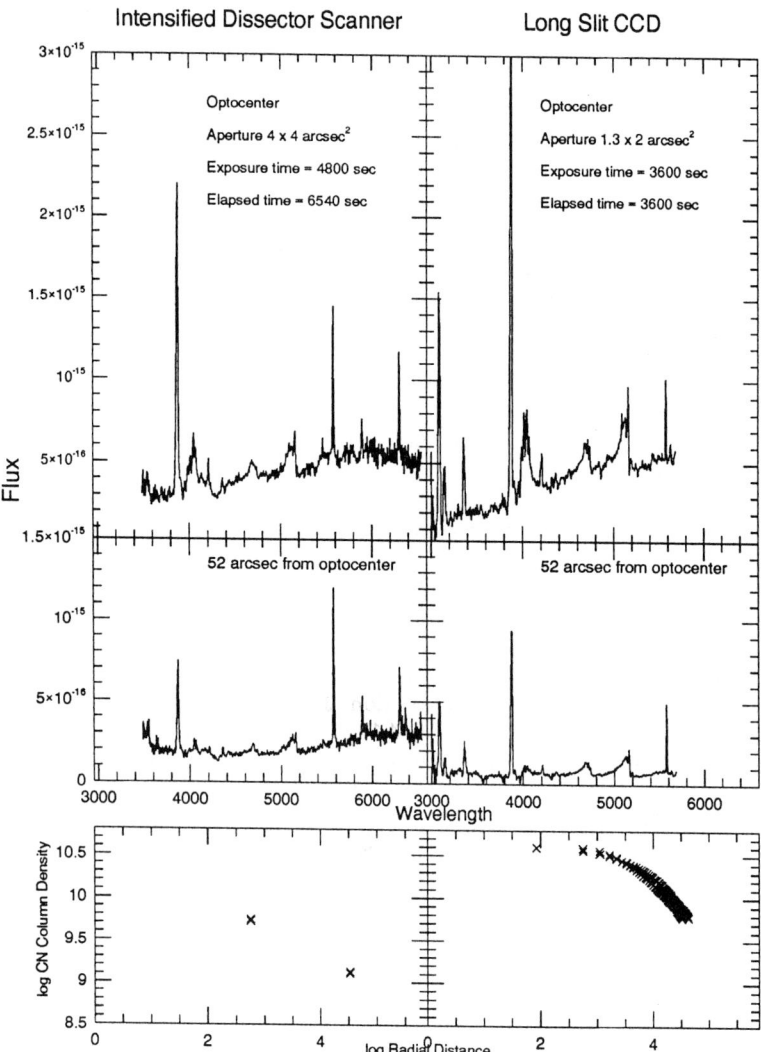

Fig. 3. Comparison of Intensified Dissector Scanner spectra and CCD long slit spectra. The IDS spectral information is the left-hand panels and the CCD is the right-hand panels. From top to bottom, we show optocenter spectra, spectra 52 arcsec from the optocenter and column densities measured. See text for more information.

spectrum is much higher signal/noise than the comparable IDS spectrum. The bottom panel shows the real strength of the CCD for cometary spectroscopy. Instead of four column density points in 6540 sec elapsed time, a single spectral image of 3600 sec yields 120 column densities. It is obvious that one can define well the gas distribution profile from the CCD data but not from the IDS data.

Since we believe that comets are remnants from the early solar nebula which have undergone very little change since their time of formation, understanding the isotopic abundances of their constituents is important to provide constraints to solar nebula models. Unfortunately, isotopic studies are quite difficult since even the brightest comets have relatively low surface brightnesses.

One of the nicest examples of isotopic studies of comets is that of Kleine (1994; Kleine et al. 1994, 1995a, b). Using the Photon Counting Array on the coudé spectrograph (R ~ 75,000) at the Mt. Stromlo 74-inch reflector, this group observed one periodic comet (Halley) and three long-period or non-periodic comets (Levy, Austin and Okazaki-Levy-Rudenko) to determine the $^{12}C/^{13}C$ abundance via the R-branch of the $B^2 \Sigma^+ - X^2 \Sigma^+$ band of CN. The carbon isotope ratios determined were between 85 and 95, indicating that the ^{13}CN lines are much weaker than the ^{12}CN lines. To determine the isotopic ratio, Kleine (1994) and Kleine et al. (1994, 1995a, b) had to model the complete P, Q and R branch of CN to predict the position and strength of the isotopic lines. The overlap of the various branches makes it easy to identify the wrong weak lines as isotopic lines. Fig. 2.14 of Kleine (1994) shows the model and the Halley data illustrating the problem of weak P-branch lines in the region of the strong R-branch lines. In addition, the Swings effect changes the ratio of strengths of the various lines (see Fig. 2.1 of Kleine (1994)).

Once the data were modeled, Kleine (1994) was able to identify and measure the various lines present. This resulted in the identification of some very weak ^{13}CN lines on the wings of stronger ^{12}CN lines of the R-branch. Fig. 4.11b of Kleine (1994) shows the R(0) through R(7) lines and the accompanying isotopic lines. The isotopic abundances which were determined in this manner are quite important for determining if comets have interstellar $^{12}CN/^{13}CN$ ratios of a few or solar system values around 90. The results of this study of four comets shows that comets have solar system $^{12}CN/^{13}CN$ ratios, and that this ratio is consistent from comet-to-comet.

Another nice feature of the observations of Kleine (1994) and Kleine et al. (1994, 1995a, b) is that the Mt. Stromlo coudé spectrograph employs a long slit. Thus, not only was it possible to determine the isotopic abundance of carbon, but it was possible to study the spatial distribution of the gas at high resolution. Fig. 2.6 of Kleine (1994) shows an example of spatial information for Halley. The data are scaled to the R(8) line (the strongest) and the figure shows that the ratios of the strengths of various R-branch lines changes with position. This is the Greenstein effect, or differential Swings effect, and has important uses for determining outflow velocities of the coma.

7. SUMMARY

The examples given in this paper are but a few of the ways in which modern array detectors have increased the scientific information on solar system bodies. Good quantum efficiency has allowed studies of asteroids and comets to push to fainter limits. Spectral coverage has allowed for in-depth, high-spectral-resolution coverage of brighter objects. High signal/noise has allowed very weak features to be detected for the first time. Large dynamic range has allowed for searches for very weak features, such as lunar sodium and potassium, near a bright source. The two dimensional format of arrays has allowed for spatial coverage well suited to the extended nature of solar system bodies such as Jupiter and comets. The nature of array detectors not only improves upon the types of data which has been collected in the past, but allows for new types of data, new questions, and new insights in our study of the origin and evolution of the solar system.

ACKNOWLEDGEMENTS

I thank all the planetary scientists who supplied figures for my presentation at the IAU symposium and who allowed me to reference work in progress. In particular, thanks are due to Drs. R. Binzel, W. Cochran, B. Flynn, S. Sawyer, J. Spencer, P. Wehinger and to their collaborators on these data.

REFERENCES

Bell, J. F., Hawke, B. R., Owensby, P. D., and Gaffey, M. J. 1988 Lunar Planet. Sci. Conf. XIX, 57
Binzel, R. P. and Xu, S. 1993 Science 260, 186
Binzel, R. P., Xu, S., Bus, S. J., Skrutskie, M. F., Meyer, M. R., Knezek, P. and Barker, E. S. 1993 Science 262, 1541
Chapman, C. R. and Gaffey, M. J. 1979 in Asteroids, T. Gehrels, ed., The University of Arizona Press, Tucson, p. 1064
Chapman, C. R., Morrison, D. and Zellner, B. 1975 Icarus 25, 104
Kleine, M. 1994 Carbon Isotope Abundance Ratios in Comets, Ph.D. thesis, Arizona State University.
Kleine, M., Wyckoff, S., Wehinger, P. A. and Peterson, B. A. 1994 ApJ 436, 885
Kleine, M., Wyckoff, S., Wehinger, P. A. and Peterson, B. A. 1995a ApJ 439, 1021
Kleine, M., Wyckoff, S., Wehinger, P. A. and Peterson, B. A. 1995b ApJ 439, 1021
McFadden, L. A., Bell, J. F. and McCord, T. B. 1980 Icarus 44, 410
Potter, A. E. and Morgan, T. H. 1988 Science 241, 675
Sawyer, S. R. 1991 Ph.D. thesis, The University of Texas at Austin
Spencer, J. R., Calvin, W. M. and Person, M. J. 1994 JGR Planets, submitted
Tholen, D. J. 1984 Ph.D. thesis, Univ. of Arizona
Vilas, F. and McFadden, L. A. 1992 Icarus 100, 85
Woodman, J. H., Cochran, W. D. and Slavsky, D. B. 1979 Icarus 37, 73
Xu, S., Binzel, R. P., Burbine, T. H. and Bus, S. J. 1994 Submitted to Icarus

DISCUSSION

WEST: A very important effect of the new detector technologies for solar system research, not mentioned by the speaker, is the probability to observe extremely faint and distant, minor objects, e.g. the new trans-Neptunian objects and long period comets on their way out. This has in fact opened up an entirely new field in the solar system research.

COCHRAN: I agree entirely. However, most of these new objects are too faint to be observed spectroscopically, at least with today's technology. This talk concentrated entirely on spectroscopic applications of arrays, but imaging applications are certainly opening up new opportunities.

WATSON: Are the terrestrial NaD emission lines present in the high resolution spectrum of the linear atmosphere you showed?

COCHRAN: No, the terrestrial emission lines are too faint to show up. Flynn and Stern moved the telescope back to the same region of the sky in which they had observed the moon and obtained a comparable length exposure of the sky. At this resolution and the short exposure times, no terrestrial NaD is observed.

"VA-ET-VIENT" ("BACK-AND-FORTH") CCD SPECTROSCOPY: A NEW WAY TO INCREASE THE LIMITING MAGNITUDE OF VERY LARGE TELESCOPES

G. Soucail, J. C. Cuillandre,

Observatoire Midi-Pyrénées, Toulouse

J. P. Picat and

Observatoire Midi-Pyrénées, Bagnères-de-Bigorre

B. Fort

DEMIRM, Observatoire de Paris

1. INTRODUCTION

Over the last decade, the quantity of scientific results brought by the observations of very faint objects has been quite spectacular. In particular, they concern the photometry of faint galaxies up to B = 27 (Tyson 1988) or K = 22 (Cowie et al. 1994). The consequences of these observations are the detection of a large population of faint galaxies more numerous than any prediction given by standard galaxy evolution and probably a new vision of the distant universe. For faint object spectroscopy, the most recent surveys of field galaxies reach a magnitude range of 23 - 24 (Colless et al. 1990, 1993, Lilly and Cowie 1993, Tresse et al. 1993) with a reasonable S/N ratio that allows a redshift measurement from absorption-line identification. But in this magnitude range, the sky background flux is dominant with respect to the source, being at least 10 times brighter. It is well known that in that case, with a low readout noise detector and a "perfect" spectrograph, the S/N ratio scales as:

where D is the telescope diameter, T is the exposure time and R the resolution of the spectrograph. In practice, the S/N does not increase as expected and tends to saturate towards an upper limit depending on the characteristics of the instrument but not on the telescope size. The main limitations come from residuals in the flatfielding process or from a bad sampling of the strong sky emission lines resulting in a poor sky subtraction. As a figure, this limit is typically of a few hours on a four-meter telescope with a low resolution of R = 300. Extrapolating to the next generation of eight to ten-meter class telescopes, it is clear that these limitations will become crucial as people will require to reach fainter magnitudes. Moreover, even now, faint object spectroscopy at very low resolution (such as R = 100) is hardly possible despite the scientific interests, so what will happen with the VLTs?

In this framework, we examine more quantitatively the main limitations which can occur in faint objects spectroscopy and propose a new mode of observations which we call "Va-et-Vient" ("Back-and-forth", Cuillandre et al. 1994). It is based on several shifts of the charges on the CCD related to a shift of the telescope from the object to an empty region of the sky, the detector being read only once at the end of the total exposure. We have tested this method in a laboratory experiment and proved its advantages over standard long slit spectroscopy in some specific cases. The implementation of the "Va-et-Vient" on a telescope is rather simple and is briefly described at the end of the paper as well as some future investigations of the method.

2. QUANTITATIVE ESTIMATES OF THE S/N RATIO AND ITS LIMITATIONS

To understand the main limitations which occur when observing faint objects with a spectrograph, let us write formally the S/N ratio of the observed data. We call Φ_O the true object signal per second recorded through a spectral element of the instrument and Φ_S the sky signal. Both of them can be written as:

$$\Phi \propto \frac{D^2 \eta}{R} g \, \varphi_\lambda$$

where η is the quantum efficiency of the detector and g is the geometrical factor which depends on the pixel size, the slit width, the possible pixel binning, etc. (see Picat et al. 1994, for more details). φ_λ is the spectral flux at the entrance of the telescope. After sky subtraction, the signal of the object on the detector will be:

$$S = [\Phi_O + (\Phi_S - \Phi_{S'})] T$$

The sky residuals are considered as sources of noise which can be written

$$N = \sqrt{\sigma_{CCD}^2 + \sigma_{photons}^2} + \mu_{sky} + \mu_{flat} + \mu_{lines}$$

where the first two terms are the standard sources of noise (readout and photon noises) which add quadratically. The other ones are additive because they are systematic and occur always on the same pixels.

In the perfect case, only the photon noise has to be taken into account provided we neglect the readout noise of the detector (this will be assumed throughout the rest of this paper). This noise term can be written as $\sigma_{photons}^2 = (2)\Phi_S T$ with a factor (2) depending how the sky is subtracted (pixel-to-pixel or average of a large number of sky rows). The S/N per pixel is therefore:

$$(S/N)_{perfect} = \frac{\Phi_O T}{\sqrt{(2)\Phi_S T}} \propto \frac{\Phi_O}{\Phi_S} \sqrt{\frac{D^2 \eta}{R} g T}$$

as mentioned at the beginning of the paper. But in the "real world", one has to explicit the additive terms of the noise estimate:

$\mu_{sky} \sim \beta \Phi_S T$ represents the temporal and/or the spatial variations of the sky below the object.

$\mu_{\text{flat}} \sim \epsilon \Phi_s T$ corresponds to the flatfield errors remaining after the sky subtraction.

$\mu_{\text{lines}} \sim \omega \Phi_s T$ is due to sampling errors, grating misalignment or slit width variations, mostly sensitive in the spectral range dominated by the strong sky emission lines.

All these factors are simply proportional to the sky flux and the exposure time, if we assume that the object is faint enough to be negligible with respect to the sky. In that case, the S/N will be:

$$(S/N)_{real} = \frac{\Phi_O T}{\sqrt{(2)\Phi_S T + (\beta + \varepsilon + \omega)\Phi_S T}}$$

At high flux level (or long integration time), it will reach a limit

$$(S/N)_{limit} \sim \frac{\Phi_O}{\Phi_S} \frac{1}{\beta + \varepsilon + \omega}$$

which does not depend anymore on the diameter of the telescope D, the resolution of the instrument R nor the integration time T. Increasing the diameter of the telescope or the total exposure time, with everything else kept constant, will not allow to reach fainter limiting magnitudes, but the limiting S/N will be reached faster!

3. WHAT IMPROVEMENTS FOR THE LONG SLIT SPECTROSCOPIC MODE?

This concern is not new, and some improvements can be expected mainly from the high quality of the optical design of the instruments. This means a reduction of the optical distortions and mechanical flexures, and a careful inspection of the slit quality and its alignment. In the long slit spectroscopic mode, the sky and the object signals are integrated at the same time, but not at the same position on the detector. The temporal sky fluctuations which are known to be important for long exposure times are then minimized. Moreover, with two-D detectors the underlying sky can be interpolated from the sky spectrum on each side of the object to minimize the spatial variations of the sky spectrum. But in any case flatfield and sky lines residuals will dominate the noise structure for very faint objects. In this mode the S/N will be limited to:

$$(S/N)_{limit} \sim \frac{\Phi_O}{\Phi_S} \frac{1}{\varepsilon + \omega}$$

and the "saturation" time (the integration time necessary to reach 1/2 of the limiting S/N) can be expressed as:

$$T_{sat} \sim \frac{1}{\Phi_S (\varepsilon + \omega)^2}$$

As a numerical example, one can compare two similar spectrographs in use at ESO, namely EFOSC on the 3.6-m telescope and EMMI on the NTT, in their low resolution mode (R = 400). In the spectral region around 5000 Å where the sky spectrum is dominated by the continuum, the saturation time of EFOSC is about five hours for a object-to-sky ratio of 0.1 and the S/N limit is about ten per pixel of 0".675 size, while for EMMI it corresponds to 28 hours for the same object but with a pixel size of 0."44. In the red part of the spectrum around 8000 Å where the sky spectrum is dominated by emission lines of atmospheric molecular bands, the saturation time of EFOSC is shorter than 30 minutes, while for EMMI it is of about two hours, both for an S/N limit of two per pixel. There are two major differences between the two spectrographs which can explain these numbers: first, the CCD in use on EFOSC is a thinned Tek CCD with a higher quantum efficiency than the thick coated Thomson CCD EMMI (it has changed now.). Second, the pixel sizes of the two spectrographs are quite different and the spatial and spectral sampling are quite better in the second case. Undersampling and high efficiency are consequently the two major points which favor an easy detection of the limiting effects in low resolution spectroscopy. One can also note that a VLT (D = eight-m) with a pixel size of 0."3 and a blue thinned CCD will be equivalent to EFOSC in the estimate of the saturation time!

Some alternatives such as beam switching mode have already been proposed to solve this problem and increase the sensitivity of the spectrographs. It allows the removal of the spatial errors, as the sky and the objects are recorded through the same optical path (or the same fiber). But as they are not anymore recorded at the same time, the temporal fluctuations of the sky brightness can be large for long exposure times and they are difficult to monitor accurately.

Another solution may come from a technique similar to the "shift-and-add" one introduced by Tyson (1988) for very deep imaging. Individual exposures are recorded with random spatial shifts and the spatial defects on the detector do not correspond to the same position with respect to the objects and can be removed by median filtering. In spectroscopic mode, this has been attempted in a few cases (Cowie and Lilly 1989), but it requires a long individual exposure time to reach the photon-noise regime even with a low readout noise CCD, and the shifts are not easy to produce inside the slit. Moreover, to obtain a significant gain in the S/N, the number of shifts must be large (typically larger then ten), and this is hardly compatible with the previous requirements of a long enough individual integration time. So this method can be efficient only in a very few specific cases!

4. PRINCIPLE OF THE "VA-ET-VIENT" MODE

Because of these remaining S/N limitations, we propose to implement a new mode of observation that we call "Va-et-Vient". It is like a beam switching method but instead of reading the CCD at the end of each elementary exposure, the charges are shifted as described in Fig. 1. After each exposure, the telescope is moved from the object to a sky reference so that the object and the sky are integrated through the same pixels and recorded on two different zones on the CCD. The shifting period can be chosen as small as necessary to freeze out the temporal sky fluctuations, a typical value being three to five minutes. Finally, there is only one readout at the end of the sequence. Note that this idea of chopping technique was already used in the 70's in spectroscopy with photon counting detectors (Shectman and Hiltner 1976), and tentatively implemented in an imaging polarimeter (McLean et al. 1981).

In the "Va-et-Vient" spectroscopic mode, the sky subtraction must be done pixel by pixel and the photon noise in the remaining object spectrum is twice that of the sky flux, instead of only once in the case of a long slit with a sky modeling:

$$\sigma^2_{photons} = 2\Phi_S T$$

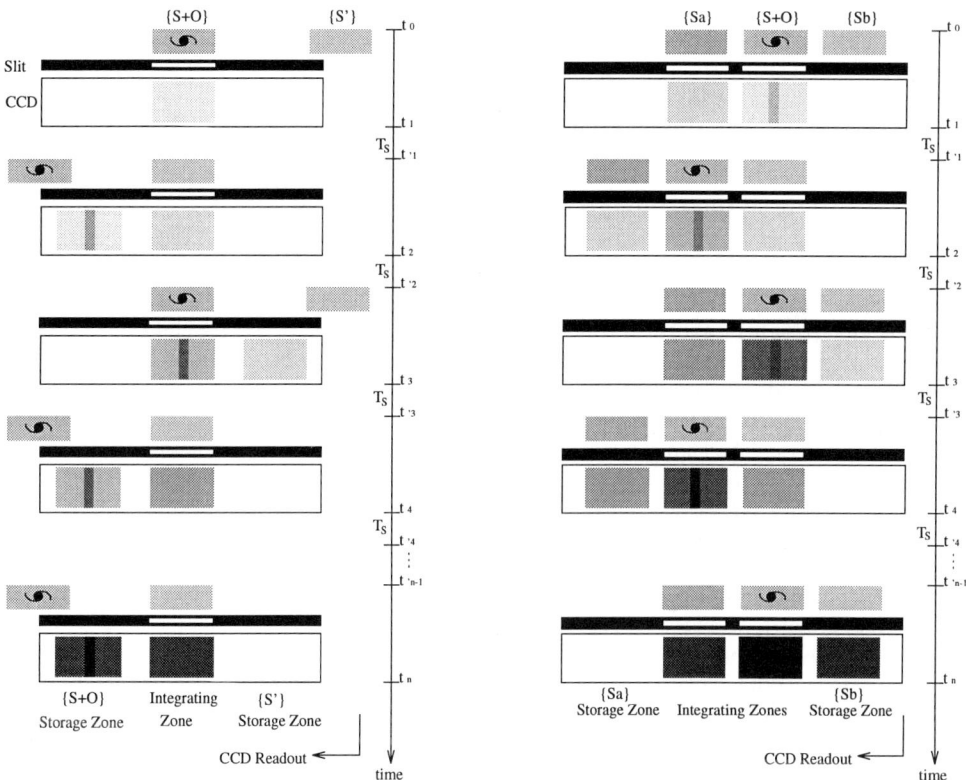

Fig. 1. Configuration of the CCD and the telescope during a "Va-et-Vient" spectroscopic exposure. Relative positions of the telescope pointed on the object and on the adjacent sky appear on the figure as the corresponding charges are moved on the CCD. The projection of the slit on the CCD determines the physical integration zones. In order to avoid any mixing of the data, a slitlet is needed to mask the light from the sky in the storage zones. Only 50% of the total integration time is spent on the object, while the other 50% is spent on the sky. (right) Same as before but in order to optimize the telescope time, two slitlets are used and the shifts of the pixels on the CCD are related spatially with the shifts of the telescope. In that case, the object in integrated 50% of the time in one slit and 50% in the other slit, both in the same strip on the CCD, while two strips of sky spectrum are recorded on the CCD. In the final image, three strips are produced for one object spectrum (from Cuillandre et al. 1994).

A major advantage of the method is that the flatfielding residuals apply only on the object flux, and there is no more sky lines sampling errors (the sky is nearly perfectly subtracted). The S/N is therefore:

$$S/N = \frac{\Phi_O T}{\sqrt{\sigma_{CCD}^2 + 2\Phi_S T + \beta \Phi_S T + \varepsilon \Phi_O T}}$$

It is mainly limited by temporal sky fluctuations (β). For an elementary exposure time short enough, "Va-et-Vient" spectroscopy will work in the photon noise regime, even at high flux when working with long integration time T, large telescope and small resolution. To test more quantitatively the ideas developed above, we have implemented a laboratory experiment to simulate the observing conditions in faint object spectroscopy.

5. THE LABORATORY EXPERIMENT AND ITS RESULTS

The implementation of the "Va-et-Vient" mode on the CCD is rather straightforward as the physical structure of the CCD allows charge shifting along the columns in both directions. The parameters which must be included in the controller software are: the shifting period, the number of periods for one exposure and the number of lines to be shifted. In our experiment we used a backside illuminated 512 x 512 Thomson CCD driven by a controller developed at ESO. The sky spectrum was simulated by a combination of two lamps, a halogen lamp for the sky continuum and a Neon spectral lamp for the emission lines. For the galaxy spectrum, we also used a halogen lamp, but with a blue filter introduced in the beam. The last beam was directly focussed on the entrance of the spectrograph, while the two other beams were combined through a beam splitter, giving a uniform illumination of the slit (Fig. 2). The spectrograph was made of two conjugated lenses with a low dispersion grism in the parallel beam region, giving a dispersion of 25 Å/pixel on the detector. The entrance is a slitlet, with a width of about four pixels, so that the resolution was 100 Å. To simulate the telescope offsets from the object to the sky, the "galaxy" lamp was alternatively switched on and off, with the sky one always on. The lamp fluxes and the exposure times were adjusted to give a signal per pixel equivalent a one-hour exposure on a 4-m telescope. The details of the experiment can be found in Cuillandre et al. (1994).

The first results were aimed at showing if there was a degradation of the output signal in "Va-et-Vient" mode. More quantitatively, we measured an additive noise lower than one electron rms, and no degradation of the MTF (60 shifts of 100 lines) as we used a high quality CCD with a very small number of traps. Charge Transfer Efficiency (CTE) losses were only detectable for more than 1000 shifts of 100 lines. A by-product of this method is a measure the CTE with a value quite similar to the one measured with other methods.

The data reduction procedure for the spectra was nearly the same as in long slit spectroscopy, except a few changes in the sequence of operations (see table below). In particular, the sky subtraction in "Va-et-Vient" mode must be done pixel-by-pixel before any other process to keep exactly the same sampling between the object and the sky.

The experimental results on long slit and "Va-et-Vient" methods are displayed in Fig. 3. They clearly show the saturation of the S/N with long slit spectroscopy, while the S/N still increases with flux in the "Va-et-Vient" mode. Both sets of data were obtained in the same

Fig. 2. Picture of the optical bench used in the experiment.

TABLE 1

Major steps in the data reduction procedure, both in long slit and in "Va-et-Vient" spectroscopic modes.

Long slit	"Va-et-Vient"
bias subtraction	bias subtraction
flatfielding	sky subtraction (pixel-by-pixel)
sky subtraction (optimal)	flatfielding
object extraction	object extraction
wavelength calibration	wavelength calibration

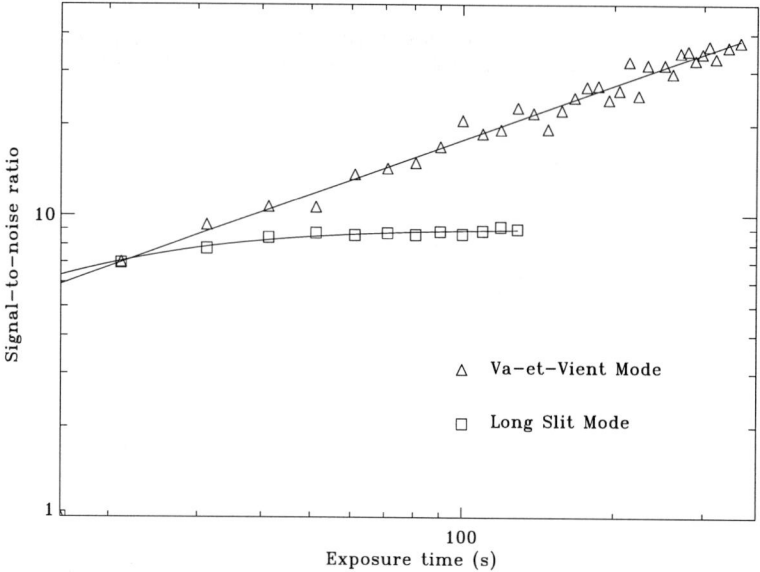

Fig. 3. S/N versus integration time for the long slit and the "Va-et-Vient" modes acquired and reduced in the same conditions, from the laboratory experiment. The gain in S/N in "Va-et-Vient" is mainly due to the better subtraction of the sky emission lines.

experimental conditions, with an object-to-sky ratio of 0.2 approximately. The gain in magnitude can reach 0.6 magnitudes in the continuum for an object-to-sky ratio of 0.1, and can be as large as 2.5 magnitudes in spectral regions dominated by strong emission

6. TELESCOPE IMPLEMENTATION AND EXPECTED PERFORMANCES

Preliminary tests on a telescope were successfully attempted on the ESO NTT. We chose an elementary exposure time of three minutes and the dead time (shift of the telescope and shift of the charges) between them was less than five seconds. Moreover, to save time during the shifts, we kept the telescope guiding on the object position while it was tracking on the sky because of the long recentering time on the guide star. As no link was implemented between the CCD controller and the Telescope Control Software (TCS), we worked in a semi-automatic way, the shifts of the telescope and the starting of individual exposures being manually controlled.

Because we did not have enough time allocated on the experiment with the spectrograph available on the NTT, the tests of the "Va-et-Vient" capabilities were not obtained in the optimal conditions discussed above. In particular they were performed in a too high dispersion mode (typically 5 Å/pixel) to be the most efficient, compared to standard spectroscopy. The total integration time was then not sufficient to reach the limiting regime in the long slit mode. However we found that the sky subtraction of strong emission lines was significantly improved.

The next step in the implementation of the "Va-et-Vient" will be to use a fully automatic procedure with a link between the TCS and the CCD controller at the Canada-France-Hawaii Telescope in the Pegasus software environment, following the observing sequence detailed below (the steps written in italics are repeated n times, where n is the number of sequences in the total integration):

- Initialize the CCD

(Telescope is guiding)
- *Open the shutter*
- *Individual exposure (t)*
- *Close the shutter*
- *Move the charges on the CCD*
- *Move the telescope*

(Telescope is tracking)
- *Open the shutter*
- *Individual exposure (t)*
- *Close the shutter*
- *Move back the charges*
- *Move back the telescope*

- Read the CCD

The new parameters which have to be defined in the user interface are: the number of "Va-et-Vient" n, the individual integration time t, the number of lines to shift on the CCD ΔY, and the shift of the telescope $\Delta\alpha, \Delta\delta$. In these conditions, the total integration time will be $T = 2nt$. Moreover, in "Va-et-Vient" mode with one slit, ΔY will be independent of $\Delta\alpha, \Delta\delta$, otherwise they will be combined.

7. CONCLUSION

In conclusion, we have demonstrated numerous interests of the "Va-et-Vient" mode for faint object spectroscopy and many others will be explored in a near future. In particular, it has long been known that the flatfield corrections with thinned CCDs are quite difficult to apply in spectroscopy because of the strong fringes which appear in the red part of the spectra. They are generally badly removed, especially when the sky is spatially and spectrally varying on the chip. In a "Va-et-Vient" mode, they will apply on the object flux only, as we checked it in the laboratory and their contribution to the underlying sky will disappear in the sky subtraction.

Another potential interest of the "Va-et-Vient" mode will be the possibility to draw curved slits, with a much better efficiency in the data reduction as the spectral sampling residuals are removed. This was already used by Soucail et al. (1988) for acquiring the spectrum of the giant arc in Abell 370, with a MOS starplate punched with adjacent holes which followed the shape of the arc itself. The data reduction and the sky subtraction were not easy to calibrate accurately, because the transmission and the sampling were not constant and not well defined along the slit. In a "Va-et-Vient" mode, the object and the sky would be obtained through the same slit, avoiding the duplication of these very special apertures!

Finally, from the laboratory experiment, we predict a gain of one to two magnitudes in the limiting magnitude for low resolution spectroscopy (Picat et al. 1994). This could be a major gain for the next generation of very large telescopes. From our point of view, the implementation of the "Va-et-Vient" on future faint object spectrographs will be unavoidable. Even if the observing procedures are slightly more complicated, the data reduction is straightforward and does not differ very much from standard procedures. "Va-et-Vient" can be extrapolated on different purposes such as Multi-Object Spectroscopy. A double slit is used for each selected object, but the same shift between each object and a corresponding sky is required all over the field, which can be difficult to maximize the number of objects. A dedicated software will consequently have to be developed for the selection of objects in this configuration. Moreover, with the double slit system, three spectroscopic strips per object will be acquired on the CCD and the number of objects per mask will be limited to 20 typically for a ten by ten arcminute field of view. But this will be a solution to keep the efficiency of two-D detectors on the spectroscopic observations of very faint objects.

REFERENCES

Boksenberg, A. 1975 in Image Processing Techniques in Astronomy, C. de Jager and H. Nieuwenhuizen, eds., Reidel, Dordrecht, p. 59
Colless, M. M., Ellis, R. S., Taylor, K. and Hook, R. N. 1990 MNRAS 244, 408
Colless, M. M., Ellis, R. S., Broadhurst, T. J., Taylor, K. and Peterson, B. A. 1993 MNRAS 261, 19
Cowie, L. L. and Lilly, S. J. 1989 ApJ 336, L41
Cowie, L. L., Gardner, J., Wainscoat, R. J. and Hoddap, K. W. 1994, preprint
Cuillandre, J. C., Fort, B., Picat, J. P., Soucail, G., Altieri, B., Beigbeder, F., Dupin, J. P., Pourthié T. and Ratier, G. 1994 A&A 281, 603
Lilly, S. J., Cowie, L. L. and Gardner, J. P. 1991 ApJ 369, 79
McLean, I. S., Cormack, W. A., Herd, J. T. and Aspin, C. 1981 Proc. SPIE 290, p. 155
Picat, J. P., Cuillandre, J. C., Fort, B. and Soucail, G. 1994 Proc. SPIE 2198, Astronomical Telescopes and Instrumentation for the 21st Century, D. L. Crawford, ed., p. 1274
Shectman, S. A. and Hiltner, W. 1976 PASP 88, 960
Soucail, G., Mellier, Y., Fort, B., Cailloux, M. and Mathez, G. 1988 A&A 191, L19
Tresse, L., Hammer, F., Le Fèvre, O. and Proust, D. 1993 A&A 277, 53
Tyson, J. A. 1988 AJ 96, 1

DISCUSSION

TINBERGEN: This is an established technique in polarimetry. Also a low-contrast high-signal situation. Stenflo's group in Zurich has an operational solar spectropolarimeter using this technique. You may like to compare notes with them to see how far you can push this technique for the VLT.

SOUCAIL: Yes, we knew that similar techniques were used in polarimetry for low contrast detections. In the case of the Sun, the problems may be different as it is not a question of low flux but low contrast. For faint object spectroscopy, we have to combine the effects of undersampling with those of low spectral flux.

FINGER: Why does this method only work at very low resolution?

SOUCAIL: "Va-et-Vent" spectroscopy is less efficient than standard long-slit spectroscopy while the defects of the spectrograph are, in some sense, negligible (flatfielding residuals, undersampling or poor sky subtraction), which is the case when the signal is split in many spectral elements. But at low resolution they become dominant and limit the S/N of the data for long integration times.

LEACH: This method has been applied to stellar polarimetry with TI 800 x 800 CCDs with CTE not being a serious problem for polarimetry accuracy down to 10^{-3}. Do you see streaks due to charge trapping?

SOUCAIL: We have tried to work in regions of the CCD where no defects were apparent as far as possible and did not detect any strong streaks. In any case, we can limit the number of shifts to a few tens, with a few tens of lines for each shift, which minimizes the sensitivity of the method to CTE limitations.

CULLUM: In your two slit method, would it not be better to record the two object fields separately on the CCD like for the sky? (ie. use four zones on the CCD per object, two sky and two object.) By adding the charges recorded by two separate slits (and recorded by different pixels) one would expect worse subtractor of the sky.

SOUCAIL: This is not exactly the case, because in the two slit mode, you have three strips of data at the end of the exposure, the central one corresponding to the sum of the charges integrated through the two slits for the (object + sky). On the adjacent strips, you have the sky charges recorded through each slit and by simply adding them, you recover the correct signal to subtract to the central strip. Indeed this would be perfectly correct if there was no flatfield correction to apply, otherwise the correction on the object signal is not perfectly applied. To avoid this problem, the solution proposed with four strips is better, but it also takes more space on the CCD.

TINBERGEN It is suitable for all low contrast situations. The basic thing is you avoid flat fielding by using the same pixel for measuring both large quantities, whose small difference you are interested in.

SOME PROBLEMS OF WIDE-FIELD ASTROMETRY WITH A SHORT-FOCUS CCD-ASTROGRAPH

I. S. Guseva

Pulkovo Observatory

ABSTRACT: The worldwide application of CCDs in astronomy is concerned mainly with the problems of photometry and spectrophotometry. Astrometry with CCDs is limited by the small fields of view and now is used to solve some special tasks such as parallax determination of faint stars, observations of QSOs, etc. Very interesting results were obtained using a CCD with a meridian circle in a drift scanning mode to solve the problem of the extension of the reference system to stars of 16th - 17th mag and the linkage between radio and optical reference frames (Stone 1994).

Based on the preliminary results of our observations, some proposals on astrometry with CCDs are discussed in the paper. The aim is to show the use of short-focus telescopes with CCDs in the mode of traditional astrographs. Numerous tasks may be solved with such instruments more simply and correctly than with meridian instruments in a drift scanning mode.

The problems of the application of CCDs with short-focus optics are also discussed. Some effects of CCD properties on the accuracy of astrometric and photometric measurements are also considered.

1. INTRODUCTION

Since the early 1980s numerous results on the application of CCDs for different tasks in astrometry have been obtained. The majority of observations was made with large telescopes and, therefore, with small fields of view. It appears that the application of CCDs allows one to solve some very interesting new astrometric problems, but seriously limits the traditional tasks of these instruments in a photographic mode.

A general way to extend the possibilities of telescopes with CCDs is to create new, large CCDs or to use a mosaic of CCDs. It is very promising, but is an expensive and complicated way with numerous technical problems. On the other hand it may be shown that from the point of view of "rational sufficiency" it is enough to use rather small short-focus instruments with available CCDs to solve numerous traditional astrometric problems. The main reason for this approach is that the internal precision of such an instrument is quite comparable with the accuracy of reference systems, and the resulting accuracy of astrometric measurements cannot be improved significantly by improvement of the instrument. It may be shown that such instruments allows us to solve a wide class of traditional and modern astrometric problems.

This approach is based on the experience of CCD-observations, made in 1990, in context with the work on the REGATTA-ASTRO space project, and on the results of observations made at Pulkovo in 1993 - 1994 with a small short-focus astrograph, which can be regarded as a prototype of a desirable instrument.

2. EXPERIMENTS OF CCD-ASTROMETRY WITH SHORT-FOCUS INSTRUMENTS.

We started CCD observations in 1990. It was a part of the work on the REGATTA-ASTRO space project, developed by the Space Research Institute (Moscow) and Pulkovo observatory (St Petersburg). The project was aimed at creating a catalogue of 400,000 stars, including positions, proper motions, parallaxes (with accuracy of 0.01 arcsec) and WBVR-photometry (with accuracy of 0.01 mag). In the summer of 1990 we made a series of experimental observations with a very small prototype of our onboard CCD-telescope operated in the mode of the space project: fast routine reading of direct images of the sky areas crossing the field of view at a diurnal rate. The precision of the astrometric measurements (for a single exposure) was evaluated as 0.01 - 0.02 of the pixel size (with a complicated algorithm of the image center) and 0.02 - 0.04 (with a simple centroid method) for all stars in the range of V magnitudes 6 - 11. The precision of the photometric measurements was 0.02 - 0.11 mag for the same range (Alexeeva et al. 1993). In addition, the CCD-observations with the Pulkovo Large Transit Instrument were made during night and day time. In the day time we obtained direct images of stars (up to V = 7), Mars, Venus and the Sun. It allows us to conclude that traditional day-time fundamental astrometric observations are quite possible with CCDs.

In December, 1992 we began regular experimental CCD-observations at Pulkovo. The first aim was to make a complete meteorological investigation of our CCD and optical system. We use a very small refractor (Double Short-focus Astrograph) with D = 100 mm and F = 712 mm. Our CCD-camera was created in the Space Research Institute using the virtual-phase CCD ISD015A produced by "Electron" (St Petersburg, Russia) (Khmelevitskij et al. 1991). The parameters of the CCD are: 520 x 580 pixels (18 x 24 microns), sensitive area 9.4 x 13.9 mm, spectral band 0.2 - 1.0 micron. The Quantum Efficiency (QE) at different wavelengths is as follows: 15% at 0.2 microns, 28% at 0.4, 58% at 0.7 and 12% at 1 micron. The full well capacity is 220,0000 e^- and the real (working) dynamic range is 6 to 7 mag. The readout noise (nominal) is 10 e^- with high speed amplifier and 7 e^- with low noise amplifier. We use the CCD placed in a gas-filled housing equipped with a thermoelectric (Peltier) cooler, which provides a temperature difference between the CCD and that of the environment of about 40° - 50° C. The dark current (nominal at -40° C) is 12 e^-/pixel/s. Our CCD-camera was designed to be used onboard the spacecraft and works without any shutter. An 11-grade analog-digital converter is used (2048 cu, one cu ~ 100 e^-); every frame takes 603 Kbytes. A very short readout time (1.5 sec) allows the camera to work without a shutter, but this creates additional problems to be solved. The observation routine is as follows: a fast reading (1.5 s), exposure (from 0.1 s to 30 minutes), a fast reading, display on the PC/AT monitor and a store (if necessary). The focal length of our astrograph provides an angular field of 45 x 67 arcmin (the angular scale is 5.2 x 7 arcsec/pixel). Thus, the expected accuracy at the level of 0.02 of the pixel size should be 0.10 arcsec in declination and 0.15 arcsec in right ascension.

The first observations were made to evaluate the accuracy of our positional and photometric measurements. The Pleiades cluster and some other standard fields were chosen for this purpose. We have made a series of observations of the same fields with different exposure times. For the first evaluations we have used a simple and rough centroid method to determine the star image center. The internal precision of the positions obtained appears to be four to five times better than in the case of photography and varies from 0.10 to 0.35arcsec (for a single observation) for a wide range of magnitudes. By use of different exposure times (from 0.1 seconds to 10 - 20 minutes) we can observe all stars from the brightest to the stars of 16th - 17th mag with our small instrument. The precision of a single magnitude determination varies from 0.03 to 0.15 mag over the whole range. It means that from a series of 5 - 10 frames it is possible to obtain positions (with an accuracy 0.05 - 0.15 arcsec) and magnitudes (with an accuracy 0.01 - 0.05 mag) for a set of 500 - 700 stars. To realize this opportunity it is necessary to make a special investigation of possible quasi-systematic errors of the instrument at the same level of accuracy. We have made the "upper" evaluation of our accuracy by a comparison with the Pleiades catalogue by Vasilevskis et al. (1979). The mean deviations of our star positions from that by Vasilevskis do not exceed 0.2 arcsec and, thus our systematic errors are not large. A detailed investigation of all possible sources of error has been made and methods of appropriate corrections are now developed. After the completion of these investigations it will be possible to begin a wide program of observations. We are planning to observe some selected areas on the sky, minor planets and QSOs, which can be observed at Pulkovo with our small instrument. We are planning also to create an instrument with better optics and CCD, more suitable for our tasks.

3. POTENTIAL OF ASTROMETRY WITH SHORT-FOCUS CCD-ASTROGRAPHS

An important problem of modern ground-based astrometry is to extend the best reference systems, HIPPARCOS and TYCHO, to the faint stars of 15th - 17th mag. Two ways are known to solve this task: astrophotography with large telescopes and differential observations with meridian instruments equipped with CCDs. It may be shown that an optimal variant of the instrument can be created for this task and numerous other applications, when we take into account all the aspects of situation.

The application of CCDs allows one not only to reach much fainter stellar magnitudes with the same telescope in comparison with photography, but also to increase significantly the precision of positional and photometric measurements. With comparatively small CCD-astrographs one can solve the same problems as with a large photographic telescope, with a great advantage in the data processing. The optimal variant of CCD-telescope can be chosen if one takes the internal precision of positional measurements and the accuracy of reference catalogue to be equal. For ground-based observations the TYCHO catalogue will become the reference frame of choice due to the dense set of stars with precise positions. This catalogue will provide us with 25 stars per square degree (in the average) with a mean positional accuracy of 30 mas. Thus, correct differential astrometry will be possible within small enough field of view, and the main limitation will be imposed by atmospheric turbulence. It is very important that differential observations were made for all stars within the field of view simultaneously, in a mode of classical astrographs. Observations in a drift scanning mode are affected by the long-period variation of refraction and extinction, variation of the instrument position, variation of detector parameters and so on. Moreover, in this variant of observations the star image is affected by all the cosmetic defects of the CCD and variations of the distortion "gathered" through the path of charge transfer (in the case of so-called "stare mode" these errors can be detected and correctly taken into account for every pixel separately). It causes some difficulties in the data reduction and leads to a worse external accuracy of such observations in comparison with very high precision of the internal errors.

The common experience of astrometric CCD-observations shows that it is quite possible to obtain a precision of star image location of 0.02 of one pixel and better. With a telescope having a focal length of one meter and a CCD of 1k x 1k pixels (15 x 15 microns) one could obtain a field of view of one square degree, a scale of 3.1 arcsec/pixel and a precision of a single observation of 60 mas. The series of five to ten frames allows us to achieve a position accuracy of 20 - 30 mas. It is about ten times better than the accuracy of the best existing reference catalogue, PPM, and quite comparable with the accuracy of the future best reference catalogue, TYCHO. A telescope with a focal length of two meters and a CCD of 2k x 2k pixels (15 x 15 microns) provides about the same field of view and the precision of single observation of 30 mas; that is about the limiting precision for the ground-based observations in the presence of atmospheric turbulence within the field of view of one square degree.

The resolving power of short-focus telescopes is the main problem. To illustrate this problem we compiled Table 1 (on the basis of "Astrophysical Quantities" by Allen (1973). We suppose here that the minimum useful variant is F = 1000 mm, CCD 1k x 1k pixels, that provides for one square minute a set of 20 x 20 pixels of size three arcsec. We suppose also that star images occupy 3 x 3 pixels with the FWHM of approximately 1.5 - 2 pixels.

TABLE 1

Mean density of stars per one square minute (20 x 20 pixels) with magnitudes less than "m" as a function of galactic latitude "b" (Allen 1973).

m\b	0	20	40	60	90
15	0.73	0.25	0.11	0.07	0.05
16	1.67	0.55	0.23	0.13	0.10
17	3.75	1.11	0.43	0.24	0.17
18	8.78	2.36	0.77	0.41	0.29
19	17.5	4.40	1.39	0.70	0.44
20	27.8	8.8	2.2	1.1	0.7

It is easy to see that the problem of resolving close stars arises only for 17th - 18th mag near the galactic equator and is practically absent up to 20th - 21st mag at mean and high galactic latitudes (it is especially important for observations QSOs). Certainly, these are only the statistical evaluations and the real distribution of stars on the sky is irregular. But it may be shown that if two stars are located on the CCD with a separation of more than one pixel, the resolving of stars is not a problem. On the other hand, the seeing for ground-based telescopes often exceeds two arcsec, and all the stars which are "closer" than three seconds cannot be resolved. In any case, there exist some regions on the sky which cannot be resolved with any telescope (for example, globular clusters) and for every instrument the proper "resolving" limit should be accepted and kept in mind. That is why the idea to use short-focus CCD-telescopes for such work seems to be reasonable.

Numerous astronomical problems may be solved by use of such instruments. Among them are:

ASTROMETRY WITH A SHORT-FOCUS CCD-ASTROGRAPH

a) Creation of large catalogues of positions, proper motions and photometric values of stars up to 16th - 17th mag.

b) Observations of solar system bodies: planets (with their satellites), minor planets, comets, asteroids etc.

c) Observations of QSOs and the establishment of a precise link between radio and optical reference frames.

d) Ground-based supplements of space projects, for example, searching for peculiar objects and their monitoring, etc.

e) Different astrophysical problems, such as investigation of variable stars, investigation of star clusters, etc.

4. SOME PROBLEMS OF CCD-ASTROMETRY WITH SHORT-FOCUS OPTICS

Numerous effects appear to be significant in CCD-observations with a short-focus astrograph, because small linear errors in the focal plane correspond to large angular errors. Additional problems for us are connected with the properties of our CCD-camera and the conditions of observations.

Some specific problems at Pulkovo observatory are lighting and electromagnetic pollution. We are surrounded by luminous St Petersburg and neighboring towns. But the worst effect is connected with a powerful radio (or TV) broadcast interference, because the half period of these waves exactly corresponds to the image diameter of stars, and a faint star may increase its image size, vanish or shift in both directions. At present some improvements in the construction of the CCD-camera have been made which reduce this effect. On the other hand a method of mathematical filtration has been developed to diminish the errors caused by this interference.

The next problem is that we are working with a thermoelectric (Peltier) cooler without stabilization of the CCD temperature. Nominally this cooler should maintain the temperature difference between the CCD and the environment of 60° - 65° C. We have this difference varying from 41° C (at an air temperature of -20° C) to 51° C (at an air temperature of +20° C). Partially it may be explained by the construction of our special CCD-camera, by the shortcoming of the concrete CCD sample. The variation of the CCD temperature during the night creates an additional problem for the reduction of the data. Fortunately, we can measure the CCD temperature anytime by use of a thermistor. The dependence of the dark current on the temperature appears to be very stable and for every observation frame we can calculate the dark current frame, including the separate response of every pixel on the temperature variation.

The investigation of the spectral sensitivity of our chip shows that it has significant irregularities. There are large variations of the spectral response from pixel to pixel and variations with CCD temperature which need a careful flatfield study and to determine the corrections.

An important problem is connected with cosmic ray events and their accidental impact on the star images, and their effect on the precision of positions and photometry. The problem is specific for each type of CCD. We have investigated the characteristics of our chip on the basis of a series of dark current frames. We have extracted only the events with energy greater than 1000 e^- to be sure that these are real events. The results are presented in Table 2 and on the Figures. In the table one can see the statistics of five dark current frames (events, which affect one pixel, two adjacent pixels and so on with a signal exceeding 1000 e⁻).

TABLE 2

The number of events "n", which affect "m" adjacent pixels with a signal exceeding 1000 e⁻ (from five dark current frames obtained with a ten minute exposure)

m (pixels)	1	2	3	4	5	6	7	8	17
n (events)	161	59	22	11	7	5	2	1	01

On the average there are 100 "bad" pixels in each frame for a ten minute exposure with false signals, that exceed the noise by approximately a factor of ten. The distribution of these events with energy is shown in Fig. 1. Moreover, we can suppose the presence of a large number of false signals (events) with smaller energy. Very often we can find on dark current frames star-like images and tracks (like the moving NEOs). An example of such a track is shown on Fig. 2. Some conclusions follow from these results: a) it is not worthwhile to use very long exposures, because a majority of star images will be affected by radiation events; b) it is necessary to take some frames (no less than three) of the same sky area to exclude possible accidental errors caused by radiation events.

One of the important sources of error in position and brightness determinations is connected with the complicated structure of pixels. In every case the CCD pixel consists of parts with different sensitivity. Especially it is important for front-illuminated CCDs, in which a large part of a pixel may be entirely insensitive. The measured positions and photometric values depend on the location of the star image center within the pixel. The result is that periodic errors are created with a main spatial period equal to the pixel size. It may be shown that this effect depends significantly on the subpixel sensitivity, on the shape of image (point spread function, quality of guiding and atmospheric conditions) and on the method of image center determination. The errors of the position determination can reach 0.1 of the pixel size and should be taken into account, as well as the errors of photometry.

Some different ways may be proposed to determine the necessary corrections to the measured positions and brightness. The first is to measure directly the sensitivity inside a pixel in the laboratory with complicated enough equipment and to use these data in the process of image reconstruction. The second way is to use the information on the geometry of pixel structure, to create a model of the subpixel response and to reconstruct the parameters of this model on the basis of real observations. And the third way is to obtain the necessary corrections statistically on the base of a large enough series of observations of the same field on the sky with a great number of stars. It may be done if we use many frames taken with small different shifts of the CCD field of view with respect to the stars observed. This way seems to be most useful, because all the above-mentioned obstacles appear to be taken into account automatically. The preliminary and rough results of the application of this method (with a small statistics - only three frames) are shown in Fig. 3 (positions) and Fig. 4 (photometry).

ASTROMETRY WITH A SHORT-FOCUS CCD-ASTROGRAPH

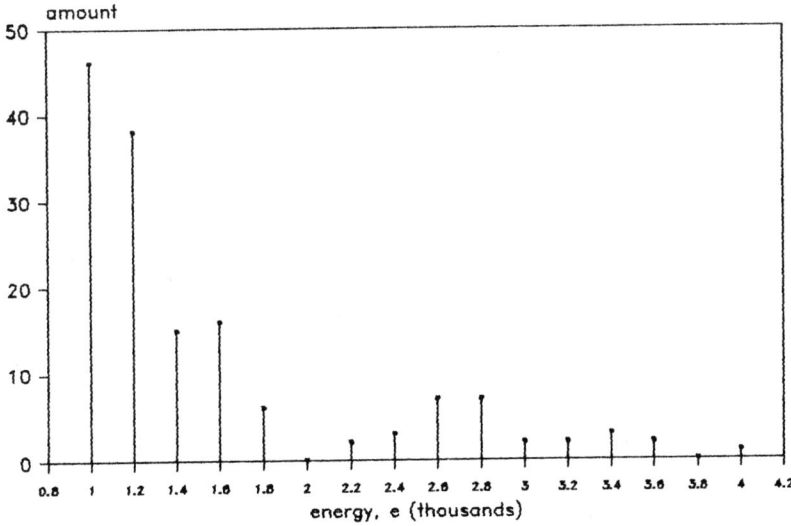

Fig. 1. Radiation events (for ten minutes) - the energy distribution.

8	5	-1	1	3	-1	3	5	2	-1	1	4	1	1
0	7	4	1	5	4	6	2	5	3	3	2	11	-1
2	-1	4	-1	3	1	7	1	7	6	1	4	23	7
3	1	2	0	0	2	2	2	1	3	0	20	5	2
5	5	3	-2	1	2	2	10	13	3	23	22	4	5
1	1	0	4	4	3	-1	1	3	15	27	7	5	-2
1	3	3	1	0	1	3	8	5	19	10	1	2	1
4	2	3	2	3	-2	5	6	16	11	2	1	1	5
2	2	0	0	3	4	6	9	18	9	1	4	5	0
5	6	0	8	2	5	4	12	11	1	3	5	3	7
1	5	4	-1	-1	2	14	11	5	5	7	-1	1	3
5	3	0	3	4	10	14	7	2	0	2	2	1	-1
1	0	7	4	5	21	7	1	3	1	5	6	-1	1
0	0	3	2	10	15	6	0	3	6	8	1	-2	1
1	3	5	14	30	6	3	2	2	2	-1	0	-1	3
5	-1	1	38	9	5	4	2	4	2	0	2	4	-3
4	2	25	8	8	1	0	0	3	4	4	7	2	0
2	13	15	9	3	3	3	4	7	0	3	3	-2	2
4	18	1	2	3	6	4	3	-2	1	0	4	0	6
1	2	6	8	3	0	-3	-3	3	4	2	0	1	-1

Fig. 2. An example of a particle track.

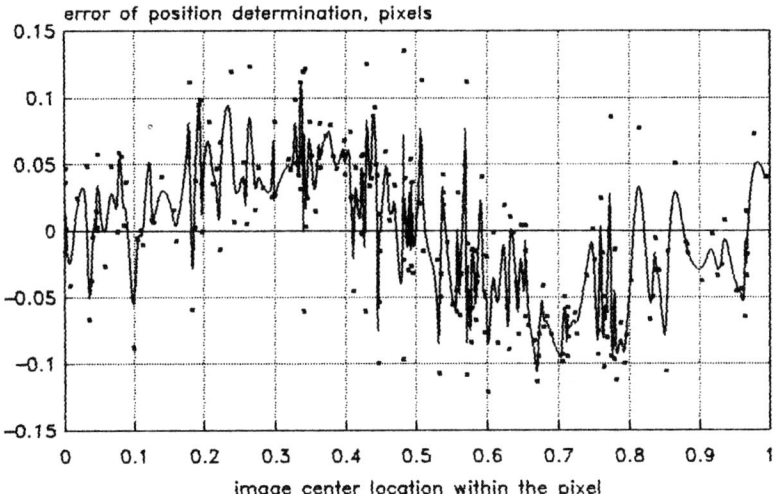

Fig. 3. Effect of the pixel structure on the position determination.

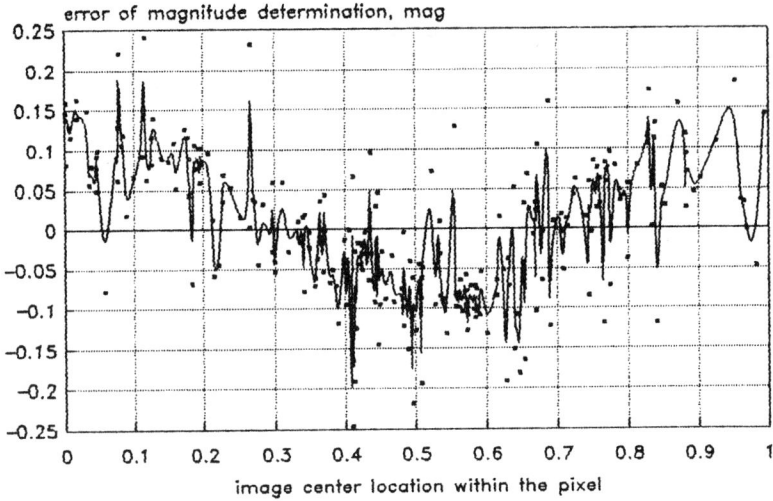

Fig. 4. Effect of the pixel structure on the magnitude determination.

The next important effect is connected with the Charge Transfer Efficiency (CTE) which can cause an error in position measurements, depending on the stellar magnitude. It is similar to the well-known "brightness equation", but this effect is quite linear, depends on the readout time and the CTE, is significantly greater along CCD columns than along rows, depends only on the star magnitude and does not depend on the position of star image with respect to the optical center of the frame. This effect may be found in two ways: by the statistical evaluation of a great number of stars with well-known coordinates and on the basis of observations of the same star field with different orientation of the CCD - direct and inverse. In the last case a difference in the positions of the same star with respect to the center of field gives a double correction to be used for data reduction. In our case (fast readout) this error is about 1 micron/mag and has to be taken into account in any case.

A very important problem is the general distortion of the instrument, which includes both the optical distortion and the distortion of the CCD. As in a previous case, the same procedure may be used to obtain the correction, but it is more useful to take many frames with different orientations of the instrumental frame with respect to the sky area. It may be done effectively if we use the North Pole region for this purpose, because a majority of instruments is able to take images of this field with any position of the instrument.

All the effects mentioned are to be carefully investigated and will be taken into account to obtain high precision positions and photometry with a CCD-telescope.

REFERENCES

Alexeeva, G. A., Andronova, A. A., Guseva, I. S., Kopylov, I. M., Novikov, V. V., Avanesov, G. A., Chesnokov, Yu. M., Kogan, A. Yu. and Ziman, Ya. L. 1993 Poster Papers on Stellar Photometry IAU Colloquium No. 136, p. 166
Allen, C. W. 1973 Astrophysical Quantities, Athlone Press, London
Khmelevitskij, A. T., Zuev, A. G., Rybakov, M. I., Kiryan, G. V. and Berezin, V. Yu. 1991 Proc. SPIE 1447, 64
Stone, R. C. 1994 AJ 108, 313
Vasilevskis, S., van Leeuwen, F., Nicholson, W. and Murray, C. A. 1979 A&AS 37, 333

DISCUSSION

PENNY: You seem to have already been studying the problem I talked about yesterday, half pixels which only have half sensitive areas. What range of magnitudes can you measure on one exposure?

GUSEVA: We can measure a range (6 - 7) magnitudes on a single exposure.

A DUAL CCD MOSAIC CAMERA SYSTEM SEARCHING FOR MASSIVE COMPACT HALO OBJECTS (MACHOs)

Kem H. Cook

Lawrence Livermore National Lab

ABSTRACT: The Macho Collaboration uses a dedicated 1.27-m telescope (The Great Melbourne Telescope) at Mount Stromlo to make photometric measurements of tens of millions of stars per night searching for the gravitational microlensing signature of MACHOs in the halo and disk of the Milky Way. A prime focus corrector and dichroic beamsplitter provide red (6300 - 7800 Å) and blue (4500 - 6300 Å) foci with one degree fields. A two by two mosaic of 2048 x 2048 pixel CCDs in each focal plane provides simultaneous images of 0.5 square degrees. By August of 1994, more than 20,000, 32 megapixel images will have been collected of fields in the Large Magellanic Cloud (LMC), Small Magellanic Cloud and the bulge of the Milky Way. We have implemented an online analysis system which produces photometric reductions of a night's data (five Gbyte of images) within 24 hours. This system allows us to identify and follow interesting events in real-time. In this search, we have identified more than 60,000 variable stars, and a preliminary analysis of their types and distribution will be presented. Microlensing events toward the LMC and the bulge have been discovered and detection efficiencies are being calculated to constrain the MACHO content of the Milky Way's halo.

DISCUSSION

HOWELL: How do you store the data? Do you send x, y and mags to LLNL each day? How do you get image data from Mt. Stromlo to LLNL?

COOK: Images are archived on eight mm exabyte tapes, and copies are distributed to the various collaboration sites. Most of the data reduction during our first two years has been done in Australia, but the analysis has been done in California. The reduced data are collected on exabyte tapes and sent to Livermore every few weeks.

GLASS: Do you propose to release data about the variable stars to the general community within a finite time?

COOK: Our collaboration has decided to release our data four years after it is obtained. We also are open to proposals to collaborate on non-Macho aspects of our data at any time.

FLORENTIN-NIELSEN: What are the approximate values of the mass of the deflecting objects that you have from your microlensing events so far?

COOK: Because of the degeneracy of mass, distance and transverse velocity of the lens, one must appeal to a particular model of lens distribution to estimate a mass. For a generic halo model (which has not been corrected for our detection efficiency) the most probable mass of our 7x amplitude event is about 0.1 M_o, but the 50% confidence limits extend from 0.03 M_o to 0.5 M_o.

WEST: If and when your alert system is running, how soon would you be able to tell other observers about an event? And if spectra could be taken (at the maximum) what could you expect to be seen in these?

COOK: We are beginning our alert system by only notifying the CTIO 0.9-m. After the system has worked well, we may distribute alerts publicly. The OGLE collaboration is currently making available alerts from their survey. As for spectra, I would expect that they would show nothing abnormal which would the case for microlensing as opposed to some stellar phenomenon.

POLOJENTSEV: What was the reason to delay of the reduction of observation with CCD?

COOK: As we began data collection, the scale of the databases containing reduced photometry overwhelmed the facilities in the dome and database construction and analysis was moved to Livermore. This process was impeded by the necessity of archiving image data utilizing the vast majority of our exabyte writing capability.

PENNY: Does the gravitational lens resolve the stellar disks?

COOK: This is a possibility, but for lens masses in the Jupiter range and larger and for giants or dwarfs in the LMC, the Einstein ring diameter is large enough compared to the angular diameter of the source that the source can be considered a point.

MULTI-FIBER SPECTROSCOPY WITH WIDE-FIELD TELESCOPES

F. G. Watson

Royal Greenwich Observatory,

ABSTRACT: Fiber-fed multi-object systems are now the preferred instruments for gathering spectroscopic data on survey scales. The technique lends itself particularly well to telescopes with fields of 30 arcminutes or more. This paper gives a broad overview of the instrumental considerations involved in its implementation.

1. INTRODUCTION

The use of low-loss optical fibers to rearrange a randomly-distributed set of target objects for multi-object spectroscopy has become a well-established technique over the last dozen or so years. It provides one of the most efficient means of marrying a telescope with a wide field of view (which may be large in linear dimensions) to an array detector.

The astronomical problems that can be addressed by a particular system are determined as much by the telescope's field as its aperture. For any telescope feeding a multi-fiber system, the greatest attainable number-density of a particular class of target object is the density at the telescope's limiting magnitude for spectroscopy, broadly determined by its aperture. In practice, this is usually greater than the number of available fibers. On the other hand, the lowest number-density at which all the available fibers can be used depends primarily on field-of-view. Thus, field is crucially important in delineating the useful number-density range over which the system can work (see Dawe and Watson 1984, Watson 1994a). It is this consideration that has driven designers of multi-fiber spectroscopy systems for four-m class telescopes towards prime focus (e.g. Jenkins et al. 1993, Gray et al. 1993).

While detector format determines the number of fibers that can be used to feed a given spectrograph, there is no intrinsic reason why multiple spectrographs should not be operated simultaneously. The limit on the number of fibers in a multi-object system is usually set by the practicalities of packing them into the telescope's focal surface. Only for the more sparsely-distributed object-classes is the fundamental limit mentioned above approached.

There is a related technique - area spectroscopy of single resolved objects using fiber image-dissectors - for which the requirements are rather different. Here, field-size becomes subsidiary to plate-scale, though it remains important for the acquisition of guide stars. Area coverage and spatial resolution are dictated by the number of fibers, which, in turn, is generally determined by spectrograph design and detector format, since multiple spectrographs are not normally contemplated for this technique.

This paper reviews some of the technical considerations associated with multi-object spectroscopy using optical fibers. Some aspects of fiber-fed area spectroscopy are also described.

2. FIBER FEEDS

2.1 Fiber Properties

It is the excellent transmission properties of fused silica fibers that have endeared them to astronomers. The high OH^- type ("wet" fibers) have very good blue performance at the expense of uniform transmission in the far-red, where deep OH^- absorption bands occur (see, e.g., Nelson 1988). Low OH^- fibers, on the other hand, have a flat, very high transmission profile out to approximately 2 μm. A new, hydrogenated "dry" fiber type promises wide-band performance, having the enhanced blue transmission normally associated with wet fibers.

Beyond 2 μm, other materials are needed. Fluoride glasses transmit reasonably well to ~ 4 μm (e.g. Levin et al. 1988, Levin et al. 1993, Dallier et al. 1993), while beyond that, chalcogenide glasses offer a window centered on 6 μm. These materials are less robust than fused silica.

A well-known property of all fiber types is focal ratio degradation (FRD), or beamspreading, due to microbends, internal scattering, applied stresses and diffraction (see, e.g., Ramsey 1988). Fused silica fibers with excellent FRD properties are now available. Though FRD affects slower beams ($\lesssim$ f/5) more than the fast ones found at prime foci, there are serious potential losses when the beam speed approaches that corresponding to the numerical aperture of the fiber ($\sim$ f/2 for fused silica).

Beamspreading can also result from the axis of the incoming beam being inclined to that of the fiber, a situation that will arise in most telescopes if the fibers are simply placed perpendicular to the focal surface. The effect is quite significant; for example, a four-degree inclination will speed an f/2.8 beam to f/2.0. Only in telecentric (e.g. Schmidt) telescopes, or systems where the fibers are aimed at the telescope exit pupil rather than the center of curvature of the focal surface, will the inclination be everywhere zero.

Related to FRD are the image-scrambling properties of fibers. Good FRD characteristics are associated with poor radial-scrambling of an image formed on the fiber input face for beams faster than f/5 (Barden et al. 1993, Watson and Terry 1993). The azimuthal scrambling is essentially perfect, so that an off-axis point source on the input face is transformed into a ring on the output face. These scrambling effects have implications for the illumination of fiber-fed spectrographs.

2.2 Matching

A number of considerations, often conflicting, govern the choice of fiber diameter for a given telescope and spectrograph combination. Clearly, if the fiber is to be used without any auxiliary optics (e.g. microlenses), its diameter is determined at the input end by the telescope plate scale, and at the output end by the detector pixel size imaged back through the spectrograph. If the spectrograph was designed to be fed directly by the telescope (as frequently occurred in early multi-fiber systems), these will be reasonably well-matched, but the collimator will normally be overfilled because of FRD, resulting in throughput losses.

The use of microlenses at one or both ends of the fiber offers greater flexibility of design. For example, the 153 μm fibers of the large-diameter (2.7 arcsec) feed for the prime-focus multi-fiber system of the 4.2-m William Herschel Telescope (WHT) are used without microlenses at their input ends, accepting light at the f/2.8 delivered by the prime-focus corrector (Worswick et al. 1994). At their output ends, 2 mm sapphire balls re-image each fiber face onto the pupil of the f/8.2 collimator, so that throughput is not dependent on FRD. Spectral resolution is FRD-dependent, however, since the effective slit is a pupil image (formed about 1 mm from the microlens surface) whose diameter depends on the speed of the beam emerging from the fiber. (This feed also features connectors to allow the fiber cables to be mounted permanently on the telescope; they are described by Worswick et al., 1994.)

Though it is not strictly a wide-field application, another example is provided by the SMIRFS experimental fiber system for the 3.8-m UK Infrared Telescope (Haynes and Parry 1994). Here, "dry" fused silica fibers are used for the (J,H) wavebands, and zirconium fluoride for K. The fibers couple the f/36 focus of the telescope to the matching focus of the CGS4 cooled-grating spectrometer, but have microlenses at each end so that the beam within the fiber propagates at f/5. A problem with this system is the alignment of the f/36 beam accepted by the fibers with the telescope pupil, which demands fine angular tolerances.

2.3 Area Spectroscopy Feeds

A parameter of interest in fiber-coupled integral-field spectroscopy is the packing fraction or filling factor, i.e., the ratio of total fiber-core area to the area covered by the array at the input end. It depends on such factors as the core/cladding diameter ratio, and mode of packing (e.g. hexagonal or rectangular). Typically it might be 50 or 60 percent. One way of improving this is by the use of lenslet arrays on the input ends; if these are used in "Fabry mode" so that the telescope pupil is imaged onto the fiber cores, the packing fraction can be dramatically increased.

Examples of area spectroscopy feeds include DensePak (Barden and Wade 1988) and HEXAFLEX (Arribas et al. 1993); several new instruments are currently planned or under construction.

3. MULTI-FIBER POSITIONERS

3.1 Fiber Alignment

Requirements for aligning a fiber set with targets in the field of a telescope are principally concerned with positioning accuracy within the focal surface. Except at very low or very high focal ratios, longitudinal and angular alignment are generally more tractable problems.

The accuracy with which fibers can be placed sometimes involves more than one component; for example, the placement accuracy of a robot within its own reference frame and the accuracy of the model relating the reference frame to the real world (see, e.g., the Autofib-2 tests reported in Watson 1994b). The other side of the coin is the accuracy with which the focal-surface positions of the target images themselves are known. This involves considerations like astrometry, telescope flexure, the distortion function of the telescope optics, and atmospheric refraction (see, e.g., Watson 1984, Donelly et al. 1993).

The various means of supporting fibers in accurate alignment with target objects are now well-known. Manual plug-plate systems like the original "Medusa" and FOCAP (Hill et al. 1982, Gray 1984) have largely given way to robotic positioners. However, plug-plates will be used for the Sloan Digital Sky Survey (Owen et al. 1994), and in the experimental infra-red SMIRFS multi-fiber system (Haynes and Parry 1994).

3.2 Multi-Actuators

Broadly speaking, robotic positioners fall into the two classes of multi-actuator and pick-place systems. Both are highly software-intensive, and demand sophisticated algorithms for fiber/target allocation, placement order, anti-collision, etc.

The prototypical multi-actuator system is MX on the Steward 2.3-m telescope (Hill and Lesser 1986); it consists of 32 computer-controlled steerable arms, each of which carries object and sky fibers at its tip. While it has the advantage of rapid field set-up, the cost is prohibitive for large numbers of fibers.

3.3 Pick-Place Systems

Focal-plane pick-place systems, in which magnetic buttons carrying the fiber ends are positioned on a ferrous field plate by a single robot, are usually known as Autofib class instruments (Parry and Gray 1986). The original Autofib was developed for the 3.9-m Anglo-Australian Telescope (AAT); the more sophisticated Autofib-2 has recently been commissioned at prime focus on the 4.2-m WHT (Parry et al. 1994, Parry Lewis and Watson 1994). Autofib-2 has the capacity to position 160 fibers, and is arranged with interchangeable fiber modules.

Since Autofib devices configure fibers sequentially, their use in the telescope's focal surface incurs a time overhead which might be substantial (about 25 minutes in the case of Autofib-2). This is avoided in an off-focus pick-place system like 2dF, the AAT's two-degree prime-focus facility (Gray et al. 1993, Taylor and Gray 1994). Here, two field plates, each with its own fiber set, are interchangeable so that one can be re-configured while the other is on the sky. If (as is usual) the reconfiguration time is less than the typical exposure time, then there is a negligible set-up overhead. The penalty for this is the need to duplicate the fiber set. 2dF has two sets of four hundred fibers, distributed equally between two spectrographs.

3.4 Hybrid Systems

One or two systems do not fit exactly into the broad categorization defined above. FLAIR II on the 1.2-m UK Schmidt Telescope (UKST) is an off-focus pick-place system that requires manual selection of target objects on the (photographic) field plate before each fiber is cemented in place by a single-shot robotic positioner (Watson et al. 1993, Bedding et al. 1993, Watson and Parker 1994). Interchangeable field-plate holders allow one to be in use on the telescope while the other is being reconfigured.

A pick-place system of unusual form is represented by the Calor Alto 3.5-m telescope's "Spaltspinne" (Pitz 1993) and its near-twin, TAUMOK (originally "Feldspinne") for the 1.3-m Tautenburg Schmidt (Pitz et al. 1993, Marx 1994). Here, there is no field plate. Instead, a steel plate surrounds the focal surface; pivoted rods with fibers on their inner tips have magnetic buttons at their outboard ends which are moved by a pick-place robot.

A focal-plane pick-place system configuring 80 fiber feeds is being developed for the FUEGOS spectrograph at the ESO VLT (Felenbok et al. 1994). The 30 arcmin (1-meter) field is patrolled by two robot heads to reduce configuration time to less than ten minutes. The "Medusa" mode of this instrument fits neither the "one fiber per object" of a conventional multi-fiber feed, nor the "one object per feed" of area spectroscopy. Rather, seven fibers per object will be fed via microlens arrays to allow a spectral resolution of 30,000 to be achieved.

4. FIBER-FED SPECTROGRAPHS

In the pioneering days of multi-fiber observing, conventional long-slit spectrographs were simply spaced back from the focal surface of the telescope to make room for the plug plate and fibers. FLAIR was the first multi-fiber system to remove the spectrograph from the telescope altogether (Watson 1986), resulting in immunity from flexure with only the most elementary mechanical construction. Because of the compactness of the UKST, this involved a fiber length of only 11 m.

While long fiber runs are entirely feasible, they do reduce blue sensitivity, and a compromise solution that is being adopted on the WHT is to place the spectrograph on one of the Nasmyth platforms, resulting in a fiber length from prime-focus of 26 m. At the AAT, blue performance has been given very high priority, so the two 2dF spectrographs are mounted on the telescope's top-end ring. One consequence of this is the use of closed-cycle coolers for the detectors rather than the conventional liquid-nitrogen dewars.

Wide-field spectrographs built especially for use with multiple-fiber feeds exhibit a range of designs. "Conventional" instruments are typified by the FISCH spectrograph for FLAIR II (Watson et al. 1993), which uses the Schmidt optical system for both collimator and camera. Likewise, the 2dF spectrographs use Schmidt-type cameras fed by off-axis Maksutov collimators (Gray et al. 1993). The disadvantage with this is that the use of the unmodified beam from the fibers drives the design of the camera to very fast focal ratios. The pupil being essentially on the grating, there is also field-dependent vignetting.

Both these drawbacks have been eliminated in the WYFFOS spectrograph for the WHT (Bingham et al. 1994), which uses the "white-pupil" design of Baranne (1988) fed by microlensed fibers. Here, the pupil of the lenticular collimator is imaged onto that of the (Schmidt-type) camera by a relay mirror; in addition, the pupil is enlarged to ease the camera design and reduce the effect of the detector central obstruction. A white-pupil design is also proposed for FUEGOS (Felenbok et al. 1994).

Other unconventional designs currently under consideration include compact, slow spectrographs fed by small-diameter fibers which are themselves fed by lenslet arrays in the telescope focus (Taylor and Parry 1994). A more radical prospect still is the development of spectrographs based on slab waveguides stacked to provide a multi-object capability (Watson 1995).

5. OBSERVING WITH MULTI-FIBER SYSTEMS

Conventional wisdom maintains that optical fibers cannot compete with multi-slit spectroscopy when working on very faint objects because of imperfect sky-subtraction. This results from inadequate sampling of the sky in the immediate vicinity of the objects, uncertainty in the transmission of the fibers, and the spatial separation of object and sky spectra on the detector. The subtraction of background scattered light is also more difficult in the multi-fiber case, particularly if the fibers are close-packed.

Investigations by, e.g., Parry and Carrasco (1990) and Barden et al. (1993), have explored the limitations of the technique, showing that better than one-percent sky-subtraction is possible. Instruments with atmospheric dispersion compensation (e.g. WHT and AAT) will extend this to higher zenith-distances. Nevertheless, the unfavorable comparison with multi-slit work still exists for the faintest objects. Full-field and high-dispersion capability remain the principal advantages of the multi-fiber technique.

Observation with fibers requires normal calibration exposures such as flat-field, dark and bias frames, and wavelength calibration spectra. In addition, fiber flat fields (to calibrate the fiber-to-fiber transmission differences and telescope/spectrograph vignetting functions) are essential. Sky exposures invariably give better results than dome flats for these.

The particular instrument and observing campaign will naturally dictate the observing strategy, but a prerequisite of all multi-fiber work is the most accurate possible astrometry. Good photometry is also desirable, since the range of magnitudes that it is possible to observe simultaneously is limited by the dynamic range of the instrument.

The end-product of a multi-fiber observing run is usually several hundred spectra; with 2dF, this will run into many thousands. As far as possible, the reduction software should be automated to cope. At the UKST, spectra are reduced using FLAIR-specific tasks within IRAF; a similar approach is being adopted at the WHT for WYFFOS data, but 2dF reductions will be carried out using specially-written software that incorporates accurate modelling of the instrument. The philosophy here is that each frame can be fully reduced while the next is being obtained.

6. CONCLUSION

Though highly selective, this survey has outlined some of the more important aspects of multi-fiber spectroscopy. The sophistication of today's instrumentation and techniques contrasts strongly with the original "Medusa". It is exactly fifteen years since that instrument was used for the very first multi-fiber observations (December 1979). Though the run yielded spectra of only eight galaxies (Hill 1988), the explosion in survey-type spectroscopic data that is now imminent will owe its existence to that first pioneering step.

7. ACKNOWLEDGEMENTS

It is a pleasure to thank Richard Bingham, Brian Boyle, David Carter, Charles Jenkins, Ian Lewis, Quentin Parker, Ian Parry, Tom Shanks, Ray Sharples, Sue Worswick and Charles Wynne for useful discussions. The forebearance of the editor of these proceedings in accepting this late contribution is also gratefully acknowledged.

REFERENCES

Note: the abbreviations FOA1 and FOA2 refer, respectively, to Fiber Optics in Astronomy, S. C. Barden, ed., ASP Conference Series, 3, and Fiber Optics in Astronomy II, P. M. Gray, ed., ASP Conference Series, 37.

Arribas, S., Mediavilla, E. and Rasilla, J. L. 1993 In FOA2, 322.
Baranne, A. 1988 In Very Large Telescopes and their Instrumentation, M. -H. Ulrich, ed., European Southern Observatory, 1195
Barden, S. C., Elston, R., Armandroff, T. and Prior, C. P. 1993 In FOA2, 223
Barden, S. C. and Wade, R. A. 1988 In FOA1, 113
Bedding, T., Gray, P. and Watson, F. 1993 In FOA2, 181
Bingham, R. G., Gellatly, D. W., Jenkins C. R. and Worswick, S. P. 1994 In Instrumentation in Astronomy. VIII., D. L. Crawford and E. R. Craine, eds., Proc. SPIE 2198, 56
Dallier, R., Baudrand, J. and Cuby, J. G. 1993 In FOA2, 310
Dawe, J. A. and Watson, F. G. 1984 In Astronomy with Schmidt-type Telescopes, M. Capaccioli, ed., D. Reidel, Dordrecht, p. 181
Donelly, R. H., Brodie, J. P. and Craig, W. W. 1993 In FOA2, 270
Felenbok, P., Cuby, J. G., Lemonnier, J- P., Baudrand, J., Casse, M., Andre, M., Czarny, J., Daban, J- B., Marteaud, M. and Vola, P. 1994 In Instrumentation in Astronomy. VIII., D. L. Crawford and E. R. Craine, eds., Proc. SPIE 2198, 115
Gray, P. M. 1984 In Instrumentation in Astronomy. V., A. Boksenberg and D. L. Crawford, eds., Proc. SPIE 445, 57
Gray, P., Taylor, K., Parry, I., Lewis, I. and Sharples, R. 1993 In FOA2, 145
Haynes, R. and Parry, I. R. 1994 In Instrumentation in Astronomy. VIII., D. L. Crawford E. R. Craine, eds., Proc. SPIE 2198, 572
Hill, J. M. 1988 In FOA1, 77
Hill, J. M, Angel, J. R. P., Scott, J. S., Lindley, D. and Hinzen, P. 1982 In Instrumentation in Astronomy. IV., D. L. Crawford, ed., Proc. SPIE 331, 279
Hill, J. M. and Lesser, M. P. 1986 In Instrumentation in Astronomy. VI., D. L. Crawford, ed., Proc. SPIE 627, 303
Jenkins, C. R., Gellatly, D. W., Bingham, R. G. and Worswick, S. P. 1993 In FOA2, 209
Levin, K. H., Tran, D. C., Kindler, E., Glenar, D. and Joyce, R. 1993 In FOA2, 295
Levin, K. H., Tran, D. C. and Mossadegh, R. 1988 In FOA1, 23
Marx, S. 1994 In IAU Colloquium No. 148, Future Utilization of Schmidt Telescopes, in press
Nelson, G. 1988 In FOA1, 2
Owen, R. E., Siegmund, W. A., Limmongkol, S. and Hull, C. L. 1994 In Instrumentation in Astronomy. VIII., D. L. Crawford and E. R. Craine, eds., Proc. SPIE 2198, 110
Parry, I. R. and Gray, P. M. 1986 In Instrumentation in Astronomy. VI., D. L. Crawford, ed., Proc. SPIE 627, 118
Parry, I. R. and Carrasco, E. 1990 In Instrumentation in Astronomy. VII., D. L. Crawford, ed., Proc. SPIE 1235, 702
Parry, I. R., Lewis, I. J., Sharples, R. M., Dodsworth, G. N., Webster, J., Gellatly, D. W., Jones, L. R. and Watson, F. G. 1994 In Instrumentation in Astronomy. VIII. D. L. Crawford and E. R. Craine, ed., Proc. SPIE 2198, 125

Parry, I., Lewis, I. and Watson, F. 1994 Spectrum, No. 4, 20
Pitz, E. 1993 In FOA2, 20
Pitz, E., Lorenz, H. and Elsässer, H. 1993 In FOA2, 166
Ramsey, L. W. 1988 In FOA1, 26
Taylor, K. and Gray, P. 1994 In Instrumentation in Astronomy. VIII., D. L. Crawford and E. R. Craine, ed., Proc. SPIE 2198, 136
Taylor, K. and Parry, L. 1994 personal communication
Watson, F. G. 1984 MNRAS 206, 661
Watson, F. G. 1986 In Instrumentation in Astronomy. VI., D. L. Crawford, ed., Proc. SPIE 627, 787
Watson, F. G. 1994a In IAU Colloquium No. 148, Future Utilization of Schmidt Telescopes, in press
Watson, F. 1994b In Wide-Field Spectroscopy and the Distant Universe, (35th Herstmonceux Conference), World Scientific, in press
Watson, F. 1995, in preparation
Watson, F. G., Gray, P. M., Oates, A. P., Lankshear, A. and Dean, R. G. 1993 In FOA2, 171
Watson, F. G. and Parker, Q. A. 1994 In Instrumentation in Astronomy. VIII., D. L. Crawford and E. R.Craine, eds., Proc. SPIE 2198, 65
Watson, F. and Terry, P. 1993 Gemini, No. 42, 32
Worswick, S. P., Gellatly, D. W., Ferneyhough, N. K., Terry, P., Weise, A. J., Bingham, R. G., Jenkins, C. R. and Watson, F. G. 1994 In Instrumentation in Astronomy. VIII., D. L. Crawford and E. R.Craine, eds., Proc. SPIE 2198, 44

PERFORMANCE OF A 2048 X 2048 PIXEL THREE-SIDE-BUTTABLE CCD DESIGNED FOR LARGE FOCAL PLANES IN ASTRONOMY

J. A. Cortiula

Thomson-CSF Semiconducteurs Specifiques

ABSTRACT: TH78997M is a 2048 x 2048 pixel full-frame CCD sensor featuring 15 x 15 micron pixels, MPP operating mode, four parallel outputs and buttability on three sides. This new device makes it possible to build very large focal plane detectors for telescopes (two x n butted CCDs) with less than a 25 dead pixel zone between adjacent sensors. Other features are high full-well capacity (typically 180 Ke$^-$) and very low dark current thanks to four phase MPP clocking which allows a very long exposure time together with a high S/N ratio.

The readout of a complete frame is achieved through four parallel outputs running at frequencies ranging from ten KHz up to five MHz per output. The four on-chip amplifiers have been designed to be completely user controlled and to meet most of terrestrial and spaceborne astronomy requirements: very low noise (less than 4e$^-$ at 50 KHz, -40 degrees C), very low power consumption (150 micro-W/amplifier at 10 KHz/output), linearity better than 0.5% together with high conversion factor (4.2 micro-V/e$^-$) over 2.5 V output range. The packaging of the device is compatible with the good flatness of the chip and with the low parasitic "cosmic event" rate thanks to specific care in selecting the packaging material.

DISCUSSION

IWERT: To avoid possible confusion, I would like to point out that the presented frontside product, THX 7837M, differs considerably from the product, THX 7397M, under development by Thomson for ESO. The frontside product, THX 7837M, has four outputs. They are arranged that way because the operation of all outputs is needed to read out the full CCD imaging area. The backside product, THX 7397M, is an independent development for ESO, taking thinning into account right from the start. This device has two outputs in the conventional scheme. There are many other differences between those two product developments, which cannot be covered in this context. However the fabrication process technology used for those two products is the same and the results on cosmetics and amplifier similarity due to a clean process are certainly encouraging.

CCD MOSAIC DEVELOPMENT FOR LARGE OPTICAL TELESCOPES

G. A. Luppino, M. R. Metzger

Institute for Astronomy, University of Hawaii

and S. Miyazaki

Institute for Astronomy, University of Hawaii and
Subaru Telescope Project, National Astr. Obs. of Japan

ABSTRACT: We outline the recent developments in CCD imager technology aimed at producing the very large format (8192 x 8192 pixels and larger) detector mosaics required for existing 4-m class and new 8 to 10-m class telescopes. The key technology areas include buttable array design and buttable element packaging, and optimization schemes for QE and readout time. As an example, we highlight the University of Hawaii effort to develop an 8192 x 8192 15 μm pixel CCD mosaic.

1. SCIENCE WITH LARGE CCD MOSAICS

The need for larger and larger focal plane detectors has steadily increased due in part to two main technical advances: a) the steady improvement of image quality at observatories worldwide, and especially on Mauna Kea, and b) the demand for physically larger focal planes required for the next-generation 8 to 10-m telescope instruments. These are discussed in Sections 1.1 and 1.2 below.

1.1 Imaging

In recent years there has been a resurgence in "purely imaging" projects. For a time it was thought that direct imaging on large telescopes would give way to spectroscopy. This has not been the case. In fact a number of key scientific projects use both small (1 to 2-m) and large telescopes (4-m class) exclusively for imaging. Some current examples of such projects include: a) the surface brightness fluctuation (SBF) method for measuring distances to early-type galaxies out to 4000 - 5000 km s^{-1} (Tonry and Schneider 1988, Jacoby et al. 1992), b) the study of the dark matter distribution in clusters of galaxies from observations of gravitationally lensed arcs and arclets (Tyson et al. 1990, Kaiser and Squires 1993, Luppino et al. 1993), and, c) the search for baryonic dark matter (MACHOs) in our galactic halo from observations of gravitational microlensing of stars in the LMC (Stubbs et al. 1993, Griest 1991, also see Cook 1995 in these Proceedings). There are numerous additional scientific projects in galactic structure and solar system studies that, like the above, share a common need for large-format, high-QE detectors with high spatial resolution. By "large-format" we mean a detector with a field of view of exceeding 20'. By "high spatial resolution" we mean a detector that can adequately sample the seeing, which, for the best sites, implies a pixel size of order 0″.2. In general, the science falls into two main categories: a) surveys, where we require observations of large numbers of objects or where we search large fields for objects with intrinsically low space density, and b) observations of intrinsically large objects.

1.2 Detectors for Large Spectrographs

Since the physical size of a focal plane scales with the size of the optical system, we can naturally expect that the focal planes required for the 8 to 10-m telescope instruments will be quite large. The spectroscopic instruments planned for the Keck I and Keck II 10-m telescopes, the Subaru 8.2-m Japanese National Large Telescope (JNLT), and the Gemini 8-m telescopes will require detectors of order 100 mm format and larger. Moreover, for imaging applications it is often more cost effective and technically conservative to tile the large telescope focal planes with CCDs and bin pixels on-chip to achieve the desired pixel scale than to build extremely elaborate reimaging optics that may be more expensive than the CCDs, and may compromise the performance of the system. Most of the low resolution spectrograph designs call for at least a 4096 x 4096 CCD with 12 - 15 μm pixels. Some of more ambitious instruments as well as the high resolution, cross-dispersed spectrographs and large imagers demand even larger formats. For example, the Keck DEIMOS spectrograph under development at UCSC requires two 8k x 8k (15 μm pixel) mosaics that are thinned, AR-coated, and mechanically flat and co-planar to within ±five μm (see Stover et al. 1995 in these Proceedings).

2. THE FIRST-GENERATION CCD MOSAICS: THE 4096^2 DESIGNS

A number of groups have developed or are developing CCD mosaics with 4096 x 4096 pixels. While there are some notable exceptions (Tyson et al. 1992, Sekiguchi et al. 1992), the majority of these mosaic designs are based on a family of two- and three-edge-buttable 2048 x 2048 imagers designed by John Geary (e.g. see Geary et al. 1991) and fabricated at Loral Fairchild. The first of these close-packed 4096^2 mosaics were successfully employed by the MACHO collaboration in their search for baryonic dark matter through observations of microlensing of stars from the LMC (Cook 1995, in these Proceedings). Similar mosaic prototypes have also been fabricated by the University of Hawaii (Luppino et al. 1992) and by NOAO (the "mini-mosaic", Boroson et al. 1994). An observer instrument for the CFHT, MOCAM (Cuillandre et al. 1995, in these Proceedings), will see first light this Fall. Many other mosaics and mosaic cameras are in the design or fabrication phase (see the list in Luppino, et al. 1994). Fig. 1 shows an example of a typical first-generation 4096^2 mosaic. This drawing shows a two by two array of two-edge-buttable CCDs mounted on modified kovar packages.

3. STRATEGIES FOR LARGER MOSAICS

While a 4096^2 mosaic is certainly useful for a variety of observational projects, for many projects such a mosaic offers only marginal improvement over what one can accomplish with a single thinned Tek 2048 CCD. A 4096^2 mosaic with 15 μm pixels is only slightly larger (60 mm) than the monolothic Tek 2048 with 24 μm pixels (49 mm). In situations where the smaller pixels are of no advantage (e.g. direct imaging on the UH 2.2-m telescope), the thinned Tek 2048 CCD is superior to an unthinned 4096^2 mosaic. Thus, the UH 4096^2 mosaic was a stepping stone to the larger mosaics that we require for wide field imaging cameras and large format spectrograph readouts.

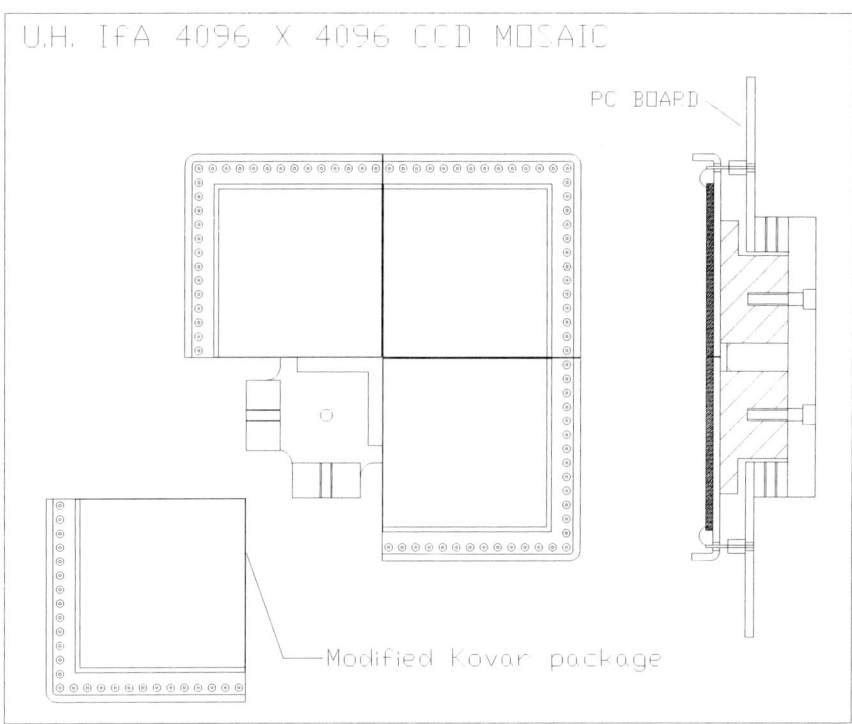

Fig. 1. The 4096 x 4096 CCD mosaic prototype design built by the University of Hawaii in 1992. This is the same design employed in MOCAM for the CFHT (see Cuillandre et al. 1995, in these Proceedings).

What approach can we take to move to the next level in mosaic design? Consider that our goal is to construct a detector with 8192 x 8192 15 μm pixels. When designing such a large CCD mosaic, a number of approaches can be taken, trading off mosaic tile size, gap size, number of amplifiers, readout time, etc. First, we address gap size. In some applications, especially wide field imaging, relatively large gaps are not considered a problem since deep images are often obtained by taking many unregistered, short exposures, thus "filling" the gaps while allowing removal of cosmic rays and improved flatfielding. For spectrograph focal planes, however, gaps can be a serious nuisance, especially for multiobject and cross-dispersed instruments. Therefore, for the remainder of this paper we consider only close-packed mosaics.

Next we consider the optimal CCD tile size. It is now routine to fabricate 2048 x 2048 15 μm pixel CCDs with excellent yield. Four devices can fit on a 100 mm wafer, and a foundry run will produce 80 devices, often with more than half the devices electrically functioning and a sizeable fraction of scientific quality. Furthermore, a variety of designs with two and

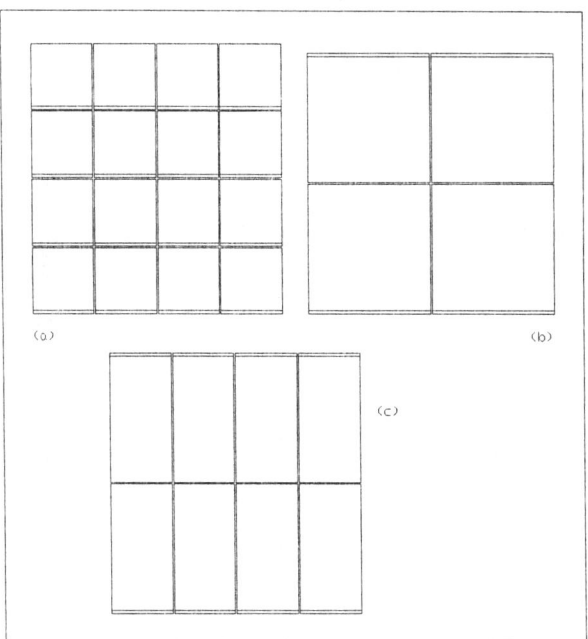

Fig. 2. Various strategies for constructing a close-packed 8k x 8k mosaic: (a) a four by four array of three-edge-buttable 2k x 2k CCDs, (b) a two by two array of three (or two) edge buttable 4k x 4k CCDs, and (c) a four by two array of 2k x 4k three-edge-buttable CCDs.

three-edge buttability and standard quad amp readout exist for these 2048 devices (Geary et al. 1991). One could assemble an 8192 x 8192 mosaic from 16 of these 2048^2 imagers arranged in a four by four array. Unfortunately, some or all of the gaps would exceed one mm, and unless four-side-buttable packaging were developed, the gaps could be considerable ($\geq$ 1 cm). A mosaic with 16 CCD elements would also be electrically and mechanically complex, requiring numerous readouts, and making alignment and flatness difficult to achieve (see Fig. 2a).

Considering the high yield on the 2048 devices, we could attempt to build a 4096 x 4096 two-edge or three-edge-buttable CCD with 15 μm pixels (Fig. 2b). This device would fill the 100 mm wafer and the yield would be considerably lower than for the 2048 CCDs. But the resulting 8192 x 8192 could be assembled from a simple two by two array of these devices in a scaled-up version of the first generation mosaics described in the previous section. Loral Fairchild has attempted on several occasions to build such a device. They have not been successful, however, in producing any scientific quality imagers. Rather than pursue this strategy, one could adopt an intermediate solution and choose a design that is midway between the buildable $2k^2$ imagers and the ambitious $4k^2$ imagers: a 2048 x 4096 three-buttable design with 15 μm pixels. Two of these devices would fit onto a 100 mm wafer, thus significantly increasing the expected yield. A close-packed $8k^2$ mosaic can be built from a four by two array of such imagers.

4. CRITICAL DESIGN ISSUES

In this Section, we touch on three key technology areas relating to large CCD mosaics: a) mosaic packaging, b) thinning and QE optimization, and c) readout time.

4.1 Mosaic Packaging

Packaging large CCD mosaics presents a considerable technical challenge. The simplest approach is to mount the CCD tiles on a common substrate, achieving small gaps, precise alignment and flatness in the mounting process. Such a scheme, however, permanently joins the CCD dies and precludes replacement of a mosaic element should it become damaged. Furthermore, CCD dies are usually selected for inclusion in the mosaic based on room-temperature wafer-probe tests, and many of these CCDs are not of scientific quality when operated at cryogenic temperatures. Thus, a mosaic that is permanently assembled from room-temperature-selected dies may contain a number of elements that do not work properly when cold, thereby destroying the entire mosaic. The preferred technique used in nearly all the first-generation, close-packed mosaics was to employ a custom-designed, two-edge-buttable package (see Luppino et al. 1992, Luppino and Miller 1992, Stubbs et al. 1993) that can be inserted and removed easily from a mosaic.

Although a three-edge-buttable package is somewhat more technically challenging than the two-edge-buttable designs, we can still adopt the same strategy for the larger mosaics as for the earlier mosaics; individual mosaic elements should be replaceable. The technical issues in the package design involve choice of materials, method of fabrication, and method for achieving electrical contact. Material choices include ceramics (e.g. aluminum nitride) and metals (Kovar, Invar, molybdenum). Aluminum nitride is an excellent choice for a package material, but it must be worked by the manufacturer. It is also non-conductive which precludes electrical contact to the CCD substrate if that is necessary. Kovar packages were employed almost exclusively in the first generation mosaics. It turns out, however, that Kovar is a poor thermal match to silicon, but was used because it is a common, commercially available microelectronics package material (Kovar's popularity in the microelectronic industry results not from any match to silicon, but because Kovar has a coefficient of thermal expansion precisely matched to the glass used to mount and insulate the pins in these packages, allowing such packages to be hermetically sealed and used over a wide temperature range). If we are to match the coefficient of thermal expansion (CTE) of silicon, we must choose an appropriate metal, such as Invar 36 (Szentgyorgyi 1993) or molybdenum. While Invar has the better CTE match, molybdenum is a far better thermal conductor (by nearly an order of magnitude), and may prove to be the better material overall.

In either case, the material can be machined to the required close tolerances using conventional (grinding) or EDM machining. An important consideration for the package design is maintaining focal plane flatness. If the large mosaics are to be used in fast optical beams (e.g. prime focus cameras of spectrographs) then the deviation from flatness cannot exceed of order a pixel size to maintain focus across the array. This specification places severe constraints on the mechanical package specifications which must be held flat to of order ± 10 μm.

4.2 Thinning and QE Optimization

It is well understood that to achieve the highest possible quantum efficiency from these CCD mosaics, the individual CCDs must be thinned, back illuminated and anti-reflection (AR) coated. Thinning exposes the bare silicon surface that can be anti-reflection coated to allow the incident photons to enter the device. Backside treatments are then necessary to ensure that the photoelectrons are actually detected and not caught in surface traps generated by the thinning process. While thinning conventional CCDs is hard enough, thinning edge-buttable CCDs is harder still. It is important that thinning techniques are developed that thin the entire device and do not leave a "lip" near the buttable edges. This may require full wafer scale thinning where the entire CCD wafer is thinned and processed before dicing the individual CCD elements.

An enormous amount of effort has been expended developing the technology for thinning CCDs. Much of the present day techniques have been pioneered by groups like Lesser's at the UA Steward Observatory, and we refer readers to the many papers in this area (e.g. see Lesser 1994 and references therein). While CCD quantum efficiency in the blue and mid visible depends on the AR-coating efficiency and the properties of the thinned backside interface, the QE redward of ~ 800 nm depends primarily on the AR-coating and the on the thickness of the thinned silicon membrane. Although the CCD has no response at wavelengths longer than 1100 nm, there is a scientifically critical area between 900 nm and 1100 nm where boosted CCD response will have enormous scientific gains. The one sure way to increase CCD response in this area is simply to build thicker devices (note, however, that we still need thinned, back illuminated devices so we can AR coat the backside for the highest sensitivity). Since we presently thin devices to around 15 μm, would it be possible to build devices with a 30 μm thick epitaxial layer (EPI) and thin to ~ 30 μm? This is certainly possible, and will result in higher red QE, but such a device built on conventional 10 - 50 Ω cm silicon will experience charge spreading since this thicker EPI will only be depleted a few μm deep, and charge generated in the "field free" region below the depletion region will diffuse laterally and contaminate neighboring pixels, resulting in a loss of image sharpness. On the other hand, we can build thicker devices and avoid the image sharpness tradeoff by building the CCDs on high resistivity silicon. The depth of the depletion region depends both on the gate voltage and the resistivity of the silicon. For a fixed gate voltage depletion depth is proportional to the square root of the resistivity. Thus to fully deplete a 30 μm membrane, we need to build CCDs on material with approximately ten times the resistivity of conventional CCDs (i.e. 1000 - 5000 Ω cm silicon). Such "deep depletion" CCDs have been pioneered by the X-ray astronomy community where high energy (5 to 10 keV) X-ray response is analogous NIR response (see Burke et al. 1991). An added advantage of thicker CCDs is the reduction of interference fringing.

4.3 Readout Time

As we attempt to build larger CCD mosaics, is crucial that we address the problem of readout time. For applications where we are detector noise limited, such as high dispersion spectroscopy or narrow band imaging, the lowest redout noise possible is essential. The lowest noise levels achieved with commercially-available CCDs are around 2.5 - 3 e$^-$. These noise levels are only possible, however, by reading out the CCD in "slow scan" mode with a pixel rate in the range of 20 - 50 kHz (kpixels/sec). For the large arrays we intend to build, however, such a slow readout speed is a severe limitation, since it leads to objectionably long readout times; nearly seven minutes for a single 2k x 2k CCD at 20 kHz. Long readout times present a serious efficiency problem in applications where one is taking many short exposures, such as broad band imaging. Most deep imaging observations are obtained using the "shift and stare" technique where one takes many short exposures while shifting the telescope slightly between the exposures, thus allowing one to build up a "flatfield" using the disregistered images. Taking many short exposures also helps with the removal of cosmic ray hits in the CCD images. When this technique is used on present-day 4-m class telescopes, the individual exposures are often limited to five to ten minutes each. On 10-m telescopes, it would be even better to use only one to two minute exposures. The limitation on the exposure time is dependent on the readout noise level of the CCD and the brightness of the night sky (we integrate until the sky photon shot noise dominates the detector readout noise).

Clearly, if we are taking many exposures, and reading out the CCD between those exposures, our observing efficiency is strongly affected by the CCD readout time. If the readout time is of order five minutes and our exposures are of order five minutes, then half the night is wasted reading out the CCD instead of collecting data from astronomical objects. One approach taken to offset this problem is to build CCDs with multiple output amplifiers, thus reducing the readout time by reading through two or four amplifiers in parallel. While this approach certainly solves the problem, it introduces other drawbacks that are only now being realized in recently constructed systems using multiple on-chip amplifiers. First, in a multiple amplifier CCD or mosaic, each separate amplifier must be calibrated. And that calibration must remain stable over time. The standard observational approach for calibrating one's data is to observe standard stars through the exact same telescope optics and atmosphere as the data were taken. But with multiple amplifier configurations, one needs to place standard stars on all regions of the CCDs to calibrate all the amplifiers. Trusting that one can calibrate with just one amplifier using previously determined relative amplifier gains is a risky business, and is not adequate for precise photometry. An additional problem is amplifier to amplifier crosstalk that can appear in multiple on-chip amplifier configurations. This can be a difficult problem in astronomical observations where objects of enormous brightness difference can be present on single CCD frames. We therefore argue that minimizing the number of amplifiers is desirable. But then how do we decrease the readout time? Clearly the best solution is to increase the readout speed, without increasing the readout noise beyond acceptable levels.

This goal is possible with present day technology. Manufacturers, however, must integrate the appropriate amplifiers into their designs. As an example of what can be done now, we show in Fig. 3 the readout noise as a function of readout speed for a MIT Lincoln Labs (MITLL) CCID-10 1024 x 2048 CCD measured in the UH IFA CCD Lab. Our CCD controller electronics uses a dual slope integrator implementation of a correlated double sampler circuit, and plotted on the x-axis of Fig. 3 is the dual slope integrator integration time. The conventional dual slope integrator integrates "up" on the reset level and "down" on the signal level, thus the minimum pixel time is twice the dual slope integrator integration time. Certain values for the corresponding pixel frequency (in kps or kpixels/sec) are also shown on the graph. This MITLL amplifier achieves a noise floor of 1.7 e^- rms per pixel at a conventional slow-scan rate of 50 kHz, but has a readout noise below five e^- at a speed as fast as 500 kHz. The readout time for a 2k x 4k CCD at 500 kpixels/sec is only 16 seconds! Clearly, amplifiers of this type should be incorporated into CCD designs that are intended for large format mosaics.

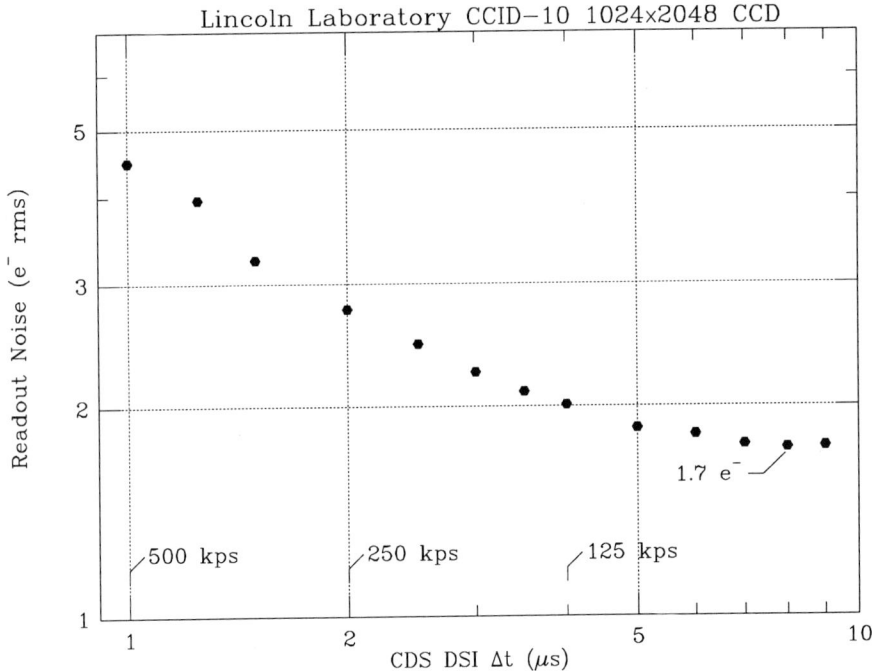

Fig. 3 This plot illustrates that low noise, high speed operation is possible with state-of-the-art CCD amplifier designs. The data were taken at the UH IFA CCD Lab with a MITLL CCID-10 CCD.

5. THE UH 8192 x 8192 CCD MOSAIC DESIGN

The University of Hawaii, Institute for Astronomy is presently building an 8k x 8k CCD mosaic camera for use on the various telescopes on Mauna Kea. In this Section, we will describe this camera as an example of a state-of-the-art astronomical CCD mosaic.

The UH 8k mosaic camera is based on a 2k x 4k three-edge-buttable CCD fabricated at Loral Fairchild (see Luppino, Bredthauer and Geary 1994). The device is designed to be three-edge-buttable with the single serial register running along a short (2048) edge. All of the bond pads are also confined to this edge. We can construct an 8192 x 8192 mosaic from a two by four array of these devices. The single serial register has a standard floating diffusion LDD amplifier at each end, and is split to allow the CCD to be read out of either amplifier or out of both simultaneously (note this is the same device design being used by NOAO in their 8 K mosaic project; Boroson et al. 1994). A single 2k x 4k mosaic element and its associated three-edge-buttable package is shown in Fig. 4. The CCD die is attached via electrically conductive sheet epoxy to a custom molybdenum package. Mounted adjacent to the CCD bonding pads is a small PC board containing a 25-pin micro-D connector. The CCD is wire bonded directly to the PC board. The CCD and package are designed so that the gaps in the resulting mosaic are less than one mm.

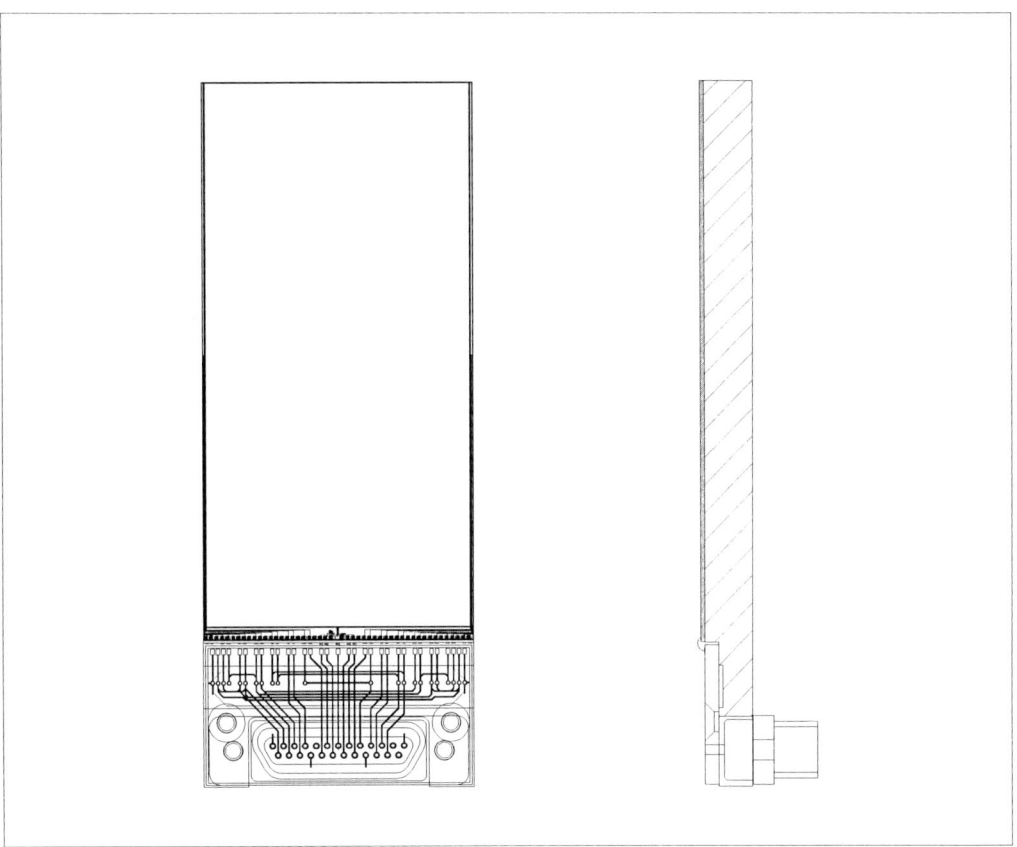

Fig. 4. The Loral 2k by 2k three-edge-buttable CCD and the buttable package design being developed at the UH IFA. The frontside illuminated CCD die is attached directly to a precision machined molybdenum carrier. A PC board containing a micro-D connector is also attached to the carrier and is wire bonded to the CCD.

The eight mosaic elements mount to a common base suspended from thermally insulated supports behind a 12 mm thick quartz window in a large LN2 cooled, vacuum cryostat. Each individual mosaic element will be precisely aligned using a microscope and precision alignment fixture. A scale drawing of the mosaic focal plane is shown in Fig. 5.

The first light instrument will be equipped with a large focal plane shutter and a two position filter slide for 150 mm square filters. Design and fabrication work is underway on these components. The lot run of CCDs at Loral Fairchild was reasonably successful, with the first 20-wafer lot (40 devices) produced 16 functional devices, 12 of which appear to be of scientific quality in room temperature wafer probe tests. We will shortly test these devices at cryogenic temperatures to select the eight best devices for inclusion in the mosaic.

Fig. 5. The University of Hawaii's 8192 x 8192 CCD mosaic. The resulting active area is 122.9 mm x 122.9 mm. The gaps are less than 1 mm.

CCD MOSAIC DEVELOPMENT FOR LARGE OPTICAL TELESCOPES

Fig. 6. A photograph of a non-functional 8k x 8k mosaic formed from eight dead 2k x 4k devices. This is shown for comparison to a 4096 x 4096 mosaic (at the left), a Tedtronix 2048 x 2048 CCD (at the top) and a Loral 4096^2 monolithic (15 μm pixel) device (at the right). Various other smaller CCDs of historical significance are also shown (TI 800^2WF/PC CCD at lower right, Fairchild 100^2 at upper left and to its right an RCA 512 x 320.

This 8k mosaic camera is a step toward future, improved versions large format detector focal planes. In its first incarnation, this camera will use frontside illuminated CCDs with a peak QE of only ~40%. Thinning is an option for future mosaics, as is deep depletion. We will explore the compromise between readout noise and readout speed, but we have not yet incorporated new on-chip amplifier designs that are optimized for low noise at higher speeds. Nevertheless, we intend to use this mosaic camera for several wide field imaging projects on the CFHT 3.6-m, the UH 2.2-m and the UH 0.6-m telescopes where the substantial increase in field of view will outweigh the compromises in other areas.

REFERENCES

Boroson, T., Reed, R., Wong, W.-Y. and Lesser, M. 1994 Proc. SPIE 2198, 877
Burke, B., Mountain, R., Harrison, D., Bautz, M., Doty, J., Ricker, G. and Daniels, P. 1991 IEEE Trans. Elec. Dev. 38, 1069
Cook, K. H. 1995 in IAU Symposium No. 167, New Developments in Array Technology and Applications, A. G. D. Philip, K. A. Janes and A. R. Upgren, eds., Kluwer Academic Press, Dordrecht, p. 285

Cuillandre, J. C., Murrowinski, R., Crampton, D., Mellier, Y., Luppino, G. and Arsenault, R. 1995 in IAU Symposium No. 167, New Developments in Array Technology and Applications, A. G. D. Philip, K. A. Janes and A. R. Upgren, eds., Kluwer Academic Press, Dordrecht, p. 213

Geary, J. C., Luppino, G. A., Bredthauer, R., Hlivak, R. J. and Robinson, L. 1991 SPIE 1447, 264

Griest, K 1991 ApJ 366, 412

Jacoby, G. et al. 1992 PASP 104, 599

Kaiser, N. and Squires, G. 1993 ApJ 404, 441

Lesser, M. 1994 SPIE 2198, 782

Luppino, G. A., Bredthauer, R. and Geary, J. 1994 SPIE 2198, 810

Luppino, G. A. and Miller, K. 1992 PASP 104, 215

Luppino, G. A., Jim, K. T. C., Hlivak, R. J. and Yamada, H. 1992 SPIE 1656, 414

Luppino, G. A., Gioia, I., Annis, J., Hammer, F. and Lefevre 1993 ApJ 416, 444

Sekiguchi, M., Iwashita, H., Doi, M., Kashikawa, N. and Okamura, S. 1992 PASP 104, 744

Stover, R. J., Brown, W. E., Gilmore, D. K. and Wei, M. 1994 SPIE 2198, 803

Stover, R. J., Brown, W. E., Gilmore, D. K. and Wei, M. 1995 in IAU Symposium No. 167, New Developments in Array Technology and Applications, A. G. D. Philip, K. A. Janes and A. R. Upgren, eds., Kluwer Academic Press, Dordrecht, p. 19

Stubbs, C., Marshall, S., Cook, K., Hills, R., Noonan, J., Akerlof, C., Alcock, C., Axelrod, T., Bennet, D., Dagley, K., Freeman, K., Griest, K., Park, H.-S., Perlmutter, S., Peterson, B., Quinn, P., Rodgers, A., Sosin, C. and Sutherland, W. (The MACHO Collaboration) 1993, SPIE 1900, 192

Szentgyorgyi, A. 1993 Harvard CfA/SAO Technical Memorandum

Tonry, J. and Schneider, D. 1988 AJ 96, 807

Tyson, J. A., Bernstein, G., Blouke, M. and Lee, R. 1992 SPIE 1656, 400

Tyson, J. A., Valdes, F. and Wenk, R. A. 1990 ApJ 349, L1

DISCUSSION

IWERT: I think it is worth mentioning that the bottleneck in large CCDs is (and probably will be) the thinning of those devices, although many small groups are working on it. We can certainly build giant mosaics of frontside illuminated CCDs, but which additional options besides the well known groups do you see for thinning in the future?

LUPPINO: There are vendors, most notably SITE, who claim they are developing thinned 2k by 2k CCDs. And of course, there is the Foundry Kesser route. But we don't have to wait for thinned devices to do good science.

TINBERGEN: To add a point to your list of science drivers: I think photometry will benefit greatly from over-sampling and slower "camera" beams.

LUPPINO: Yes I agree. I have a very "personally biased" list of science drivers.

HIGHLIGHTS OF IAU SYMPOSIUM No. 167

Sandro D'Odorico

European Southern Observatory

1. ARRAY DETECTORS: FROM THE FAR UV TO SUBMILLIMETER WAVELENGTHS

The talks at this IAU Symposium have illustrated the spectacular development which has taken place in the last decade in the field of array detectors for astronomy. Just a few years ago it was possible to speak of two-D detectors for the UV-red wavelength range only. At this meeting we have witnessed presentations on array characteristics from the extreme UV (Bonanno 1995), through the blue-visual range (D'Odorico 1995, Jorden and Oates 1995, Iwert 1995 and Luppino et al. 1995); the infrared 1 to 5 μm window (McLean 1995, Finger et al. 1995, Gilmore et al. 1995, Fazio 1995, Glass et al. 1995 and Ueno et al. 1995); the 10 - 20 μm window (Fazio 1995, Gezari 1995) and finally to an array of bolometers to operate at submillimeter wavelengths (Moseley 1995). Field imaging and spectroscopy are now possible across this entire energy spectrum and some of the first exciting astronomical results obtained with these devices have been presented here.

The infrared region between 1 to 5 μm is the one where the progress has been most impressive. Arrays of 256^2 pixels, InSb-based by SBRC and NICMOS MCT by Rockwell are now in regular operation at different observatories. Still, their operational characteristics are not yet fully investigated (see the presentation by Finger et al. (1995) for a flavor of the present status). Experience with CCDs shows that only by the use by different scientists on different programs can the detector behavior be fully explored and its capabilities well understood. The Shift-and-Add and Weighted SAA techniques to enhance the image quality discussed by Finger et al. (1995) and Ueno et al. (1995), respectively, are examples of potential, added-value applications of infrared detectors. In the spectroscopy field the user experience with the new arrays is even more limited because to our knowledge there are just two spectrometers based on 256^2 detectors which started operation in 1994 (KSPEC of the University of Hawaii and the upgraded CRSP of NOAO), although many are in the construction phase. The two principal suppliers of infrared detectors for astronomy, Rockwell and Santa Barbara Research Corporation, have also announced the availability in 1995 of 1024^2 arrays. First observations with a prototype 1024^2 HgCdTe detector from Rockwell (device name: Hawaii) have been already obtained. If these large arrays become indeed off-the-shelf products in 1995, this can lead in the infrared to the embarrassing situation that astronomers will not be able to fully exploit the capabilities of the new arrays because the development time of a matched cryogenic instrument is at least three years.

2. A SOBER VIEW OF THE STATUS OF CCDs FOR ASTRONOMY

Charged Coupled Devices remain, even in this era of expanding capabilities for infrared detectors, the devices with which the largest fraction of astronomical data are collected at ground-based telescopes. In an ideal world, astronomers would like to have for their instruments CCDs with sizes up to a 50 cm^2, pixel sizes from 15 to 50 μm, QE close to 100%, negligible read-out-noise and dark current and, last but not least, affordable prices. It is tantalizing to see that for some of these characteristics we are close to realizing the ultimate goal but there are also areas where progress has been slow. After the disappearance of RCA from the market of manufacturing chips for astronomy it has been difficult to find a stable supplier of high-efficiency CCDs. Tektronix, now SITe, has been the main source of thinned devices in the 512^2 and 1024^2 formats, but progress was slow on the 2048^2 devices. ESO had foreseen the use of a 2048^2 thinned CCD in the 1985 design document of the multi-mode instrument EMMI for the Nasmyth focus of the 3.5-m New Technology Telescope. The instrument was completed and installed in 1990 but the planned detector could be implemented in 1994 only, that is nine years after it was announced as an off-the-shelf device. Clearly one of the reasons for the slow progress by industry has been the lack of investment and coordinated efforts by the scientific community on this crucial component. We should be careful to make sure that this does not happen again on other crucial detector developments.

At the beginning of this decade a new approach to CCD procurement to circumvent both the high cost and the lack of flexibility in the CCD industry was proposed to the astronomical community. In the new scheme the astronomers and engineers responsible for detector definition and procurement actively participate in the chip design using standard CAD packages running on PCs. After the preparation of the corresponding mask, a production run (typically 20 four-inch silicon wafers) is commissioned to a silicon industrial laboratory which acts as a foundry on a best effort basis. J. Geary at the Harvard Smithsonian Center pioneered this approach and has been the main contributor to the design of new chips. Loral South (the former Ford Aerospace) has been the most effective foundry of CCDs for astronomy. The working chips are identified by on-wafer testing, again in collaboration between the final user and the manufacturer. The selected chips are finally packaged and bond-wired. They are however front illuminated devices with a low QE. The subsequent steps, thinning, surface treatment and coating, have also to be subcontracted to a specialized laboratory. M. Lesser of the Steward Observatory of the University of Arizona has set up a well equipped thinning and coating laboratory with the associated testing facilities and has been the most active in this area producing in the last five years a number of chips with very high QE. Many astronomical groups, including ESO, have gone all the way down this route with generally satisfactory results. It has to be remarked however that the quality of the chips produced on this best effort basis has been sometimes disappointing and the thinning process has encountered unexpected difficulties.

A few years down the road, one can recognize that the new approach has opened the way to new ideas and provided a number of devices which found useful applications at the telescopes, but has yet to prove itself as the reliable source of a significant number of large, high quality CCDs which the astronomical community needs. We certainly hope that the success rate will increase in the future but the community should also make sure that the industrial suppliers maintain an interest in this field. A personal view of the status of some key issues on CCDs as emerged from the presentations and discussions at this meeting is given below.

2.1 CCD Size and QE

In the last few years we have often heard about a coming boom in the size of CCDs which could be used for astronomy. The CCD mosaic projects now in operation (Macho, MOCAM and the Tokyo Camera, just to quote the ones presented here) use however front illuminated CCDs of low QE. The somewhat disappointing truth is that if you want to buy a 2048^2 high efficiency CCD off the shelf, there does not exist a single manufacturer which will accept the order and guarantee the specifications (SITe). A large number of dedicated efforts are "almost" there. M. Lesser is working intensively on the thinning of three-side buttable 2048^2 CCDs produced by Loral South and has been already successful on a few devices of comparable size. Stover et al. (1995) have presented here the work on 4096 x 2048 devices produced by Orbit which he hopes to thin in a newly set-up detector lab at Lick Observatory. ESO has a running contract with Thomson CSF for the delivery of 2048^2, 15 μm pixel, three-side buttable thinned CCDs. Negotiations for the procurement of similar devices are going on with EEV and Lincoln Lab. It is likely that in 1995 - 1996 some of these developments will be successful and the first mosaics with high performance devices will be assembled.

The results on the intrapixel variations of the QE presented by Jorden and Oates (1995) deserve also special attention. They have to be taken into account in evaluating the photometric accuracy and the positioning accuracy to be extracted from a given CCD-camera combination.

2.2 Read-Out-Noise

Values of 2-5 e^- have been achieved at different observatories using chips from SITe, EEV and Thomson. The use of the so-called skipper on-chip amplifiers introduced a few years ago should permit a further reduction of the read-out-noise by multiple readout of the pixels and subsequent averaging of the values. To my knowledge, there has not been any astronomical use of these devices (mounted on a number of Loral CCDs) yet. The sampling of the pixels required to achieve low values of the read-out-noise is relatively long. This leads to overall read-out times of largest format CCDs of the order of three to four minutes. At present this time can be reduced only by the use of multiple outputs read-out, with the disadvantage of different calibration parameters for the different sections of a device and sometimes crosstalk between the different channels. High speed, low-noise amplifiers are now becoming available as reported by Luppino et al. (1995) in his talk. They are likely to become the preferred solution to the read-out speed problem in the future.

2.3 CCD Controllers.

The introduction of 2048^2 CCDs in the standard operation of most major observatories and the prospect of even larger mosaics of CCDs has driven the development of many new controllers based on the use of transputer modules and DSP. At this meeting we had the opportunity to hear directly on the status of different projects (ARCON from CTIO, SDSU from San Diego, ACE from ESO and the controller for the Italian Galileo telescope) and to witness a lively discussion on the relative merits of transputers and DSP. Equally interesting developments for the control of infrared detectors were reported by Finger et al. (1995) (ESO system IRACE) and Glass et al. (1995) (The Rutherford-SAAO controller). As a user, my impression is that there is no unique technical solution to the problem of building a fast, flexible CCD controller and the present technology does provide all the necessary tools to do the job. At the end the relative success of one system with respect to the other will depend more on factors like friendly user interface, reliability and simplicity of operation.

3. HOT TOPICS FOR FUTURE DISCUSSION

The symposium has provided an exciting, updated view of the current development in the field of array detector applications. Optical and infrared arrays play however such a central role in so many fields of modern astronomy that it is impossible to cover all applications in a single meeting. I give below my personal list of topics which, although of relevance, have not been discussed extensively at this meeting and should be kept in mind for future gatherings.

Astrometry was discussed in poster papers only. With the introduction of large size CCD detectors and mosaics of CCDs we can expect that CCD-based astrometry will become of increasing relevance. It would be of particular interest to discuss the requirements set by astrometric programs on CCD cameras and on the specific properties of the detectors.

At large telescopes the combination of better and larger detectors and improved image quality has opened the way to the deepest surveys at optical and infrared wavelengths ever done in astronomy. Special techniques are used to collect and combine several images and to reach the faintest magnitudes. A discussion of the results obtained so far and on the sources of errors is needed to identify the parameters of the detectors to be improved and the best technique for make further progress in this field. With single optical detectors now approaching the 4096^2 format and infrared detectors at the 1024^2 size, one of the more challenging problem in astronomical instrumentation is that of data transmission, preprocessing and archiving. The constraints are particularly severe in the infrared where the detectors have usually to be read at high frequency and through multiple outputs to avoid pixel saturation. The talks of Blecha (1995) and that of Cook (1995) on the CCD mosaic used for the Macho project have given some hints on these problems but the subject deserves much more attention in the future.

There was no talk dedicated to photon counting devices as detectors for the blue-visual spectral region. It is true that the applications where the use of CCDs is advantageous have increased because of their high QE and reduced read-out-noise, but photon-counting detectors are still the only detectors to be used when high time resolution is required, such as very fast photometry and speckle imaging. This type of observation is the key to many major astrophysical problems, so photon-counting devices are not to be forgotten, at least until they can be efficiently replaced by ultra-fast read-out CCDs.

REFERENCES

All the references are to IAU Symposium No. 167, New Developments in Array Technology and Applications, A. G. D. Philip, K. A. Janes and A. R. Upgren, eds., Kluwer Academic Pub., Dordrecht.

Blecha, A. 1995 IAU 167, p. 57
Bonanno, G. 1995 IAU 167, p. 39
Cook, K. H. 1995 IAU 167, p. 285
D'Odorico, S. 1995 IAU 167, p. 9
Fazio, G. G. 1995 IAU 167, p. 93
Finger, G., Nicolini, G., Biereichel, Meyer, M and Moorwood, A. F. M. 1995 IAU 167, p. 81
Gezari, D. Y. 1995 IAU 167, p. 97

Gilmore, K., Rank, D. and Temi, P. IAU 167, p. 79
Glass, I. S., Sekiguchi, K. and Nakada, Y. 1995 IAU 167, p. 109
Iwert, O. 1955 IAU 167, p. 67
Jorden, P. R. and Oates, A. P. 1995 IAU 167, p, 27
Luppino, G. A., Mezger M. R. and S. Miyazaki 1995 IAU 167, p. 297
McLean, I. 1995 IAU 167, p. 69
Moseley, H. 1955 IAU 167, p. 95
Stover, R. J., Brown, W. E., Gilmore, D. K. and Wei, M. 1995 IAU 167, p. 19
Ueno, M., Tsumuraya, F. and Chikada, Y. 1995 IAU 167, p. 117

Section II - Poster Papers

Chairmen: Discussion Sessions

Session A:	Ian S. McLean

Session B:	A. G. Davis Philip

The Poster Paper discussions will be found at the end of Session II, starting on page 377.

THE TRANSPUTER BASED CCD CONTROLLER AT ESO

Roland Reiss

European Southern Observatory

ABSTRACT: A new controller concept based on transputer modules (TRAMs) and Digital Signal Processors (DSPs) has been developed at ESO. The Array Control Electronics (ACE) can handle single CCDs or CCD mosaics with 1 to 64 outputs and with up to 256 clocks with software programmable independent high and low voltages. The same basic concept will be used for slow scan scientific and fast scan auto guiding/wave front sensing applications. While the transputers are mainly used for data acquisition, communication and supervisory tasks, a DSP with very simple external logic performs the time-critical CCD clock pattern generation. The low noise, high speed clock driver with built-in telemetry is designed for highest operational safety. A newly designed video processor board has four video channels each with its own 16-bit ADC. ACE can easily be interfaced to various types of host computers like PCs, Unix workstations or ESO's VME-based LCUs (Local Control Units with VxWorks real-time operating system).

1. KEY FEATURES

ACE is host independent and works with PCs, workstations or VME systems. It has a multi-transputer controller and a DSP-based sequencer with reduced complexity. The DSP acts both as a co-processor for CCD timing and CCD Hardware control (DAC setup, gain switching, Shutter control). The hardware structure allows for multiple windowed readout with individual gain settings for each window. The system is built around classical microprocessor bus structure (ACEbus). The bus does not carry any CCD specific signals. The ACEbus is based on a standard VMEbus P1 backplane. No firmware, Transputer and DSP programs and CCD clock patterns are loaded from the host disk. There is an Electronic ID for each board with I2C-bus EEPROMs. The modular concept allows growth with detector arrays. The same concept supports technical applications (auto guiding, wave front sensing) by exchanging the video board. There are up to 256 (384 planned) clock phases per system and 4 - 64 video channels with separate ADCs. The data rate to the host is 1.6 MBytes/sec per transputer link (up to 25 MBytes/sec total). The power consumption is low, 25 watts for a single CCD. The system is safe, operating from 24 V DC. It is of compact size, (standard configuration) 150 x 150 x 300 mm^3 + two VME slots.

2. SEQUENCER/CONTROLLER

The sequencer/controller is a T805 with four MByte RAM. A T225 is interfaced to a 20 MHz DSP56001 with 96 KByte RAM (commercial product from Perimos, Germany). Buffered DSP data, address and control pins are available on the bus backplane. The DSP wait state logic is in one 44-pin MACH PLD and is integrated in a RS-422 interface for the host link. A single high precision (25 ppm) clock generator (5, 8, 10, 20, 40 MHz) eight opto-isolated inputs and outputs for shutter control and external synchronization. The I2C-bus controller is an IC (Philips PCD8584).

3. DC AND CLOCK DRIVER BOARD

There are eight DC outputs: four unipolar (0 V..25.5 V) and four bipolar (-12.8 V..+12.7 V). There are 16 bipolar clock outputs (-12.8 V..+12.7 V). Each output has its own eight-bit DAC (40 DACs in five ICs in total). Reference voltages (0, 1, 10 V) are software selectable. The driver can be turned off under software control for "on-the-fly" CCD change. No transients appear on the CCD lines during power-up (typical less than 100 mV). There are fast and equal (!) rise and fall times (typical 5 ns for 10 V amplitude). A built-in, 32 channel telemetry with a dedicated 12-bit ADC is connected directly to the DC and clock outputs. A hardware selectable board has addresses (0 - 15); 16 boards are individually addressable. In software selectable address mode (0 - 3) up to four boards can be cascaded to generate 64 independent clocks; up to 16 boards can operate in parallel.

4. 16/20-BIT VIDEO PROCESSOR

There area four channels with separate 16-bit ADCs (Crystal CS5101A). Data acquisition is done with a T225 transputer. The instrumentation amplifier input has software programmable gain. There is DC restoration with pixel clamp and a CDS with proven dual slope integrator design. The ADC offset is directly fed into the integrator. No extra opamp stage is required! A special 20-bit mode has two integrators in parallel with different time constants per channel connected to the two ADC inputs. The time penalty is only 2.5 sec for a 1024 x 1024 quad readout CCD. The video gain range is 1 - 128 in eight steps (1,2,4,..,128). There is an integrated RS-422 interface for an optional data link. A hardware selectable board has addresses (0..15); 16 boards are individually addressable. Under software address modes (0..3) are selectable; four groups of CDS timing are possible. There is an I2C-bus EEPROM for identification and/or setup storage.

5. AUXILIARY FUNCTIONS BOARD

This board contains shutter and temperature control with a dedicated transputer. There is temperature control with a digital control loop via 16-bit DAC and ADC. Four channels measure temperatures with telemetry readout and direct readout via voltmeter.

6. POWER SUPPLY

DC/DC converters generate all the necessary supply voltages (+5 V, ±15 V, +30 V) from a 24 V input. There is a common mode for input filters and a differential mode for output filters. There is a controlled power-up and power-down sequence and a voltage supervisory circuit with "power good" output.

7. FUTURE PLANS

We plan to expand the clock driver to 24 clocks per board. We will use faster DSP (56002) with 40 MHz, a single fiber link for multiple (four) transputer links using the TAXI chip set and a Local transputer graphics board for real-time display.

DETECTOR CONTROLLERS FOR THE GALILEO TELESCOPE: A PROGRESS REPORT

G. Bonanno, P. Bruno, R. Cosentino,

Osservatorio Astrofisico di Catania

F. Bortoletto, M. D'Alessandro, D. Fantinel,

Osservatorio Astronomico di Padova

A. Balestra and P. Marcucci

Osservatorio Astronomico di Trieste

1. INTRODUCTION

The GALILEO telescope requires a flexible detector controller that allows one to drive single and mosaic CCDs in different modes of operation. The CCD controller can be divided into two main parts: the preamplifier, located close to the cryostat and the other modules, the Sequencer, the Bias generator, the Clock driver, the Preamplifier and the Correlated double sampler, located into a rack and connected together via a CCDC-BUS.

2. MAIN FEATURES OF THE CCD CAMERA

For the TNG focal planes, we have chosen the Loral three-side buttable CCDs (2048 X 2048 pixels, 15 μm pixel size). Each CCD can be read on two outputs. The controller is able to drive 2 x 2 mosaics and the readout time is compatible with a single chip system. The image is at once reformatted to simulate a single sensor having the full mosaic size.

3. CCD SEQUENCER

The functions supported by the CCD SEQUENCER are: a clock waveform generation for CCD and signal processor; setting the time and monitoring of CCD bias levels (clocks and analog levels); handling of the local data-buffer and telemetry; file making up; commands reception and handling and shutter and temperature-controller handling.

Possible CCD readout modes are: full frame, only a set of predefined boxes with fast skip of unwanted pixels, with binning, drift scan mode. In drift scan mode, synchronization can be provided externally. The heart of the CCD sequencer is a standard TRAM module (DTM560) containing a couple of high speed processors: a 16 bit T222 INMOS transputer and a 24-bit 56001 MOTOROLA DSP. The Transputer and DSP communicate data and application SW through a set of shared memory locations. Thanks to the simple reconfigurable networking scheme embedded on the transputer, fast communication of data and commands can be easily implemented with little extra hardware.

Sequences can be generated in two DSP ports: PORT A that allows 16 independent sequencer status lines and PORT B that allows 12 independent sequencer status lines. The status and the delay loop are derived from a given table. An eight-bit counter is programmed with the upper eight bits of the DSP-24 bit word while the lower 16 bits are used as programmable sequencer lines. The counter holds the DSP in WAIT state until the counter rollover; each tick corresponds to 100 ns (the DSP clock is 20 MHz), so a maximum delay of 25.6 μs can be programmed. Longer delays, when needed, can be obtained by conventional SW loops.

4. PROGRAMMABLE BIAS GENERATOR

The board that generates the Bias levels has the following features; 32 buffered output voltages providing at least 10 mA (These are programmable through 8-bit D/A converters with serial loading (DAC 8800 PMI); 16/1 multiplexers with selection latch allowing the choice between 32 analog signals and a 12-bit ADC (ADC 674A Burr Brown) with 15 μs conversion time used for telemetry; DSP data-bus used for ADC programming and telemetry.

The board is a single euro card and is realized by using "Surface Mount Device" technology. The DSP 24-bit data bus is used for programming the DACs and for receiving data from the telemetry section. The addressing and the control logic for D/A programming and multiplexers driving are also provided. The DAC output is unable to drive other devices directly, so we utilize operational amplifiers (OP490) to buffer the output.

5. CLOCK DRIVER

The clock driver board performs the level translation and buffering of signals coming from the sequencer. Low and high levels are programmable and their values are monitored by telemetry. The circuit employed for the voltage level generation is the same as the bias generator board. Fast switches (AD7512), in conjunction with a fast buffer (LM6321) are used to drive the CCD. To introduce fixed rise and fall times a resistor-capacitor circuit is used.

6. CORRELATED DOUBLE SAMPLER

The correlated double sampler (CDS) performs the subtraction between the baseline signal and the CCD output signal. Two operating modes are allowed: a. The integrator performs either the integration, for a fixed amount of time, and the difference of both signals. The difference is converted by the ADC, b. Both signals are integrated and converted, the binary difference is performed by transputer. The ADC module, that consists in a size two TRAM format board, is plugged in the CDS board. This solution allows the replacement of different ADCs, with different resolution and conversion times.

THE RUTHERFORD-SAAO CCD CONTROLLERS AND THEIR APPLICATIONS

I. S. Glass, D. B. Carter, G. F. Woodhouse,

South African Astronomical Observatory

N. A. Waltham and G. M. Newton

Rutherford Appleton Laboratory

1. THE DEVICE

We describe a versatile array controller developed at RAL and SAAO. The original concept was due to Waltham, van Breda and Newton (1990). A Transputer-based microcomputer forms the heart of the device.

The T225 Transputer offers the following advantages:

a) Fast output of 16-bit words using a "block move" command;
b) Built-in serial communication using Transputer links;
c) Possibility of parallel processing - useful in high-backgrounds such as may be encountered in infrared work.

The following cards have been developed:

a) Transputer microprocessor
b) Clock boards with potentiometer-controlled voltage levels
c) Clock boards with digitally controlled voltage levels
d) Video boards with eight-bit A/D for acquisition cameras
e) Two types of video boards with 16-bit A/D converters
f) Filter-wheel controller
g) Temperature, shutter and communications card.
h) Balanced pre-amplifier
i) Power supply in separate unit
j) Link/PC adaptor (with Transputer buffer)

The program and CCD drive waveforms are normally stored in EPROM on the Transputer board but can be downloaded over the link. Timing is controlled by the Transputer's own cycle and 16-bit instructions can be issued every 200 ns over the data bus. The instructions are decoded on the cards into clock and A/D control information.

The controllers can be regarded as independent devices which act on simple control codes received over the link, returning the digitized CCD output as well as certain engineering information. In our applications, each controller is connected by a single bi-directional fiber-optic link to a 486 PC which the observer operates and which stores the data on disk and DAT tape. As well as this, the PC can display the data on a Super-VGA screen and do low-level professing. The Transputer is programmed in OCCAM and the PC in Gnu C (the latter allows 32-bit addressing).

2. APPLICATIONS

64 x 64 Philips HgCdTe Infrared Camera: This detector is sensitive from 1 to 4 microns. Exposures in the L' band are read out twice per second and summed in the PC. The chip is of sandwich construction. Because of the chip's problems with high dark current in many pixels, poor pixel geometry, threshold effects and low quantum efficiency, this camera is only competitive in the L' band where its well depth of 10^7 electrons is very useful.

RCA 320 x 512 CCD: This liquid-nitrogen cooled CCD was originally controlled from a hard-wired controller and Nova minicomputer and was the main imaging instrument at SAAO. It has been upgraded to use one of the new controllers. Its noise ($85e^-$) is completely dominated by the chip. It is now semi-retired.

Tek 512 x 512 CCDs: These liquid-nitrogen cooled chips have replaced the RCA chip for most routine imaging at SAAO. Their read noise is 12-13 electrons for a complete read time of 20 sec. A camera using a Tek 1024 x 1024 chip is under development.

Acquisition TVs: The acquisition cameras make use of frame-transfer CCD detectors from the EEV CCD 02 - 06 range, having 385 x 288 pixels and are cooled to 228° K by 3-stage Peltier devices. Coated blue-sensitive chips have replaced the original commercial-grade sensors because of complaints from blue star observers. The VGA display is highly interactive, offering movable cursors and features such as seeing-size measurement. Sensitivity is altered by changing the exposure time and/or the color look-up table. Images can be saved to disk. Autoguiding can be performed on stellar images. There is no shutter and timing is accomplished by making use of frame transfers to define the start and stop times of the exposures.

PtSi Infrared Camera: A large format PtSi camera has recently been commissioned. This requires two clock cards and has four independent readout channels. It is described in the contribution of Glass et al. (1995) at this conference.

REFERENCES

Glass, I. S., Sekiguchi, K. and Nakada, Y. 1995 in IAU Symposium No. 167, Advances in Array Technology and Applications, A. G. D. Philip, K. A. Janes and A. R. Upgren, eds., Kluwer Academic Pub., Dordrecht, p. 109

Waltham N. R., van Breda I. G. and Newton G. M. 1990 Proc. SPIE 1235, 328

CCD IMAGERS WITH ENHANCED UV SENSITIVITY FOR INDUSTRIAL AND SCIENTIFIC APPLICATIONS

G. I. Vishnevsky, M. G. Vydrevich, L. Yu. Lazovsky, V. G. Kossov and
S. S. Tatautshchikov

"Electron" Research Institute, St. Petersburg

ABSTRACT: A family of virtual phase (VP) CCD array image sensors for various industrial and scientific applications has been designed, fabricated and tested. All share the common concept of a "2.5-phase" photosensitive cell, combining the advantages of known "1.5-phase" VP devices (increased quantum efficiency, especially in the UV, and radiation hardness) with the simpler fabrication process and extended functionality of three-phase devices.

1. PHOTOSENSITIVE "2.5-PHASE" CELL

The cell structure is similar to a three-phase CCD where a third level of polysilicon is replaced by a structure formed by N-type implant ("virtual well") followed by a shallow P-type implant forming a "virtual gate". Comparing the manufacturing process of this structure with that of the "1.5-phase" VP CCD, two important moments should be noted: it takes only two additional implants to form the "2.5-phase" device, compared to four necessary for "1.5-phase"; and both implants are self-aligned, with polysilicon electrodes acting as a mask.

2. VP CCD SENSORS FOR SLOW-SCAN TV SYSTEMS

These devices are intended for applications where a wide spectral range, uniform photoelectric response, high resolution, and low noise performance are essential. They have been designed primarily for operation at low readout frequencies and low temperatures (-40° C and lower), however, they can readily operate at room temperature and readout frequencies up to 10 MHz. All these devices have a peripheral drain surrounding photosensitive area to protect it from stray charge generated at the chip periphery, and a built-in active load of the first stage of two-stage on-chip preamplifiers that can be switched off to prevent light emitting effects during integration. The basic design features of the family are presented in tables on the next page.

3. CCD COOLING SYSTEMS

The laboratory develops and manufactures various kinds of CCD cooling systems for both ground and space applications. The CCDs can be supplied with coolers providing the required thermal conditions for the devices installed into equipment, meeting the ultimate requirements to weight, size and the power consumption of the cooler (heat tubes, heat accumulators and so on). We are experienced in the design of the CCD imagers with Peltier coolers, cryostats, open space heat radiators.

TABLE 1

Design Parameters

CCD Type	ISD 017A	ISD 015A	ISD 011A	ISD 034A	ISD 048A	ISD 049A
# of Vertical pixels	1040	520	512	258	290	576
Hor.	1160	579	512	256	386	576
Format	FT		FF		FT	
Pixel Vert.	16	18		16		22
size Hor.	16	24		16		22
# Registers	2		1			
Numb. clocked phases						
Image Sect	2		2		2	
Storage Sect	2		-		3	
Register	2		2		3	
Number readout ampl. in every register		2			1	

TABLE 2

Typical Performance Characteristics

Parameter	Units	Value
Saturation signal	mV	500
Full well	ke	130 (ISD011A,17A,34A)
		220 (ISD015A,48A,49A)
Optical response non-uniformity	%	3
Sampled optical response non-uniformity	%	1.2
Dark signal @ -40° C	e$^-$/pix/s	10
Charge transfer inefficiency in any direction (by Fe55 source)		1x10^{-5}
Readout noise @20 kHz		
HS output	e$^-$ rms	10
LN output	e$^-$ rms	7

THE LARGE-FIELD BRIGHT-STAR HIGH-PRECISION CCD PHOTOMETER OF BAO

Shiyang Jiang

Beijing Astronomical Observatory, Academy of Sciences

1. INTRODUCTION

Time-series high-accuracy photometry is very important for research in stellar variability. For a long time photometry made by a photomultiplier was the only instrument for high precision stellar photometry. To overcome the atmospheric variation and instrumental problems, we must choose at least one stable star as a comparison star and move the telescope quickly between the targets. So the real efficiency is very low and one only can do it on photometric nights. To overcome this limitation, since 1989 we began to cooperate with the team of the STEPHI network. We used the Chevreton four-channel photometer which can observe the variable, two comparison stars and a chosen sky background simultaneously. The multi-channel photometer is much better than normal single channel photometer as we can see from the several STEPHI results. Now the very high quantum efficiency CCD becomes more and more popular, so we are trying to change to use CCDs. Here we give some general description of a large field high accuracy bright star CCD photometer being prepared for the Beijing Astronomical Observatory (BAO).

2. THE PRIME FOCUS OF THE 60-CM REFLECTOR OF THE XINGLONG STATION OF BAO

The 60-cm reflector of the Xinglong Station has a prime focus of F/4. But the original correctors were not good enough, hence we asked Dr. M. L. Yi of Nanjing Astronomical Instruments Research Center to redesign and rebuild a new corrector. It is a four element, K9 glass made, lens system. The largest spread image size is less than 0".8 within a field of view of one degree. Just this August we finished the installation of this new corrector system and got some CCD frames with a SBIG ST6 CCD as a test.

3. THE CCD SYSTEM

The CCD camera we chose is a TC-215 type thermoelectric cooling system made by Dr. Z. W. Zhao at the National Observatory of Japan when he was a visitor there between 1992 to 1993.

For fast photometry and a smaller data base, we will add window readout type software. That means we will make a square box for each object which we want to measure and only read those boxes and save them to the disk. In this way the read out time can be limited to 40 ms and the data base is only 0.1% of the full read of the chip.

4. SOME CALCULATIONS

a) The signal to noise ratio and the limiting magnitude: If D is the diameter of the telescope, t is the light efficiency of the telescope, a is the light transparency of the atmosphere, f is the light efficiency of the filter, c is the quantum efficiency of the CCD, T is the integration time, w is the wave band width, V is the magnitude of the objects, I is photons per angstrom per square cm per second of a zero magnitude star, then we have:

$$\text{Signal} = (\pi/4) D^2 (tafcwI)T \times 2.512^{-V}$$

$$\text{Noise} = \sqrt{\text{Signal} + (\pi^2/4) D^2 (tafcwI)TM^2 \times 2.512^{-22.0} + 64M}.$$

Here M denotes the total pixels in each box. To get a high S/N we defocus the stellar image to about 11" or M = 78. Under this condition, for V = 15, on a moonless night, with T = 100 s we can have a S/N better than 100, so it is the limiting magnitude for high accuracy photometry.

b) The dynamic range: When we defocus the stellar image to 11", we can get about 4.5 magnitudes range for high accuracy photometry. That is to say we can work on V = 5 - 9.5 with T = 1 s or V = 10 - 14.5 with T = 100 s or somewhat in between.

c) How many stars are there in each frame to a limit of V = 15? As we know generally, in the average each square sphere degree on the sky, will have one or two stars in the V band brighter than 8 mag, our CCD on the sky will cover about 0.09 square degrees, so there will be only a 9% chance to find a star brighter than 8. But for finding a star brighter than 12 mag, the chance is larger than 500%. Within the sky area we cover, one can have more than 100 stars in V brighter than 15. So for high accuracy photometry of bright stars, in most situations we can find good comparison stars.

5. RESEARCH PROJECTS

a) δ Scuti stars and asteroseismology research: Due to the low quantum efficiency and short integration time, in most situations, we only work on stars brighter than V = 10. By using a CCD, we can have a two to three magnitude gain. Hence it is very useful if we choose some clusters as a target, so that for variable stars we can not only easily get some frequencies for known variables but also we will have many chances to find new variables.

b) Faint Be star observations: Up to now we have worked only on bright Be stars, for which it is very difficult to find comparison stars within short distances. We can not get a high enough photometric accuracy. Using a CCD, one can work on clusters, and find many new faint Be stars. For these objects we can rather easily find comparison stars in the same field and get photometry of high accuracy

c) Supernova survey: For the supernova survey, the CCD will be used. The two pixel image is 2.14" which matches well with the normal seeing disk of our site. For a 300 s integration, the detected limiting magnitude can be as faint as V = 22. This is very useful for super nova surveys among galaxies.

EFFECTS OF SHUTTER TIMING ON CCD PHOTOMETRY

D. Galadí-Enríquez, C. Jordi and E. Trullols

Dept. d'Astronomia i Meteorologia, Universitat de Barcelona

CCD-shutters take some time to open and close. This results in a difference between the recorded and real exposure times. The pixel-to-pixel pattern δ(x,y) in which this difference varies depends on the characteristics of the device. These timing errors affect photometric measurements in two ways: directly, since the recorded exposure time is not the real one, and indirectly, via the flatfielding process, because the shutter effect is mixed up with the flatfield pattern. The direct effect may be negligible when exposure times are long. The indirect effect is present irrespective of the exposure times of the frames, if flatfield exposure times are short.

Generalizing Stetson (1989), to calculate the function δ(x,y) we take a set of n flatfields, $f_i(x,y)$, with recorded exposure times τ_i, n and another flatfield, F(x,y), with recorded exposure time T, all of them obtained under the same illumination conditions. Defining the ratio:

$$R(x,y) = \frac{\sum_{i=1}^{n} f_i(x,y)}{F(x,y)}$$

we can compute the shutter map as

$$\delta(x,y) = \frac{R(x,y)T - \sum_{i=1}^{n} \tau_i}{n - R(x,y)}$$

Shutter effects can be removed from any given image Z(x,y) with recorded exposure time t_n, by performing the calculation

$$\hat{Z}(x,y) = Z(x,y) \frac{t_n}{t_n + \delta(x,y)}$$

which yields the corrected image $\hat{Z}(x,y)$. Another approach to shutter-timing errors was proposed by Surma (1993).

Our study of the CCD-shutters on the 1.23-m and 2.2-m telescopes of CAHA at Calar Alto (Almeria, Spain), shows that shutter-timing influence is not negligible. For a 10 s exposure, δ(x,y) yields a difference of 0.010 mag between the most and less exposed pixels, and a global error of 0.037 mag due to the general over-exposure of all the frame.

REFERENCES

Stetson, P. B. 1989 Highlights of Astronomy 8, 635
Surma, P. 1993 A&A 278, 654

ASTRONOMICAL APPLICATIONS OF CCDs IN HUNGARY - THE FIRST STEPS AND FUTURE PLANS

G. Szécsényi-Nagy

Eötvös Loránd University

1. INTRODUCTION

The first century of non-visual astronomical image detection in Hungary was devoted exclusively to photography. The most famous Hungarian astronomers and astrophysicists of the last century were the pioneers of astrophotography too. The very first exhaustive description of photography as a scientific method was the one by Konkoly-Thege about astronomical photography in 1887. In his book one can find some interesting images printed of the plates taken by his friend and colleague, the enthusiastic astrophotographer J. Gothard, who was the first to demonstrate that the well-known bright planetary in Lyra (M 57 or the Ring Nebula) contains an extremely faint central star. His discovery, which opened new fields for astrophysics, was contested for a long time because his foreign contemporaries were unable to repeat this feat of arms.

Astrophotography remained, since then, the number one method in collecting observational evidence for scientific studies. World-known results of the Hungarian astronomical programs range from discoveries of minor planets, comets and flare stars to the first observations of novae and extragalactic supernovae. Photospheric and chromospheric features of the Sun were also amongst the targets of Hungarian sky-photographers during past decades.

2. THE ADVENT OF ELECTRONIC IMAGING

The first step into the field of electronic imaging was done during the late 1980s. The capabilities of photographic and photoelectric image detecting techniques were compared and as a result of it the introduction of CCD technology was suggested (especially to spectroscopic investigation and H-alpha photometry of dK and dM stars (Szécsényi-Nagy 1990 and 1994). Political changes made possible then the purchase of state-of-the art cooled CCD cameras, at least in principle. The budget of astronomical institutes remained the last restriction in planning this step of modernization and unfortunately this limit has been fixed at a very low level. Consequently it is no wonder that the first cameras were simple cheap commercial products. These systems have TV-image resolution and very low dynamic range (the A/D conversion offered is normally eight bits). As has been found at the Baja Observatory e.g., the uncooled EDC1000HR CCD camera fits the needs of astronomical demonstration and education. Possible targets are the Moon, the Sun and the planets or a bright comet (Hegedüs 1994). The University of Szeged (JATE) bought a chilled ST-4 CCD Star Tracker/Imaging Camera. This device is so light that it could be mounted on a 63/840 mm visual refractor (Telemator made by Carl Zeiss Jena). The ST-4 contains a very small chip (its active area is only 2.5 mm x 2.5

mm) and its short focal length is definitely advantageous when the program needs an extended field of view. The power of the system has been tested on globular clusters, planetary nebulae and galaxies. The photometry of the SN in M 81 and that of a faint mira (DQ Vul) has been performed by this device also (Szatmary 1994). Their next step is to purchase a more sophisticated version of that product an ST-6. The JATE team intends to use this new system for the photometry of pulsating variables.

Our institute, the Department of Astronomy of the Eötvös University, received a thermoelectrically cooled eight-bit CCD camera too. Unfortunately the chip of the device had been constructed for TV cameras. The greatest problem is that the CCD has an interlaced structure and their pixels are not square. Furthermore its spectral sensitivity is negligible in the violet and UV range relative to the dark noise of the device. Otherwise it is a reliable system. For comparison we were able to record an image of the M 57 with an exposure which is shorter than 1% of the photographic exposure that Gothard had to apply a century ago. Another drawback of the camera is the lack of any optical filters. This also limits the scientific use of the system. Our present efforts are aimed at ordering a more sophisticated camera with at least 12 bit A/D conversion and a larger chip (1 Megapixel), better cooling and a computer controlled filter wheel with a set of UBVRI, H-alpha and H-beta filters.

ACKNOWLEDGEMENT

The author acknowledges the partial support of this research by the NSRF (grants: OTKA-T 7595 and OTKA-U 15909).

REFERENCES

Hegedüs, T. 1994, private communication
Szatmary, K. 1994, private communication
Szécsényi-Nagy, G. 1990 Star Clusters and Associations, Balazs, B., and Szécsényi-Nagy, G., eds., ELTE, Budapest, p. 55
Szécsényi-Nagy, G. 1994 Astronomy from Wide-Field Imaging, MacGillivray, H. T. et al., eds. Kluwer Academic Pub., Dordrecht, p. 61

A SIMPLE CCD-SYSTEM FOR SECONDARY ALIGNMENT OF THE SPECTRUM-UV SPACE TELESCOPE.

L. V. Didkovsky, N. V. Steshenko, P. I. Borzyak and A. I. Dolgushin

Crimean Astrophysical Observatory

ABSTRACT. A CCD-system is proposed to maintain the alignment of an optical system of a large space telescope during ground tests and after launch. The system is able to detect alignment errors resulting from displacements of the main and secondary mirrors of the telescope and also from the tilts of these mirrors. Computer simulations have been performed to determine the sensitivity of the system to the displacements and/or the tilts. Different sources of errors have been also explored on a working stand; and the main requirements to the elements of the system have been determined. The system is to be installed on the space telescope "Spectrum-UV". It can be also used in ground-based telescopes in a quasi-real time mode.

1. INTRODUCTION

A new space Ritchey-Chrétien telescope "Spectrum-UV", which is currently being designed, will be equipped with spectrometers installed at a significant distance (50 mm) from the center of the field of view. Therefore, any displacement or tilt of the secondary mirror (SM) may strongly affect the quality of the image in the spectrometers, even if the disturbances do not change the quality of the image significantly in the center of the field of view. Our task was to develop a system which would maintain the alignment of an optical system of a telescope, the Secondary Alignment System (SAS).

2, INSTRUMENTATION

The SAS system must detect different breaches of alignment and separate those connected with decentering of the SM from breaches caused by tilts of the SM. The optical scheme of the SAS (Fig. 1.) consists of two measuring channels. The first channel includes the source of light S1, placed on the main mirror (MM), the flat mirrors M1 and M3 manufactured right on the SM as one of its parts, and the spherical mirror M4, which is manufactured as a part of the MM and serves for focusing the optical beam into the CCD. This channel is practically insensitive to any transversal motions of the SM. It detects only tilts of the SM. The second measuring channel consists of the source of light S2, placed on the MM, the spherical mirror M2 manufactured as a part of the SM, and the same CCD. The second channel is sensitive to both the displacements and the tilts of the SM. We have used a CCD camera with 256 x 288 array and 8-bit ADC. Using the CCD as a sensor, we have been able to determine the position of the center of the image with high precision in a rather wide range of displacements. It seems also to be important that the positions of the optical beams can be stored in the form of the X, Y coordinates of the beam rather than as electrical levels, that makes maintaining of the alignment more reliable.

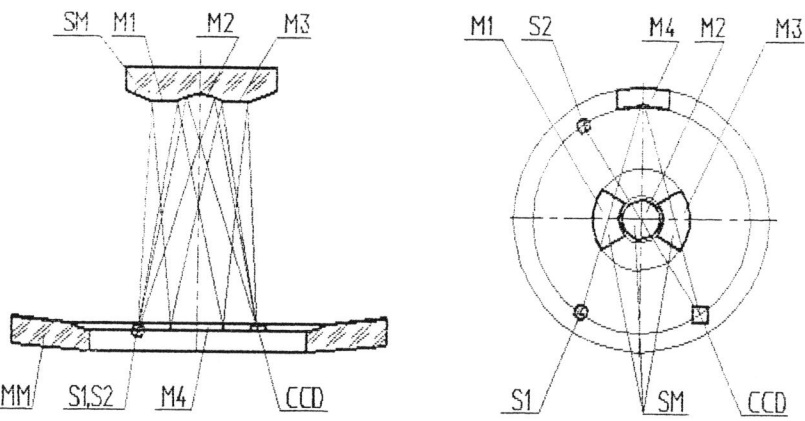

Fig. 1. Optical scheme of the Secondary Alignment System.

3. MODELING AND TESTING

We have performed computer simulations of the optical scheme of the SAS studying responses of the system in the whole possible range of the tilts and displacements of the SM. All technical tolerances for the components of the system have been taken into account; and all necessary requirements for manufacturing have been set up. We have also analyzed the main factors which affects precision of the determination of the alignment breaches. The sensitivity of the system, which is defined as minimal measurable displacement of a lightspot on the CCD, has to be about 10 μm to provide the required quality of image. However, our laboratory measurements have demonstrated that the real sensitivity of the system is significantly higher even under ground conditions.

4. CONCLUSION

We have developed a Secondary Alignment System for the space telescope "Spectrum UV", which will enable us to follow every breach in co-axiality between the main and the secondary mirrors with settled precision. The information about any alignment breaches will be used during ground-based testing of the telescope. The system gives also ability to restore the ground-made alignment and to reach the required image quality for the whole width of the field of view of the telescope after its launch.

THE MODERNIZATION OF THE PULKOVO PHOTOGRAPHIC (PHOTOELECTRIC) VERTICAL CIRCLE BY A CCD ARRAY

G. A. Goncharov, B. K. Bagildinsky, E. V. Kornilov, D. D. Polojentsev,
K. V. Rumyantsev and V. D. Shkutov

Pulkovo Observatory

The Zverev photographic vertical circle (PVC) of the Pulkovo observatory is in the process of modernization. The features of the vertical circle are:

a) Maksutov mirror-lens optical system with small aberrations and wide passband: aperture: 20 cm, focal length: 200 cm, focal scale: 103 arcsec/mm.

b) Very compact instrument: 140 cm total length, 60 cm - tube.

c) Wide field: 25 x 25 mm = 40' X 40'. Wide-field imaging can be combined with meridian observations.

d) Easily-reversible instrument: reversing takes less than 30 seconds.

e) Two divided vertical circles of glass. Photoelectric circle reading microscopes.

f) Photographic micrometer in focal plane. This will be changed with a CCD micrometer.

The Photoelectric Vertical Circle makes possible, as do usual vertical circles, the determination of stellar declinations in an absolute manner. The observed zenith distance is a linear combination of micrometer, circle and level readings before and after the reversing of the instrument. Therefore all zero-points are eliminated and the final value of the zenith distance is free from variations of the instrumental system orientation between observations.

There are three versions of the new CCD micrometer for the Photoelectric Vertical Circle:
a) An array of three CCDs arrays, 1040 x 1160 pixels, pixel size of 16 x 16 μm, covers a field 27 x 31 arcmin. In this case two arrays with filters register in two bands from U to I and a third array registers in a wide passband.

b) An array of two such CCDs in which the first is covered by two filters and the second works in a wide passband.

c) One array registers all bands with changed filters.

The CCD array allows us simultaneously to observe in the three bands (during one passage of an object). With one CCD we have to make the observations during two passages (nights).

We intend to observe ALL stars and other compact objects (extragalactic, asteroids, planet satellites and artificial satellites) from 7th to 18th, up to the pole in declination.

The aims are:

a) Absolute and differential determination of both coordinates will be made by classic and/or modern methods which use CCD abilities (Hoeg 1995, Goncharov 1995). A determination will be made directly with respect to an extragalactic reference frame. A link of Hipparcos/Tycho stars with quasars is possible.

b) Photometric determinations will be made of color indices for the purpose of elimination of chromatic refraction and for astrophysical studies.

The expected accuracy of the differential coordinate determinations is 50 mas and of the photometric determinations is 0.01 - 0.1 magnitude. A special light mark near the array and equipment on the tube will allow us to control instantly the flexure of the tube. Improvements in accuracy are expected by means of network of environmental and instrumental sensors of temperature, pressure, humidity, currents, voltages and signals.

TABLE 1

Some results of the PVC's modernization.

	Current PVC	Upgraded PVC
Observed Quantity	DECLINATION	BOTH COORDINATES AND PHOTOMETRY
MAGNITUDE LIMIT	8	18
ACCURACY, MAS	180	50
OBJECTS IN FRAME	1	1500
OBJECTS: TOTAL	2000	300,000,000
OBSERVATIONS/YEAR	3000	70,000,000

ACKNOWLEDGEMENT

This research was carried out with financial support from the Russian Foundation of Fundamental Research, grant # 93-02-3056.

REFERENCES

Hoeg, E. 1995 in IAU Symposium No. 166, Astronomical and Astrophysical Objectives, Kluewer Academic Pub., in press

Goncharov, G. 1995 in IAU Symposium No. 166, Astronomical and Astrophysical Objectives of Sub-Millisecond Optical Astrometry, Kluwer Academic Pub., Dordrecht, in press

A NEW OCULAR MICROMETER FOR THE MAHIS

T. R. Kirian,

Pulkovo Observatory

V. S. Korepanov and

Institute of Precise Mechanics and Optics

V. M. Grozdilov

Pulkovo Observatory

A working model of an ocular micrometer has been designed for the MAHIS (Meridian Automated Horizontal Instrument by Sukcharev). An optical scheme of the micrometer includes the following devices:

a) Artificial light marks in the focal plane of the objective. The marks have an increased sharpness and have a stable scale factor under defocusing. An amplitude-phase grid illuminated by coherent light from the second main point of the objective is used to make the marks..

b) The observed stars and artificial marks are imaged on the CCD chip by means of an additional objective. This objective also corrects chromatic aberations of the main objective.

c) A concentric meniscus is used to compensate for the chromatic refraction of the atmosphere. The meniscus center of curvature coincides with the center of the system's pupil image. In this case the compensation is equal at all points of the field of view.

d) The possibility of measuring the normal attitude of the flat mirror relative to the main instrumental plane during each observation is discussed. For these purposes there needs to be a holographic grid on the mirror surface, an artificial zenith or nadir horizon and an autocollimated source of light in the focal plane.

e) Laboratory investigations of a working model of a CCD camera are being carried out and the basic software is being developed. This work is planned to be finished in the autumn of 1994. The CCD matrix ISD 011A (NPO "Electron", St. Petersburg, Russia) 512 x 512 pixels, with pixel size of 16 x 16 mkm is used. The noise is 5 electrons/pixel/sec when the temperature is -40° C (thermoelectric cooler).

The basic CCD software is being adjusted in the following way. First the CCD control and information processing algorithms are to be debugged with help of an IBM PC (the software is written in the C programming language with the use of floating point codes). Then, after a real CCD image has been obtained, the algorithms are optimized for the specialized signal processor TMS320C3x (also with the use of a C-compiler). Then, if necessary, they are transformed into fixed-point TMS' codes. The DSR-adapter used with its TMS320C31 processor and with the 8-Mbyte RAM allows real-time work for the two chosen modes of the micrometer. The first one is the momentary readings of the CCD frames and the second one is so called the "time delay and shifts" mode (the TMS320C25 signal processor allows only the fixed frame mode). More detailed information concerning the new MAHIS CCD camera will be available after tests and refinements of the CCD software.

PbS AND CCD ARRAY AUTOCOLLIMATION MICROMETERS FOR THE INFRARED MERIDIAN CIRCLE

V. N. Yershov

Pulkovo Observatory

A new infrared meridian instrument is being developed at Pulkovo Observatory. The main purpose of the instrument is to extend the fundamental coordinate system to the K-infrared waveband and to faint stars at visual and I-wavebands. The instrument has a 30-cm primary mirror made from astrositall. An intermediate focal plane is used to introduce luminous reference marks. One can obtain autocollimated images of the marks at the intermediate focal plane with the use of a polished chamber located around the central hole of the primary mirror. The secondary mirror of the telescope forms images of the marks and of their autocollimated counterparts and passes them to the plane of a photodetector (Fig. 1.). The luminous marks give a reference frame for the measurements. These measurements are not affected by displacements of any optical unit placed after the intermediate focal plane or by displacements of the detector. The measurements are done relative to the coordinates of the average between positions of the luminous mark and its autocollimated image. Any small constant difference between the center of curvature and the optical axis position can be determined in the laboratory.

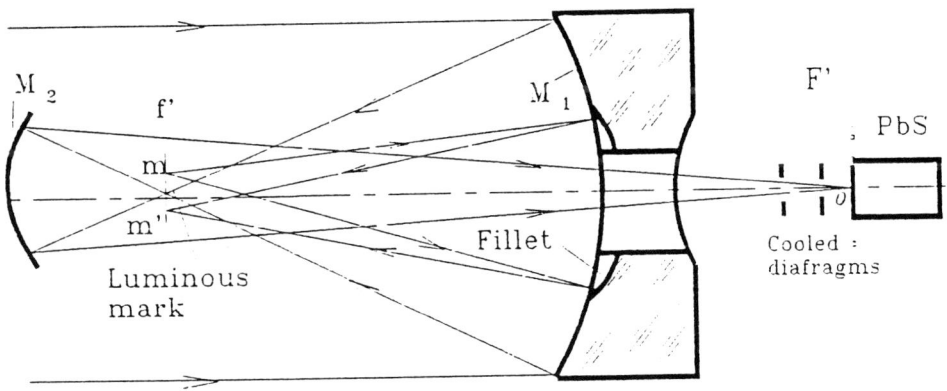

Fig. 1. The autocollimation micrometer with the intermediate focal plane.

The circle reading system of the instrument is rigidly connected to the optical unit. Two long-focus collimators measure the orientation of the sitallic unit relative to the unmovable optical axes by means of autocollimation from the lateral flat surfaces of the unit. So, both zenith distances and transit times of stars are determined relative to these fixed optical axes.

The intermediate focal plane allows one to use conventional construction for the infrared detector with its liquid nitrogen cryostat and cooled diaphragms and filters. Moreover, one may create additional focal planes in the micrometer by splitting the light beams into different colors. Two wavebands are planned to be used in the micrometer to obtain parameters of the atmosphere to be taken into account during the observations.

The visual and the near infrared wavebands (I) will be covered by the front illuminated CCD array with its 768 x 580 pixel array with square 27 micron pixels. The device will be cooled by Peltier thermoelectric elements. The sensitive area of the CCD is 21 x 16 square mm, and the total registration time will be 60 seconds. With the use of a "time delay and shift"" mode of signal accumulation one may expect the limiting magnitude of about I = 16 - 18.

The infrared K-waveband will be registered by a PbS 64 x 64 array (developed by the Electron Research Institute, St. Petersburg). The detector is in a coordinate addressing mode with a continuous layer of thin lead sulfide film. It is made by a process of chemical precipitation of poly-crystal, lead sulfide, onto a silicon substrate having the address bus.

Information is read by switching on part of the photoresistant layer, at the intersection of the selected column on the row, and by registration of the current passing through this part of the layer. For the PbS detector the time delay is a few dozens of milliseconds. The detector is working in a frame integration mode. Such a construction gives high sensitivity to the detector and a wide dynamic range (60 db). The spectral range of the detector is 1.0 to 3.0 micron and the working temperature is 90 K. The quantum efficiency is about 30%. The threshold level of the detectable signal is $(0.5 - 1.0)E^{-13}$ W/pixel (under the assumption of a S/N ratio of about one, and an integration time of 10 ms). The pixel size is about 50 x 50 micron and the pixel period is 60 microns. The total registration time is about ten seconds for these parameters of the detector and the telescope focal length is 5 m. The limiting K magnitude is estimated to be K = 5 - 10, depending on the star's spectral class (S/N = 3 is assumed).

This work was supported by the Russian Fund of Fundamental Investigations (92-02-17095)

THE NOAO CCD CONTROLLER - ARCON

Alistair R. Walker

Cerro Tololo Inter-American Observatory,

The Array Controller (Arcon) developed at CTIO is a flexible and extendable system for acquiring astronomical data from a wide variety of optical and IR detectors. It has been designed from the outset to support expansion to more demanding applications such as large CCD mosaics. In its minimum configuration it can read a quad-readout CCD through all four amplifiers simultaneously.

To provide real time performance and portability, most of the software executes within the Arcon, rather than the host, in a network of parallel processors (Inmos Transputers) which is easily extended as more demanding applications arise. Transputers (rather than DSPs) were chosen since they provide excellent hardware support for real time programs (e.g., rapid scheduling instructions, timers, eight DMA driven I/O channels), a simple, efficient and uniform inter-process and inter-processor communication mechanism, unified software development and debugging tools for the entire processor network, and a wide range of standardized, compact and relatively low power hardware subsystems from a variety of third party vendors.

The entire controller is compact enough to be housed in a single small enclosure permanently attached to the CCD dewar, connected only to AC power and a pair of optical fibers carrying commands, status and data to a small network of transputers ("Tram box") located near a SUN workstation. These commercial transputer modules are interfaced to the SUN by a single Sbus card. Most of the Arcon software runs in an Inmos transputer in the Tram box. This processor is dubbed the "Nexus" since it is the communications hub of the system. It accepts high level requests from the host and generates the appropriate subsystem commands and status requests to set-up the detector, run waveforms, and read back image data. It assembles header information in FITS format and stores the incoming image in a 2.5 MB software FIFO, plenty large enough to absorb the worst case Unix latencies. Another transputer serves as an intelligent display controller which presents a 1280 x 1024 image at full resolution as it is read out, performing extensive real time processing to allow all quadrants to be viewed with the same intensity scaling, while removing the overscan from the center of the image, orienting it to match finding charts and marking saturated pixels in red.

The IRAF-based user interface in the Unix workstation seamlessly integrates data acquisition and reduction. Observational parameters are set using the familiar IRAF parameter editor (epar). Data acquisition, telescope and instrument commands may be mixed with normal IRAF commands in CL scripts. Scripts have been provided which automate observing operations such as focusing and exposure sequences through various filters, as well as performing CCD characterization and controller diagnostics. Status information is divided into several windows so the observer can tailor the information displayed as desired. Both status windows and control windows may be opened or duplicated on consoles or terminals anywhere on the network. This versatility allows for efficient use of the system by single or multiple observers, and enables remote diagnostics by staff not present at the telescope.

A Waveform Definition Language (WDL) compiler and plotting software have been developed, which make it practical to customize waveforms for each detectors. WDL allows run time control of characteristics which were formerly considered to be "hard wired", such as output amplifier selection. It also allows the automation and acceleration of engineering tests which have formerly been prohibitively time consuming.

Other unusual features of the hardware are very low power consumption, DC coupled video to improve overscan behavior, the widespread use of surface mount technology, built-in test signal and data generators and extensive self-diagnostics, and a modularization which allows board replacement without manual adjustment.

Arcon controllers are now responsible for taking the majority of CCD data at CTIO, with implementations onto Thomson 1k, Tektronix 1k and 2k CCDs and a Loral 3k CCD. Most of these CCDs have four output amplifiers; as an example, with a Tektronix 2k CCD it takes 30 seconds from initiation of readout to having the image stored on disk, in IRAF format, with 4.5 e^- read noise. Experience has shown that the Arcon controllers are reliable, robust and immune to interference, with scientific performance greatly superior to that of CTIO's previous generation of controllers. The system has proven popular and easy to learn. We are presently converting our remaining CCDs to Arcon controllers. Meanwhile the existing controllers at KPNO are being upgraded with the Arcon software and digital hardware, and the full Arcon has been adopted for new applications such as the WIYN telescope, and the Gemini 8-m telescopes. An expanded version of the controller will be used to operate the 8k x 8k mosaic array (eight CCDs, 16 amplifiers) presently under development at NOAO.

The building of Arcon controllers has been a team effort: Tom Ingerson (project manager), Alistair Walker (project scientist), Roger Smith, Rolando Rogers, Eduardo Mondaca, Andy Rudeen (hardware), Dan Smith, Gary Webb, Jim Hughes, Pedro Gigoux and Steve Heathcote (software) have all made major contributions. But principally, the Arcon controller owes much of its success to the vision, the technical proficiency, and the driving enthusiasm of Roger Smith.

INTENSIFIED ELECTRON-BOMBARDED CCD IMAGES FOR INDUSTRIAL AND SCIENTIFIC APPLICATIONS

I. Dalinenko, G. Vishnevsky, V. Kossov, L. Lasovsky, G. Kuzmin and A. Malyarov

"Electron" Research Institute, St. Petersburg

1. INTRODUCTION

Electron-Bombarded CCD (EB-CCD) image intensifiers, based on the primary photoelectron multiplication effect in the CCD substrate, provide a number of advantages over routine intensified CCDs, optically coupled with the output of conventional image intensifiers. Despite this fact, there are no commercially available intensified EB-CCD imagers, though many CCD and image intensifier's manufactures put great efforts to design devices of such kind. This paper briefly reviews the results of more than a decade of work in the field of intensified EB-CCD imaging devices in the "Electron" National Research Institute, Russia's leader in optoelectronics.

2. THE DESIGN AND MANUFACTURE OF THE EB-CCD IMAGING TUBES IN RUSSIA - STATE OF THE ART

2.1. ICCD-16 - the Basic Model of the EB-CCD Imaging Tubes

As the result of prolonged efforts with the semi-industrial technology of the intensified EB-CCD, an image tube (ICCD-16) has been created with 520 x 580 pixels. This device can operate in the faceplate illuminance range from 50 down to 10^{-4} lx, providing spatial resolution not worse than 350 TV lines. When coupled with the output of Gen 1 image intensifier, this EB-CCD imaging tube provides spatial resolution not worse than 350 TV lines in the faceplate illuminance range down to 5×10^{-6} lx, permitting such "coupled" device successfully to compete with Gen 2 MCP-based intensified CCDs.

Along with the improved fabrication technique, the 780 x 580 CCD version of the initial device model has been designed, produced and tested. This EB-CCD imaging tube provides ultimate spatial resolution not worse than 600 TV lines in the faceplate illuminance range down to 5×10^{-4} lx and outperforms the MCP image intensifier because it allows a signal-to-noise level comparable to MCP tubes but without the MTF degradation on the high spatial frequencies.

3. THE DEVELOPMENT OF THE EB-CCD FAMILY IN THE NEAR FUTURE

The totality of experience in design and manufacturing of the EB-CCD Imaging Tubes permits us to proceed in this work in the direction of creating new devices of this family, which could widen the variety of applications. The design process of the devices proposed has already begun and some of them are scheduled to be manufactured in the near future.

3.1. Intensified EB-CCD Imaging Device With a 1024 x 1024 Full-Frame CCD.

This device is based on the scientific-oriented image-converter tube "Yasen'" and is intended for the slow-scan or pulse mode of operation in scientific fields of applications, such as high-energy physics, astronomy, high-speed photonics etc. Demagnification of the input image size and the photoelectrons energy increase up to 15 keV together with the low-noise CCD readout mode provide the photon counting mode with the signal-to-noise ratio not worse than five and spatial resolution not worse than 40 lp/mm. The EB-CCD used is a full-frame, 1024 x 1024 pixels, 13μm x 13μm, device with two (low-noise and high-speed) output amplifiers.

3.2. Intensified EB-CCD Imaging Device With Input Photo-Cathode Diameter 40 mm and With Input Image Demagnification

The proposed device is intended to employ a frame-transfer CCD and, thus, could operate in the standard TV mode. It will be designed on the base of the scientific-oriented image-converter tube PM-031, designed and produced by the Research Institute of Opto-Physical Measurements (Moscow) Demagnification of the input image size together with the photoelectron energy increase provide the threshold sensitivity of the device as low as 5 x 10^{-7} lx and ultimate resolution 600 TV lines, thus making this device competitive with the Gen2 image intensifiers with a cooled CCD. The device is intended for the night-vision TV systems, day-and-night TV cameras as well as for the wide variety of industry and scientific applications.

3.3 Electromagnetically-Focused Intensified CCD Imaging Device With Solar-Blind Photo-Cathode For UV And EUV Imaging.

The device is intended for the detection of the low-intensity light sources in the UV and EUV spectral ranges. This device is designed with the use of the basic ICCD-16 model construction, except for some specific differences.

With the CsI or CsTe photocathodes, this device provides a threshold sensitivity of about 5 x 10^{-16} - 5 x 10^{-17} W/pixel and Solar blindness factor > 10^{-10}. With availability of appropriate financing, all these new devices could be designed and manufactured in approximately a year.

4. CONCLUSIONS

As the result of prolonged efforts in the field of producing Intensified Electron-Bombarded CCD Imaging Tubes, a whole family of such devices was created for a wide range of different industrial and scientific applications. Along with the improving of the EB-CCDs fabrication techniques, some novel devices could be designed and manufactured in the near future.

THE CCDs AT ESO: A SYSTEMATIC TESTING PROGRAM

T. M. C. Abbott

European Southern Observatory

It is incumbent upon ESO to ensure that its CCDs perform according to advertised specifications (Abbott 1994). We describe a systematic, regular testing program for CCDs which is now being applied at La Silla. These tests are designed to expose failures which may not have catastrophic effects but which may compromise observations.

At the time of writing we at ESO offer 12 CCDs for use by visiting astronomers (Abbott 1994). Supporting all of these CCDs poses some unusual problems. ESO serves a very broad community and the astronomers who use our CCDs range in ability from those quite new to the field to those with many years of experience in the use of modern, state of the art detectors. To protect the former and assist the latter, we must make a concerted effort to regularly investigate the quality of the data delivered, whether or not any problems are known. To that end, we have instigated a systematic program of standard CCD tests at ESO, La Silla.

Currently, we test one CCD each week. These tests are not intended to be as thorough as be performed in a specialized CCD lab; instead, they should expose as many problems as possible with minimal technical intervention and under the simple setups available at the telescope.

For each test, we obtain the following information:

a) A map of hot pixels, from a set of nine bias frames, and a map of traps and other defects, from a set of nine low-count-level images (of order a few hundred electrons per pixel). A 16-point transfer curve (Janesick et al. 1987), from a set of 16 pairs of flatfields with exposure levels ranging over the full digital dynamic range and using a stable light source.

b) The mechanical shutter delay and two 16-point linearity curves, expressed as count rate versus true exposure time, from the same data as the transfer curve.

c) The bulk CTE (from the EPER method, (Janesick et al. 1987)

d) A map of the dark current.

e) The amplitudes and frequencies of any interference patterns, from a Fourier analysis of bias frames.

f) A map of the shutter pattern across the CCD, obtained from analysis of a flatfield obtained with multiple shutter cycles.

g) The current bias and clock voltage settings.

All images include bias overscan regions in both dimensions, cover the entire light-sensitive, unbinned area of the CCD and are collected under the same circumstances as normal observing.

The light source used to obtain the flat fields may be either a preflash LED or a beta light. To avoid possible radiation hazards, we are in the process of replacing the beta lights with compact packages of battery-powered LED regulated by feedback from a photo-diode. These are small enough to fit within a normal filter wheel in most La Silla instruments and exhibit a flux variation of 0.2% per degree C.

We extract the required information from the test data set using IDL or MIDAS. We use IDL to develop algorithms and for free-form investigation of the data when necessary. The most common reduction algorithms have been incorporated into the MIDAS CCD context (MIDAS 1994) to provide standard methods of data reduction both within ESO and at other institutions. The raw and reduced test data are to be made available to the community online (ESO 1994).

ACKNOWLEDGEMENTS

S. Deiries of ESO, Garching designed and built the stable LED light source. R. Warmels of ESO, Garching incorporated the test data reduction routines into MIDAS. We are grateful to the ESO, La Silla Astronomy Department and CCD group for their cooperation in collecting the test data necessary for the success of this project.

REFERENCES:

Abbott, T. 1994 ESO CCD Catalogue
Janesick, J. R., Elliot, T., Collins, S., Blouke, M. M. and Freeman, J. 1987 Scientific
    Charge-Coupled Devices, Optical Engineering, 26, p. 692
MIDAS User Manual 1994
ESO Document 1994 OSDH-SPEC-ESO-00000-0002/2.0, EMMI/SUSI calibration plan for an
    on-line calibration database

DEVELOPMENT OF A 7000 x 4000 PIXEL MOSAIC CCD CAMERA

Nobunari Kashikawa, Masafumi Yagi, Naoki Yasuda, Sadanori Okamura, Kazuhiro Shimasaku, Mamoru Doi and

University of Tokyo, Bunkyo-ku

Maki Sekiguchi

National Astronomical Observatory of Japan

1. CCD CHIPS, MOUNTING PROCEDURES, MECHANICAL & ELECTRIC DESIGN

The CCD we use is TC215 manufactured by TI, Japan. The pixel size is 12 microns square. It is a virtual phase CCD which has a peak QE of 60% at 700 nm and 15% QE at 350 nm. It is commercially available in a package, which is too big to meet our requirement for CCD spacing. We therefore put the CCD in a specially made compact package. Each chip is mounted on a machined ceramic spacer whose thermal coefficient is matched with that of the CCD package. We glue each CCD chip on the spacer under a microscope to measure YHE x - y position and height. Then we screw the CCD chip + spacer on a copper motherboard with the help of the gauge which has a planned grid with a good accuracy. The alignment accuracy we can get with such set-up procedures is an order of 5 microns (< 0.5 pixel) in the horizontal direction and 10 microns in the vertical direction.

The CCDs are put in a dewar of the size 30 cm x 30 cm x 10 cm (WDH) with a window optimized for an eight by four array (See Fig.1.). A large focal plane shutter and a filter holder are attached in front of the dewar. It takes about eight seconds for the shutter to fully open/close. The shutter has two bellows on each side, and the exposure time is kept uniform over the chips by moving the two bellows in appropriate turns. We have made four large (20 cm x 20 cm) filters whose responses are close to those of Johnson-Mould's B, V, R and I bands. The CCDs are cooled down to -80° to -90° by both a Stirling-cycle cooler and liquid nitrogen. The vibration caused by a compressor of the cooler is confirmed to be negligible for a 12 micron pixel. A liquid nitrogen tank is necessary, not only as a cooler, but also as a cryogenic pump. We use a system called MESSIA2 (Modularized Expandable SyStem for Image Acquisition) for CCD control and data acquisition. Four columns, each consisting of seven CCDs, are read in parallel, while in each column seven CCDs are read out one-by-one serially. The total CCD readout time, i.e., the time necessary to read CCDs, digitize, and store the data in the VME memory boards located near the camera, is about 100 seconds. Once the data are stored in VME memory boards, the next exposure can be started. The data are transferred to hard disks of a control computer during the exposure.

3. SOFTWARE, PERFORMANCE AND PROSPECTS

The data produced by the camera amount to about 1 GB per night. Software which

Fig. 1. The dewar of the mosaic CCD camera developed by the Univ. Tokyo-NAOJ group. 7 by 4 = 28 CCDs are mounted. A liquid nitrogen tank is on the left and on the right is a Stirling cooler.

reduces this vast amount of data has also been developed. The data reduction system consists of several modules, which do specific tasks such as flatfielding, sky subtraction, find and measure objects, measure seeing size, match objects in overlapped regions, star-galaxy separation, and crude morphological classification of galaxies. The system also has functions to plot variety of statistical quantities in many kinds of graphs for an easy and quick monitoring of whether or not the system is processing the data in a proper way. The best performance of the system is realized by two-pass processing. The first pass is to fine-tune several critical parameters that are used in the second pass, which produces the final catalog of objects with photometric parameters together with parameters necessary to produce one contiguous image from 28 x 4 = 112 frames.

We have not yet completely finished evaluating all the performances, which are expected to be the same as or better than those of MCCD1 (Sekiguchi et al. 1992). Our primary motivation for these cameras is observational cosmology based on wide field surveys. We need a wide-field telescope at a good site in order to take full advantage of MCCD2. In collaboration with Carnegie Institution we used MCCD2 in May 1994 with the 40-inch telescope at Las Campanas Observatory to take multi-color images of southern clusters of galaxies. MCCD2 with the Swope telescope give an image scale of 0.35 arcsec/pixel and a sky coverage of 74 x 43 arcmin with four exposures. More observations are planned in Oct. - Nov. 1994. Our goal is to expand the array to eight by eight CCDs, which would be a sort of prototype of the prime-focus camera for our Japanese 8-m telescope (SUBARU).

REFERENCE

Sekiguchi, M., Iwashita, H., Doi, M., Kashikawa, N. and Okamura, S.. 1992 PASP 104, 744

THE FEASIBILITY OF A CCD FOR AN ASTROMETRIC REFRACTOR

A. R. Upgren, Alice Morales, Jose Herrero, J. W. Griese, III, J. M. Vincent

Van Vleck Observatory

and John T. Lee

Yale University Observatory

The 0.5m refracting telescope of the Van Vleck Observatory has been active in the determination of trigonometric parallaxes since its first observations in 1922. Its lenses were ground by C.A.R. Lundin of the Alvan Clark Co. for photographic use. Coma was minimized across the field and vignetting was also kept to a minimum. Partly as a consequence the focal curve is very steep in the blue and green regions of the spectrum, as is shown in Fig. 1. A Wratten No. 12 minus blue filter is used to filter out all wavelengths to the blue of about 5200 Å. The region between 5200 Å and 6000 Å is very flat with the focal plane varying over a range of about one millimeter. Towards the red region it steepens, although not enough to impair images on photographic plates of emulsion types 103a-D and IIIa-F, the two in widespread use in recent years.

A CCD of 2048 by 2048 pixels with sensitivity limited to the flat yellow spectral region should allow the telescope to reach stars brighter than visual magnitude 14.5, the present photographic limit, yet should provide a sufficient reference frame. Table 1 shows the manufacturer's specifications for five typical CCD's of this size. The dimensions of the pixels, the linear and angular sizes and the area covered are shown in the first four columns. The final two columns give the average number of reference stars for the field areas shown, that are brighter than 14.5 for fields in the galactic plane and poles, respectively. It would appear that a judicious choice among the CCD's should be made to assure rich reference frames for all star fields.

TABLE 1

List of Most Probable CCD's (2048^2 array)

Pixel size	Linear size	Angular size	Field Size sq degrees	Average # Stars b = 0°	b = 90°
09.0 μm	18.4 mm	07.5'	0.016	10	2
15.0 μm	30.7 mm	12.6'	0.044	26	5
13.5 μm	27.6 mm	12 3'	0.042	25	5
12.0 μm	24.6 mm	10.0'	0.028	17	3
24.0 μm	49.2 mm	20.0'	0.111	67	13

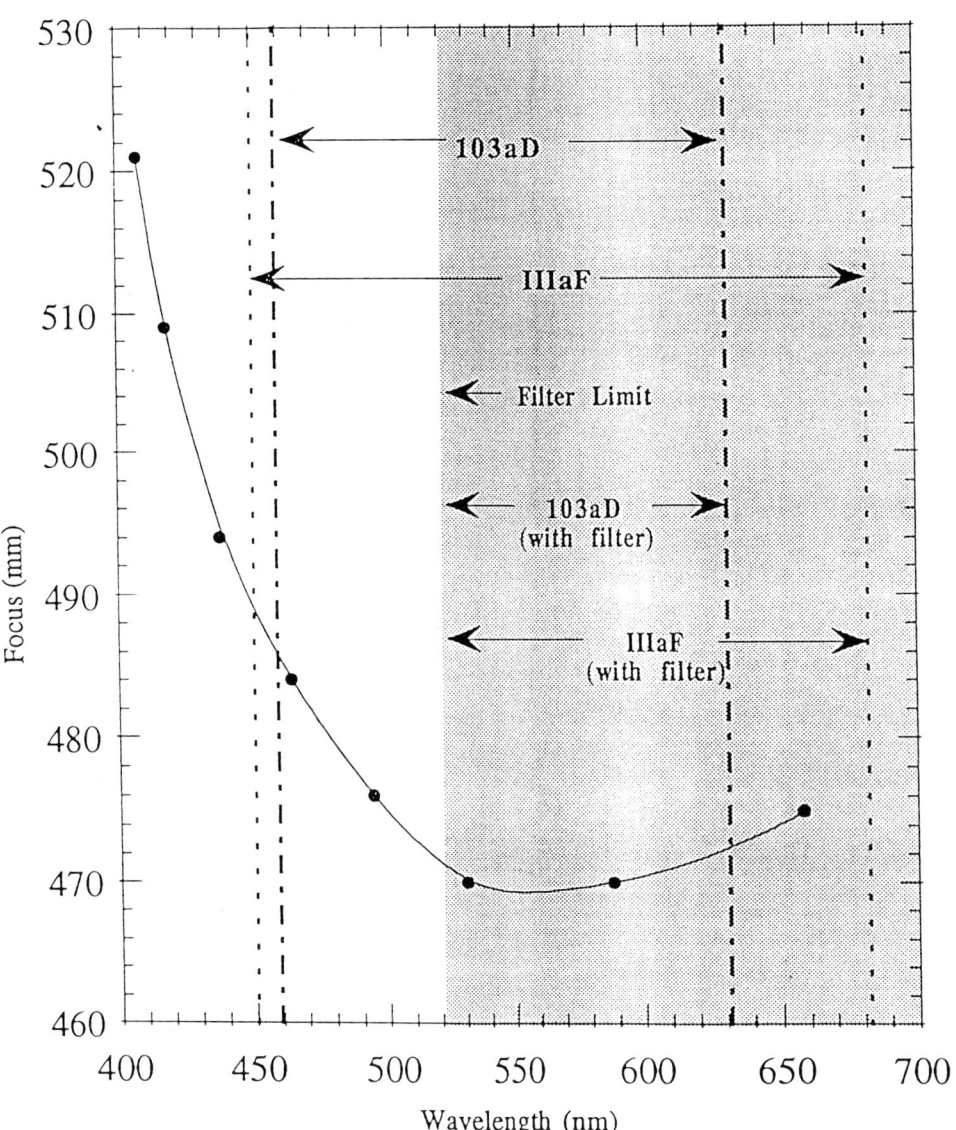

Fig. 1. The focal curve for the 0.50-m refractor. The unshaded portion corresponds to the spectral region blocked by the Wratten filter. Within the shaded portion the wavelengths covered by two emulsions are shown.

This study is supported by Research Grant AST-9218605 and amendments from the National Science Foundation.

STELLAR POSITIONS FROM CCD IMAGES

A. R. Upgren,

Van Vleck Observatory, Wesleyan University

C. Abad and J. Stock

CIDA, Merida

The CIDA visual refractor of 65-cm aperture and focal length 10.5 m, has been used extensively for position determinations on photographic plates. The combination of Kodak D plates and a yellow filter permit an almost perfect adaptation to the focal curve of the telescope. It appeared of interest to test whether the telescope could be used for astrometric purposes with a CCD detector. As is well known, the spectral sensitivity of these detectors extends well into the infrared where the images formed by the telescope optics will be far out of focus. The blue spectral region where this would also be the case can easily be cut off by a yellow filter. There are no filters which would produce a similarly sharp cut-off towards the red region. On the other hand, given the small field covered by a CCD, the displacement of the red out-of-focus image with respect to the center of the visual image might be negligible. Recently obtained accurate positions in the area of the Perseus Double Cluster made this field suitable for the test of this possibility.

A number of methods has been developed to determine the center of a stellar image on a CCD exposure. All require that an image extends over a number of pixels. Basically two different procedures can be used, namely a) to adapt a given profile such as a Gaussian distribution to the image, or b) to construct an image model in tabular form and adapt it to the actual images. We have followed both procedures.

Usually a two-dimensional Gaussian image profile is considered the most adequate mathematical representation of true stellar images. This leads to six unknowns of which one describes the central intensity, one the orientation of the major axis of the image, two describe the length of the major and the minor axes of the image, and two more the coordinates of its center.

A new dimension can be added to the formation of artificial image profiles by making use of different functions. We found that the use of the hyperbolic tangent function offers a new flexibility in the profile adaptation. We shall demonstrate its possibilities first in the case of a single dimension. An expression of the form

$$d = a \, \exp[\, -(x-x_o)^2/b^2] \qquad\qquad 1$$

represents a Gaussian profile, where d is the intensity, x is a coordinate, x_o is the coordinate of the center of the profile with maximum intensity, a, and a constant, b, which determines the width of the profile. It is obvious that no matter which values are used for the different parameters, the resulting profile will always have the shape of a Gaussian distribution. From the expression

$$d = \{\tanh[p(x/c + 1)] - \tanh[p(x/c - 1)]\}/(2 \tanh p) \qquad 2$$

a much larger variety of profiles can be derived. They can in fact simulate a Gaussian at one extreme, and flat-topped profiles with steep edges at the other. The parameter p determines the slope of the edges of the image, while the parameter c determines its width. This flexibility is obtained by the addition of only one free parameter. In fact, if one would be satisfied with a given form of the profiles, i.e. with one constant value of p, the number of free parameters would be identical to those of Gaussian profiles. Naturally, the profile given in equation (2) can readily be extended to two-dimensional elongated images in the same way as is done for Gaussian profiles. Here we should mention that the arctangent function can equally well be used for this purpose. Again, if the two-slope parameters are kept constant, the number of free parameters is reduced to six, as is the case for the Gaussian model.

In the case of CCD images, one may assume that all image profiles are based on one and the same underlying master profile, unlike photographic images. The first step for any CCD exposure should be to find the master profile. For that purpose one may use the image of a bright star, provided one is certain that it is not a double star or an object surrounded by luminous nebulosity. It is safer, though, to co-add the images of several stars, taking care that they are well centered, one on top of the other, before adding. The co-adding process requires some interpolation since the image centers will hardly ever fall right in the middle of a pixel. This is best done by the creation of intermediate pixels which we found to be beneficial for the centering programs. Some smoothing can be done at this point if that is considered to be practical.

The next step is to determine whether an artificial profile created by the mathematical procedure explained above can be satisfactorily adapted to the master profile. We found no appreciable difference in the wings of the profiles for stars of different colors, as might be expected on the basis of the focal curve of the telescope. However, we must admit that the range of colors of the stars investigated is rather small.

The centering is done by an iterative process, starting out with approximate values for all of the parameters involved including the coordinates, and calculating by least squares their most probable corrections. These are applied, and the process is repeated until the corrections become negligible. Naturally, the process makes use of the partial derivatives of the function in use with respect to all of its parameters. This necessity is in fact the reason why it is practical to have a mathematical expression which describes the image profile.

The master profile may not always be fitted by any of the expressions mentioned above. The centering can then be obtained by a strictly numerical process. One must first find a scale factor which will scale the master profile down to the amplitude of the image that is being analyzed. Then by shifting the master profile back and forth in x and in y one can find the position of the latter with respect to the real image for which the rms difference becomes a minimum. No significant differences in the positions for bright objects appear upon use of either method. For the faintest stars we found non-systematic differences up to a few hundredths of a pixel. One pixel corresponds to approximately 0.5 arcseconds.

ON THE ACCURACY OF CCD AND PHOTOGRAPHIC OBSERVATIONS OF ASTEROIDS AND THEIR CURRENT ORBIT DETERMINATIONS

O. P. Bykov

Pulkovo Observatory

An application of the Classical Laplacian Method and new Pulkovo AMP-method for current asteroid orbit determinations is given. The CERES software package created at the Institute of Theoretical Astronomy (Russia) was applied to calculate (O-C)-differences for 200 numbered minor planets observed irregularly and quasisimultaneously in 1993 by CCD as well as by photographic techniques at 25 observatories (ESO, SERGA, Kitt Peak etc.). The accuracy of the observations was estimated by means of the standard error of the average (O-C) differences for each type of observation obtained by each telescope. As a whole the CCD-observations of the numbered minor planets are considerably more precise in comparison to the photographic ones. Some results are given in Table 1.

The efficiency and high accuracy of the CCD-observations allow one to solve the problem of orbit determination for any celestial object moving in the field of reference stars. We use the Classical Laplacian Method and the Pulkovo Method of Apparent Motion Parameters (Bykov 1989). All the calculations can be directly executed in real time during the CCD-observations. The author believes that the CERES software, developed by V. L'vov, V. Shor et al. at the Institute of Theoretical Astronomy (Russia) and the software (LAPLACE) made at the Pulkovo Observatory, are very useful for this purpose. Both software packages have been used for the current identification and classification of many asteroid orbits. An example of one orbit's calculation is given in the Table 2. The orbit obtained is not very good in comparison with the real orbit of the numbered minor planet 2699 but it is a single orbit which was obtained using the supershort arc of the positional observations. Gauss's and Laplace's methods cannot give precise orbital elements using these data only.

REFERENCE

Bykov, O. P. 1989 Opredelenie orbit nebesnikh tel pryamimi metodami, Problemi postroenia koordinatnikh sistem v astronomii, GAO, Leningrad, p. 328

TABLE 1

Observations of minor planets

MPC-code and name of observatory		Telescope	Number of minor planets	nights	positions	Average error for a single observation of coordinates RA	DEC
\multicolumn{8}{c}{CCD observations}							
691	Kitt Peak	0.9 Shm	6	14	42	0.38	0.18
801	Oak Ridge	1.5 rfl	7	17	33	0.30	0.23
689	Flagstaff	1.5 rfl	1	9	38	0.05	0.05
413	Sid.Spring	1.0 Shm	3	10	21	0.45	0.40
664		0.8 rfl	2	4	31	0.56	0.39
587	Sormano	0.5 rfl	3	5	12	0.54	0.36
657	Victoria	0.5 rfl	2	4	12	0.50	0.20
596	Collev.	0.3 Shm	4	9	19	1.02	1.04
107	Cavezzo	0.4 rfl	3	11	23	0.72	0.78
\multicolumn{8}{c}{Photographic observations}							
809	ESO	0.4 Astr	30	86	256	0.77	0.56
675	Palomar	0.5 Shm	7	18	33	0.70	0.96
10	Caussols	0.9 Shm	7	18	43	0.92	0.96

TABLE 2

An Example of the orbit determination for Minor Planet 2699, Kalinin by the AMP-method

	\multicolumn{7}{c}{Positional observations by means of CERES, Nov, 1993}						
		RA	(O-C)		DEC	(O-C)	
1	16.22484	$3^h\ 34^m\ 47^s.474$	0.36	$14°\ 08'\ 11''.90$	-0.28	801	CCD
2	16.23670	3 34 46.648	0.45	14 08 15.22	-0.13	801	CCD
3	16.80346	3 34 06.932	1.06	14 10 47.80	0.91	107	CCD
4	16.81871	3 34 05.809	0.29	14 10 51.51	0.53	107	CCD
5	17.22258	3 33 37.506	0.52	14 12 39.29	-0.02	801	CCD
6	17.23487	3 33 36.641	0.48	14 12 42.51	-0.10	801	CCD

```
CALCULATED PULKOVO ORBIT        CERES for NMP 2699 (ephemeris)
   d =    1.0705                              1.35202
   r =    2.0574                              2.338
   a =    2.44598                             2.63778
   e =    0.18803                             0.16925
   i =   11.66548                            16.1306
   N =   67.040                              64.4479
   w =  308.658                             294.2558
   M =   26.574                              40.6563
```

ON THE PROBLEM OF STANDARD FIELDS FOR CCD ASTROMETRY

I. S. Guseva

Pulkovo Observatory

At the start of CCD observations one must investigate the metrological properties of the complex instrument: optics + CCD. One needs a densely spaced set of stars with precise coordinates and magnitudes. We first used the Pleiades catalogue by Eichhorn et al. (1970). Experimental observations were started at Pulkovo in 1993 using a very small refractor (D = 100 mm, F = 712 mm) equipped with a CCD ISD015A (520 x 580 pixels, 18 x 24 microns). The focal length of our astrograph provides an angular field of view of 45 x 67 arcmin (angular scale is 5.2 x 7 arcsec/pixel). With different exposures we can observe all stars with magnitudes from 2 to 16. The first observations were made to evaluate the accuracy of our positional and photometric measurements. Unfortunately, it appears that the catalogue of the Pleiades by Eichhorn is not good enough for this purpose because its epoch is very far from that of our observations and proper motions were not provided for the majority of the stars. The internal precision of our measurements (0.1 - 0.3 arcsec) allows us to determine the corrections to the stellar positions by Eichhorn et al. (1970)

After a significant improvement to our CCD camera a new series of observations was obtained of different standard fields. It appears that we cannot find any appropriate standard field catalogue to investigate the general distortion of our instrument. That is why we have started with a metrological investigation of the "optics + CCD" by use of North Pole observations with different positions of the instrument. This way does not allow us to determine the scale of the instrument only.

Thus, the problem of the creation of standard fields seems to be very important now, especially for large telescopes. The necessary parameters of standard field stars are not only their positions but their proper motions also. CCDs allow us to increase dramatically the precision of positional measurements, and all the standards without proper motions will yield imprecise results. It is very important to give positions of standard stars in a correctly defined global reference system, because any differential positions given relatively to an arbitrary chosen group of stars will not allow us to obtain correct proper motions, and the precision and use of such standards will vanish with time.

The next important obstacle is that increasing positional accuracy needs a very exact knowledge of photometric values of stars, because chromatic refraction becomes the main limiting factor for differential observations in a small field of view. On the other hand, any modern photometric standard field requires precise positions and proper motions of stars, because it is not realistic now to identify numerous faint stars without computers. Thus, stars in standard fields should have both astrometric and photometric parameters known. It is especially important to make photometric measurements with an R filter (of the UBVRI system or with corresponding filters of other systems in the red part of spectrum) because the majority of CCDs has the maximum quantum efficiency in this band. Moreover, this range of the spectrum is very useful for astrometry due to the small influence of refraction. This range is preferable also when it is necessary to make observations in daytime or in twilight. Unfortunately, this band of the spectrum is often absent in many catalogues.

Standard fields for CCD calibration should be chosen without bright stars, because they can produce additional problems for the investigation. It is necessary to avoid possible overlapping of star images on CCDs. That is why globular clusters are not very suitable to be taken as standard fields. Even with high resolution (for the ground-based observations it does not exceed 0.5 arcsec) we never know if a star is a double or not (really or optically). The accuracy of positional measurements with CCDs is much better than 0.5 arcsec, that is why open clusters without bright stars or analogous regions on the sky are preferable to be taken as standard fields.

It seems to be very useful to create standards in the polar zones. The main advantages of these regions are: a) These areas can be observed at any time. b) They can be observed for any desirable time without limitations connected with the Earth's rotation. c) The zenith distances of stars in this region vary very slowly; that simplifies the refraction problem. d) The majority of telescopes may observe these regions in any position of the instrumental CCD frame relative to the sky area. It allows us to investigate the distortion of "optic + CCD" system without exact knowledge of the star positions.

We have made some series of North Pole observations with our small instrument. Observations of the Pleiades, open clusters M 35, M 38 and others have been made also. It is important that these clusters have been used as standard fields, and it is possible to obtain proper motions for the stars for which first epoch plates have been taken. After investigation of distortion and other possible errors of our instrument a preliminary version of catalogues of these clusters will be prepared. The final versions of these catalogues will be prepared after a photometric investigation of the clusters and the correct reduction of positions, taking into account chromatic refraction.

It seems to be very useful to join our efforts with the efforts of other astronomers on the creation of precise standard fields.

REFERENCE.

Eichhorn, H., Googe, W. D., Lukac, C. F. and Murphy J. K. 1970 MNRAS 73, 125

COMBINED VISUAL AND NEAR-IR DIGITAL PHOTOMETRY: THE VERY YOUNG CLUSTER WESTERLUND 2

M. D. Guarnieri, M. Gai, G. Massone,

Osservatorio Astronomico di Torino

M. G. Lattanzi,

Osservatorio Astronomico di Torino, ESA at STScI

U. Munari and

Osservatorio Astrofisico di Padova

A. Moneti

European Southern Observatory, La Silla

ABSTRACT: We discuss CCD UBVRI and NICMOS3 JHK observations of the young open cluster Westerlund 2, with emphasis on both reduction techniques and absolute calibrations.

The optical images were carried out with the 1-m telescope at the Sutherland station of the South African Astronomical Observatory using the RCA 512 x 320 CCD camera. IR data were obtained at the 2.2-m MPI/ESO telescope at La Silla (Chile) using the new NICMOS3 256 x 256 IR camera IRAC2. The reduction of the IR images will be discussed in the light of the instrumental problems which were identified during the observing run, when the camera was still under test. Flat fielding will also be addressed. It is shown that, despite the problems, the photometric performance of the new camera was quite good, which provided us with accurate infrared photometry.

This multi-band data set allows us to derive accurate values for age and distance to the cluster. We can also study in detail the cluster's stellar content, providing mass and temperature of individual stars. For the optical images, the standard procedure of de-biasing and flatfielding was adopted to get uniform pixel response.

The IR data were reduced by subtracting from the object images a linear combination of the corresponding skies and dividing the results by a flatfield obtained on the morning or evening sky. This technique has the advantage of removing automatically bias and dark current, but any star in the sky frames will appear as a hole in the final frame. The flatfields were prepared by subtracting a low-signal image from a high-signal image obtained with the same integration time. This procedure was adopted (with the K filter in particular) to deal with certain structures in the raw images that are a function of integration time but not of incident flux. The flatfield images were then normalized to unity in order not to alter the intensity of the science images during the division process. Both sets of data have been reduced using the package for crowded fields photometry DAOPHOT II (Stetson 1987), running in MIDAS.

Magnitudes were measured on each frame by fitting stars with the best choice PSF (Daophot routine ALLSTAR). For both optical and IR data, the most isolated stars were used to link the aperture-photometry magnitudes to the instrumental ones obtained from the fitting, to calculate and remove the aperture correction. Finally, we discarded all stars with an internal magnitude error, as provided by ALLSTAR, of 0.18 mag.

Figs. 1 and 2 show the V, (B-V) and the K, (J-K) CMDs respectively. The optical data are comparable with, or better than, those presented by Moffat et al. (1991). The absolute calibrations in both the optical and the near infrared are quite good and will allow us a thorough investigation of W2, which will include photometric membership classification based on the reddening to each individual star, age and distance through isochrone fitting, luminosity function and initial mass function (including correction for the contribution of the pre-main sequence stars). The whole set of diagrams, the details concerning data acquisition and reduction, and the full astrophysical discussion comprehensive of the comparison with Moffat et al., will be published in a forthcoming paper (Guarnieri et al., 1994).

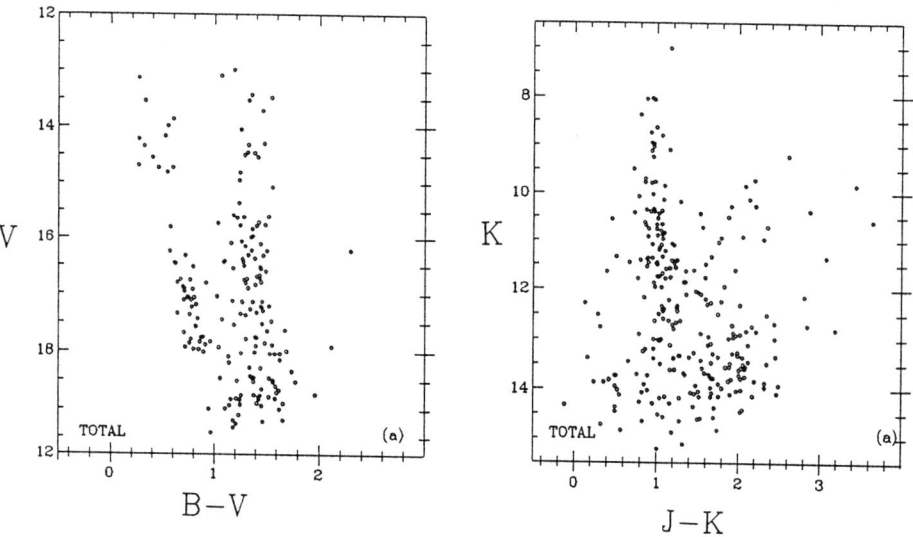

Fig. 1. (V, B-V) CMD of Westerlund 2.   Fig. 2. (K, J-K) CMD of Westerlund 2.

REFERENCES

Guarnieri, M. D., Lattanzi, M. G., Massone, G., Munari, U., Gai, M. and Moneti, A. 1995 in preparation for A&A
Moffat, A. F. J., Shara, M. M. amd Potter, M. 1991 AJ 102, 642
Stetson, P. B. 1987 PASP 99, 191

A DUAL CCD MOSAIC CAMERA SYSTEM SEARCHING FOR MASSIVE COMPACT HALO OBJECTS (MACHOs)

Kem H. Cook

Lawrence Livermore National Lab

ABSTRACT: The Macho Collaboration uses a dedicated 1.27-m telescope (The Great Melbourne Telescope) at Mount Stromlo to make photometric measurements of tens of millions of stars per night searching for the gravitational microlensing signature of MACHOs in the halo and disk of the Milky Way. A prime focus corrector and dichroic beamsplitter provide red (6300 - 7800 Å) and blue (4500 - 6300 Å) foci with one degree fields. A two by two mosaic of 2048 x 2048 pixel CCDs in each focal plane provides simultaneous images of 0.5 square degrees. By August of 1994, more than 20,000, 32 Megapixel images will have been collected of fields in the Large Magellanic Cloud (LMC), Small Magellanic Cloud and the bulge of the Milky Way. We have implemented an online analysis system which produces photometric reductions of a night's data (five Gbyte of images) within 24 hours. This system allows us to identify and follow interesting events in real-time. In this search, we have identified more than 60,000 variable stars, and a preliminary analysis of their types and distribution will be presented. Microlensing events toward the LMC and the bulge have been discovered and detection efficiencies are being calculated to constrain the MACHO content of the Milky Way's halo.

JHK PHOTOMETRY OF EXTRAGALACTIC SOURCES USING AN INFRARED ARRAY CAMERA

I. S. McLean, T. Liu and H. Teplitz

Department of Physics and Astronomy, UCLA

ABSTRACT: The UCLA two-channel IR camera has been used to perform crowded-field photometry in two distinct classes of extragalactic objects. DAOPHOT was used to reduce images of M 33 and provide JHK color-magnitude diagrams of over 1,700 stars. Galaxy photometry in crowded clusters, e.g. Abell 370, was obtained in JHK and analyzed using FOCAS. In both cases, extensive reductions were performed using synthetic aperture photometry for comparison. Methods and results are summarized.

1. INTRODUCTION

The development of infrared array detectors for astronomy has had an enormous impact (see McLean 1994 for reviews). These detectors are now sufficiently large (256 x 256 pixels) and sensitive (QE > 60%; read noise < 50 electrons) to find a wide range of photometric applications. Two examples are used to illustrate both the performance and the potential of IR arrays: JHK photometry of the core of the Local Group galaxy M 33 and JHK photometry for two galaxy clusters (Abell 370 and CL0024+1654). Full details of the two-channel camera used are given in McLean et al. (1993) and McLean et al. (1994).

2. OBSERVATIONS OF M 33

The observations were made in two sets; (J, K') in 10/93 and (H, K') in 12/93; K' refers to the Mauna Kea K-filter (1.95 - 2.30 microns). A three by three overlapping mosaic was used to obtain a field of 7.6 x 7.6 arc minutes with total integration times of 315 s at J, 300 s at H and 300 s at K'. The K' frames from October and December were compared but not combined. Conversion to standard K magnitudes was done using the relationship given by Wainscoat and Cowie (1992). Raw frames at each wavelength were first processed by subtracting a sky frame constructed from the median of nearby empty sky fields and then divided by a dark-subtracted and normalized dome flat field.

Using DAOPHOT, over 1,700 stars were measured reliably. The uncertainties in the photometry are as follows:

J: 0.17 mag at J = 16 to 0.70 mag at J = 19
H: 0.23 mag at H = 16 to 0.50 mag at H = 18
K: 0.27 mag at K = 16 to 0.53 mag at K = 18.

The DAOPHOT artificial star experiment was used to determine these error estimates and the completeness limits by recovering "fake" stars.

Photometry was performed on the final, registered mosaic using an average PSF determined from various positions in the mosaic after careful comparison of results obtained with locally determined PSFs showed excellent agreement. Aperture photometry was applied to the bright and less crowded stars to calibrate the PSF photometry.

Color-magnitude diagrams and luminosity functions have been derived. A well-defined giant branch is seen which is brighter than globular clusters and Baade's Window giants in our galaxy, but similar to M 31 and M 32. The fraction of luminous stars seems enhanced over that of M 31 and M 32. Based on the large number of OB associations in M 33, the brightest stars are probably supergiants from recent star formation.

3. OBSERVATIONS OF ABELL 370 AND CL0024+1654

JHK observations of these two clusters of galaxies at $z = 0.37$ were obtained on October 24 and 25, 1993. The K' band was observed continuously in the long wave channel, with the J and H observations "nested" within the total K' integration period. For CL0024 the K':H:J exposure times in minutes were 135:90:45 and for Abell 370 the integration times were 90:45:45. Observations were made in five minute segments and the telescope was dis-registered or dithered between each integration. A running median "sky flat" was derived from the cluster frames themselves and applied to each frame in turn. All the flatfielded frames were then registered and a median of these was formed.

Since most of the objects are slightly extended, identification was done using the FOCAS software package which produces a complete list of objects found. Synthetic aperture photometry was performed using a standard aperture size of 4.2 arc seconds for each object. Completeness was estimated by using IRAF routines to introduce "fake" stars into the frames and recover them. For Cl0024+1654 the limit is K' = 18.6 (five sigma) and for Abell 370 it is 17.9. Over 150 galaxies are found.

(J-K)' versus K' color-magnitude diagrams were formed for both clusters. Interestingly, both clusters exhibit gravitational lensing arcs. JHK photometry for Arclets A1, A3 and A4 in CL0024+16 and arcs A5 and 62 in Abell 370 reveals that all three arclets in CL0024+16 have approximately similar colors, with the average value of (J-K') = 1.33 (range is 1.2-1.5), whereas for Abell 370 the (J-K') colors are 1.4 and 2.0 for arcs 62 and A5 respectively. Since both of the arcs in Abell 370 have known redshifts of $z = 0.7$ and $z = 1.3$, it is plausible that all three arclets in CL0024+16 come from the same background galaxy at approximately $z = 1$.

REFERENCES

McLean, I. S., Becklin, E. E., Brims, G., Canfield, J., Casement, L. S., Figer, D. F., Henriquez, F., Huang, A., Liu, T., Macintosh, B. and Teplitz, H. 1993 Proc. SPIE 1946, Infrared Detectors and Instrumentation, A. M. Fowler, ed., p. 513

McLean, I. S., Macintosh, B., Liu, T., Casement, L. S., Figer, D. F., Teplitz, H., Larson, S., Lacayanga, F., Silverstone, M. and Becklin, E. E. 1994 Proc. SPIE 2198, Instrumentation in Astronomy VI, D. L. Crawford and E. R. Craine, p. 457

McLean, I. S., ed. 1994 Infrared Astronomy with Arrays: The Next Generation, Kluwer Academic Publishers, Dordrecht

Wainscoat, R. J. and Cowie, L. L. 1992 AJ 103, 332

PRECISION CCD PHOTOMETRY OF THE HORIZONTAL BRANCH

A. G. Davis Philip

Union College and ISO

The Strömgren four-color system is well suited to the measure and analysis of horizontal-branch stars. The increased accuracy of the CCD photometric system and the ability to measure fainter stars in very crowded regions make the combination of the CCD system and four-color photometry an excellent one to study horizontal-branch stars in globular clusters.

Horizontal-branch stars have been found to occupy different regions in the four-color diagrams. One group follows the ZAHB line in the y, (b-y) diagram and these stars are assumed to be stars that are evolving to the blue from their initial positions on the ZAHB. Then there is a group of stars found approximately 0.2 mags above the first group. These stars are assumed to be those stars that have evolved past the turn-around point and are now heading redward, eventually to the asymptotic giant branch. In M 15 and M 92 stars have been found in the central regions which are on the HB, but which are somewhat fainter than the ZAHB stars.

In Fig. 1. the y vs (b-y) diagram for M 92 is shown. These obeservations were made at Kitt Peak National Observatory using the T2ka chip and the No. 2 0.9-m telescope. Nine of

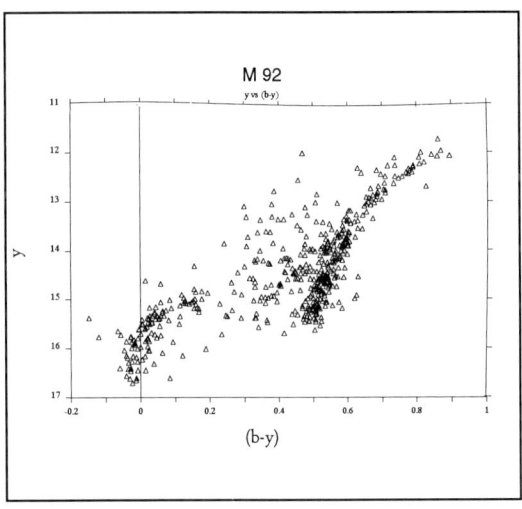

Fig. 1. Four-color CM diagram for M 92 (CCD Observations).

the Sandage (1969) M 92 BHB stars had been set up as standards by observations made with a single channel photometer on the University of Arizona's 90-in telescope. Three frames were taken in each of the four filters. Dome flats were obtained for each filter. The data were reduced at the Dominion Astrophysical Observatory using Peter Stetson's programs, DAOPHOT and ALLSTAR (Stetson 1990). The internal rms errors of the y, b, v and u magnitudes was ±0.016, ±0.014, ±0.017 and ±0.018 for stars between y = 13 and 17 mag.

In previous studies, using the smaller RCA chip, Philip (1990) showed that stars on the blue horizontal branch could be divided into two main groups. In a plot, combining observations of BHB stars in four different globular clusters, it was found that there was a lower envelope of stars, with a small scatter, running from right to left. Scattered above this relation (0.2 mag or more) were a number of stars of higher luminosity. An analogy was made to a diagram from Sweigart (1987) in which Log L was plotted vs Log T_{eff} for various values of Y and Z. Stars evolving to the blue all fell along the zero-age horizontal branch. Depending on the mass of the BHB star, the evolutionary track turned upward and to the right. The tightly grouped stars which form the lower envelope are assumed to be stars on the lower evolutionary track, evolving blueward. The more scattered group of stars above is assumed to be those stars which have evolved away from the ZAHB and are now tending toward the asymptotic red giant branch. In the M 92 observations (where the field of view included the entire cluster) it has been found that the stars fall in common groups in all the four-color diagrams, thus they form natural groups. ALLSTAR permits reductions of stars in the central region of the cluster and many new BHB stars have been found. There is also a group of stars that falls somewhat below the normal horizontal branch. Stars from both the inner and outer regions fall in this region.

A search has been made for an independent luminosity indicator that might confirm the photometric data. Gray (1991) has created an IBM PC based program, SPECTRUM, using the Kurucz (1991) models which allows one to enter a discrete set of temperatures, gravities and metallicities and the computer then plots the spectrum (from λ 3900 Å to λ 4500 Å) at a spacing of 0.02 Å. A set of model grids was computed with SPECTRUM with parameters matching those of horizontal-branch stars and two possible luminosity indicators were found in the region of the Hγ line (Philip 1994). In cooperation with Saul Adelman and Chris Aikman time for spectroscopic observations has been given at Kitt Peak National Observatory and the Dominion Astrophysical Observatory to test these luminosity indicators with observations of FHB stars on our observing lists.

REFERENCES

Gray, R. O. 1991 BAAS 23, 1382
Kurucz, R. L. 1991 in IAU Symposium No. 149, The Stellar Populations of Galaxies, A. Reis, ed., Kluwer Academic Pub, Dordrecht
Philip, A. G. D. 1990 in CCDs in Astronomy. II., A. G. D. Philip, D. S. Hayes and S. J. Adelman, eds., L. Davis Press, Schenectady, p. 107
Philip, A. G. D. 1994 in The MK Process at 50 Years, C. J. Corbally, R. O. Gray and R. F. Garrison, eds., ASP Conf. Series, 60, p. 148
Sandage, A. R. 1969 ApJ 157, 515
Stetson, P. B. 1990 PASP 99, 191

CCD PHOTOMETRY OF THE M 67 CLUSTER IN THE VILNIUS PHOTOMETRIC SYSTEM

R. Boyle,

Vatican Observatory Group, Steward Observatory

V. Straižys,

Institute of Theoretical Physics and Astronomy, Vilnius

F. Vrba,

U.S. Naval Observatory Flagstaff Station

F. Smriglio and A. K. Dasgupta

Istituto Astronomico, Universita La Sapienza

The Vilnius photometric system, consisting of seven passbands at 345, 374, 405, 466, 516, 544 and 656 nm makes it possible to determine spectral classes (or temperatures), absolute magnitudes (or surface gravities), metallicities and peculiarity types for stars of all spectral types in the presence of interstellar reddening (Straižys 1977, 1992a,b). This makes the system very useful for the determination of the physical parameters of stars which are too faint to be studied by spectroscopic methods. The system is especially effective when used with CCD detectors which combine a wide field, high sensitivity and high photometric accuracy.

For the first time, the Vilnius photometric system was successfully used with a CCD detector in 1986 and 1987 on the 90-cm telescope of the Kitt Peak National Observatory (Boyle et al. 1990a, b, 1992, Smriglio et al. 1991). It was shown that CCD Vilnius photometry allows one to obtain two-dimensional classification of stars down to 17th mag with exposures of the order of about 20 min for two ultraviolet filters and about 5 min for other five filters.

However, the number of stars in a 5' x 7' area down to 17th mag is insufficient for the statistical investigation of stellar populations. CCD chips with 2048 x 2048 pixels offer a possibility to increase the field size and the number of stars considerably. An especially large field (23' x 23') can be obtained with the CCD camera operating on the 1-meter Ritchey telescope of the Flagstaff Station of the US Naval Observatory. We decided to test a possibility of using this camera for photometry of stars in the Vilnius system. For this, the field of the open cluster M 67 was chosen.

The cluster area was observed on two nights, February 18/19 and March 25/26, 1993. The exposure lengths were: 15 min for U and P, 5 min for X, Y and V and 3 min for Z and S. The filters give an unvignetted field of 20' diameter. The standard routines of the IRAF software package were used in the reductions.

For transformation of the instrumental CCD magnitudes and color indices to the standard Vilnius system we have used five standard stars in M 67 measured photoelectrically. The average magnitudes V and color indices (U-V), (P-V), (X-V), (Y-V), (Z-V) and (V-S) for 338 stars brighter than V = 14th mag were obtained. From an intercomparison of the February and March observations we estimate that their accuracy is of the order of ±0.02 mag.

A CM diagram V, (Y-V), has been constructed for M 67 stars down to V = 14th mag. The index (Y-Y) is one of the best temperature indicators in the Vilnius system and is very close to (b-y) of the Strömgren system. The zero-age main sequence, adjusted to the true distance modulus V - M(V) = 9.6 and color excess E(B-V) = 0.05, the expected values for M 67, is shown. This diagram is not very different from the V, (B-V) diagram of the UBV system (see Montgomery et al. 1993). The photometric data have also been used to classify the stars in spectral and luminosity classes by using several methods developed at the Vilnius Observatory and at the Rome Astronomical Institute.

The CCD photometry of the cluster M 67 in the Vilnius system shows that even with short exposures of the order of 15 min for ultraviolet filters and 3 - 5 min for other filters useful photometric results can be obtained. Several hours of CCD observations replace tens of nights of usual photoelectric photometry. The results obtained for M 67 combined with the future CCD photometry of other open and globular clusters will form a data-base for the calibration of two-color and reddening-free diagrams of the Vilnius system in terms of evolutionary tracks and ages.

REFERENCES

Boyle, R. P., Smriglio, F., Nandy, K. and Straižys, V. 1990a A&AS 84, 1
Boyle, R. P., Smriglio, F., Nandy, K. and Straižys, V. 1990b A&AS 86, 395
Boyle R. P., Dasgupta A. K., Smriglio F., Straižys V. and Nandy K. 1992 A&AS 95, 51
Montgomery, K. A., Marschall, L. A. and Janes, K. A. 1993 AJ
Smriglio, F., Nandy, K., Boyle, R. P., Dasgupta, A. K., Straižys, V. and Janulis R. 1991
    A&AS 88, 87
Straižys, V. 1977 Multicolor Stellar Photometry, Mokslas Publishers, Vilnius
Straižys, V. 1992a Multicolor Stellar Photometry, Pachart Publ. House, Tucson, Arizona
Straižys, V. 1992b Baltic Astronomy, 1, 107

ESTIMATION OF THE ERRORS INVOLVED IN THE INTENSITY MEASUREMENT OF LOW S/N RATIO EMISSION LINES

Claudia Rola and Didier Pelat

DAEC, Observatoire de Meudon

We present a detailed analysis on the extent of the errors in the intensity measurements of low signal-to-noise narrow emission lines. Our first goal is to determine a model for the probability distribution function (p.d.f.) associated with the measured intensities of a line characterized by its signal-to-noise ratio. Our final purpose is to provide an error domain: - for the measured signal-to-noise ratio of an observed line; - for the ratio of two lines in terms of their signal-to-noise ratios, and eventually (with the knowledge of the noise energy) to get errors on the corresponding intensities and intensity ratios.

To reproduce a real emission line intensity measurement process, we have designed a program based on a Monte-Carlo simulation procedure. For the purposes of this simulation, the knowledge of the true line parameters (the Intensity - imposed by the true signal-to-noise $(S/N)_{true}$, the position, and width) and the noise characteristics were necessary. Each spectrum was modeled by the sum of an emission line plus a continuum of a known constant level, this sum being considered as the mean of a certain stochastic process. The line was a Gaussian profile. To this model was added a non-correlated noise ("white noise") following the Student distribution.

Our simulation was divided in two main parts: the first, called detection, tells if a line is suspected to be present in a specific segment of the spectra; the second tries to measure the detected line, that is, to give estimates of the intensity, position and width of the suspected line. Once a line is suspected in one of the samples, the program proceeds to a fitting algorithm. The line model is also a Gaussian, the parameters of which are known only to lie within reasonable bounds. The adjustment is made through a standard X^2 minimizing routine. Once the minimum is attained a X^2 rejection test is performed on the residuals of the fit. If the test is satisfied, a Gaussian line is detected.

We made at least 1000 simulations for each $(S/N)_{true}$. For each signal-to-noise $(S/N)_{true}$, we calculated the intensity of each line in the detected lines sample. We call it "observed" intensity and denote it by S_{obs}. These intensities are then normalized to the known true intensity, S_{true}, and we call this normalised intensity, i, i.e., $i = S_{obs}/S_{true}$. Fits made for each $(S/N)_{true}$ led us to consider the log-normal distribution as a possible model for S_{obs}/S_{true}, which proved to give very reasonable results. We computed, for each normalised sample, a few statistics, the "true" signal-to-noise ratio, the mean of the sample (normalised to the true intensity), its variance and the square-root of the mean quadratic error calculated relative to one (the true normalised intensity value). Our results show clearly that for $(S/N)_{true}$ up to three, there is a strong bias (> 50%) towards overestimation of the values of the "observed" intensities.

In practice, the observer's interest is to determine from the knowledge of $(S/N)_{obs}$ a plausible range where $(S/N)_{true}$ may be found. In other words, one would like to know what are the limits which contain S_{true} with probability of γ. These error bars can be easily calculated since we know for each value of $(S/N)_{true}$ the distribution followed by $(S/N)_{obs}$. By a change of variable it is easy to derive the distribution followed by $(S/N)_{obs}$, which is also a log-normal. This parameter has the advantage of being a non-dimensional universal one. By integrating the $(S/N)_{obs}$ p.d.f. corresponding to each $(S/N)_{true}$ up to a confidence level γ we obtain the chart presented in Fig. 1, where the continuous thick line delimits the confidence intervals for γ = 0.683 (or one σ), and the thin line, the one for γ = 0.954 (or two σ). This chart was constructed horizontally, but must be used vertically with the following reasoning: an observer measures a line of $(S/N)_{obs}$ and obtains vertically the range on the $(S/N)_{true}$ values determined with confidence level γ. This chart shows clearly that, for example, at the one σ confidence level (or γ = 0.683), a line measured with $(S/N)_{obs}$ = 4 is compatible with $0 \leq (S/N)_{obs} \leq 5$. One cannot then rule out the possibility that a noise fluctuation was mistaken for a line.

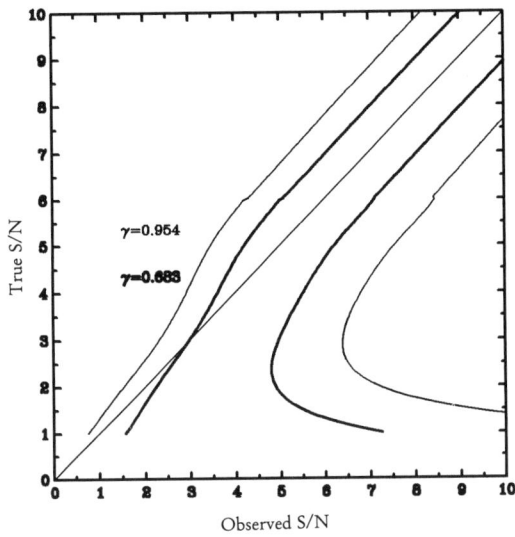

Fig. 1. Chart for the interval estimation of the $(S/N)_{true}$, given an observed value $(S/N)_{obs}$.

Furthermore, we calculated that a ratio of measured intensities or signal-to-noise ratios also follows a log-normal distribution and as such is subject to strong biases. Additionally, we determined the error domain on the ratio of a signal-to-noise ratios, which can be converted in an intensities ratio with the knowledge of the rrespondent noise energy ratio.

EEV AND ELECTRON CORP. VIRTUAL PHASE CCDs IN THE NEAR IR REGION, He I λ 10830 Å

A. G. Shcherbakov, Z. A. Shcherbakova,

Crimean Astrophysical Observatory

I. Ilyin and I. Tuominen

Observatory, University of Helsinki

He I, λ 10830 Å observations of late-type stars provide valuable information about the chromospheric and low chromospheric-coronal transition regions. High resolution measurements of the helium λ 10830 Å line profile offer a convenient way to survey the coronal emission of the Sun as a point source, as well as a variety of stars (O'Brien and Lambert, 1986, Shcherbakov and Shcherbakova, 1991).

The He I λ 10830 Å survey of solar and stellar activity commenced in 1985 at the Crimean Astrophysical Observatory, using CCD cameras for spectral observations in the near infrared region. As is well known, the sensitivity of the CCD in the near infrared region is only a few percent of the maximum sensitivity. To increase sensitivity in the infrared region, keeping the noise to a moderate level, the working temperature of the CCD should be increased to an optimal value.

Fig. 1 represents our investigations of the CCD camera, manufactured by Astromed Ltd., Cambridge, England, with an EEV P88200 CCD. The tests were made with the high resolution echelle spectrograph SOFIN at the Nordic Optical Telescope, La Palma. The various parameters of the CCD were tested as contributors of the total noise: $\delta^2_{total} = \delta^2_{stat} + \sigma^2_{spn} + (\sigma_{dark}/C)^2$, where δ_{total} is the relative error of the output signal, δ_{stat} is the photon noise of the accumulated signal, σ_{spn} is the spatial noise in the pixel-to-adjacent-pixel scale and σ_{dark}/C is the ratio of the dark current noise and accumulated signal. It was assumed that the readout noise of the CCD was included in the dark current noise value. The value of the spatial noise gives information about the photometric quality of the CCD. The noise is almost removable by a flatfield. Dark current noise does not contribute seriously to the accumulated signal of about 10000 e⁻ per one hour at temperatures near to 170 - 180° K, compared with 120 - 140° K. On the other hand, the sensitivity is increased 2.5 times at λ 10830 Å when the temperature is raised from 120° K to 180° K.

To study the activity of late-type stars the Astromed Ltd. system [with EEV CCD P8600, as well as the ASTRO-550 system by Ista Ltd., St.Petersburg, Russia (Berezin et al., 1991), with a virtual phase CCD (developed and produced by Electron Corp., St.Petersburg, Russia)] were used with the coudé spectrograph of the 2.6-m telescope of the Crimean Astrophysical Observatory. Both systems make it possible to reach an S/N ratio in the range of 30 - 100 with a spectral resolution about of 30000 for a 5 - 6ᵗʰ magnitude star in the near infrared. Both EEV

P88100 and P88200 are used with the SOFIN spectrograph at the Nordic Optical Telescope and allow a high S/N ratio of 90 - 100 with a spectral resolution up to 60,000. The ASTRO-550 system has been applied also to the observations of the Sun-as-a-star (with a resolution about of 200,000) with the 0.5-m Solar Telescope of the Crimean Astrophysical Observatory.

Fig. 2 shows the level of chromospheric activity of late-type stars as seen in the He I λ 10830 Å line profile. It is easy to see from the comparison of the line profiles for the Sun as a star, and for the active region on the solar disk, that a deeper and wider line is observed in more active stars.

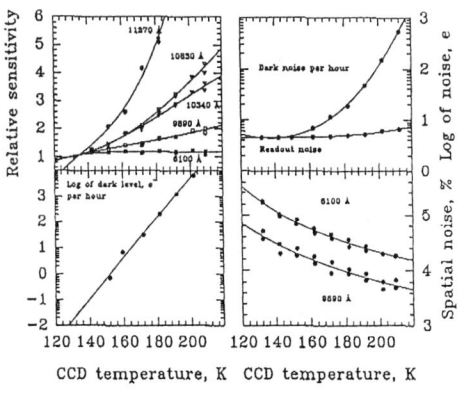

Fig. 1. Behavior of various parameters as a function of temperature for the EEV P88200 CCD.

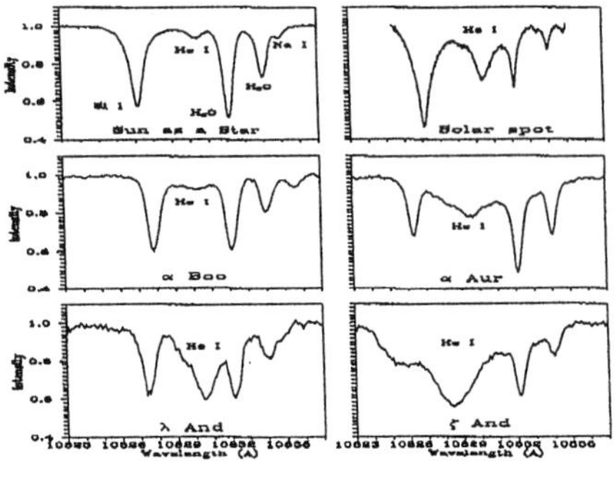

Fig. 2. He I λ 10830 Å line profile of the Sun and some other late-type stars

REFERENCES

Berezin, V. Yu., Zuev, A. G, Kiryan, G. V., Rybakov, M. I., Hvilivitsky, A. G., Ilyin, I. V., Petrov. P. P., Savanov, I. S. and Shcherbakov, A. G. 1991 Letters to Soviet AJ 17, 953

O'Brien, Jr. and Lambert, D. 1986 ApJS 62, 899

Shcherbakov, A. G. and Shcherbakova, Z. A. 1991 IAU Colloquium No. 130, The Sun and Cool Stars: Activity, Magnetism, Dynamos, I. Tuominen, D. Moss and G. Rudiger, eds., Lecture Notes in Physics, Springer, p. 252

STELLAR PHOTOMETRY WITH A PERFECT CCD

Bjarne Thomsen and Frank Grundahl Jensen

Institute of Physics and Astronomy, Aarhus University

We have developed algorithms for the construction of a simulated globular cluster, as it would appear on a frame obtained with a perfect CCD. This has allowed us to compare the absolute errors in the photometry of two different programs, DAOPHOT II (Stetson 1991, 1994) and SPS (Janes and Heasley 1993), with a special emphasis on criteria for selecting stars with the best photometry.

1. THE SIMULATED GLOBULAR CLUSTER

We assume (Thomsen and Frandsen 1983) that the shape of the Optical Transfer Function (OTF) can be represented by

$$OTF(u,v) = aH(\omega)\exp(-2\pi\theta)$$
$$\theta = \alpha u + \beta v$$
$$H(\omega) = \exp(-(\omega/\gamma)^P)$$
$$\omega^2 = u^2 + v^2 + a_2(u^2 - v^2) + 2b_2 uv$$

The location of the PSF is determined by (α, β) and (a_2, b_2) is a measure of the direction and size of the ellipticity of the PSF. The adopted FWHM of the PSF was 3.5 pixels.

The stellar positions in the simulated frame were created as the projection of a Plummer model (R_c = 100 pixels) for the spatial density of stars. The stellar magnitudes were obtained from a two-segment luminosity function, with one segment representing the giants and the other the main-sequence stars. The parameters defining the luminosity function were estimated from data in Bergbush and VandenBerg (1992). A total of 200,000 stars, between instrumental magnitudes 10 and 23, were added to the frame. The gain and readout noise were set to 3 e$^-$/ADU and 7.5 e$^-$, respectively. Subsequently we added a constant night-sky value, a random Poissonian photon noise, and finally a random Gaussian readout noise.

2. DISCUSSION

Our main result from the comparison is that the two programs seem to deliver rather similar results when comparing the bulk of the photometry. Both programs find roughly the same number of stars, and the photometric accuracy is also similar. The faint-end bias reported by Schechter et al. (1993) seems to have been removed by the addition to ALLSTAR of a repeated sky determination, as suggested by Parker (1991).

However, in both cases it is evident that a large fraction of the detected stars has a relatively poor photometric accuracy. Approximately 30% deviate more than 0.2 mag from the true value. In the world of real photometry we do not know the true magnitude of a star, and thus we need some criteria for selecting the fraction that has the best photometric accuracy.

ALLSTAR offers various quality indicators. SIGMA is the estimated standard error of the magnitude. CHI is the ratio of the observed pixel-to-pixel scatter of the fitting residuals to the expected scatter. SHARP is a measure of how well the observed profile of a given star matches that of the PSF. CHI should be close to one, and SHARP is expected to be close to zero. In addition we have implemented two new shape parameters, A_2 and B_2. They measure the ellipticity along the coordinate directions and the diagonal directions, respectively.

Through cuts in (SIGMA, CHI, SHARP, A_2, B_2)-space we isolated stars with the best photometry. After a few experiments we arrived at the following useful limits on the ALLSTAR parameters: CHI < 1.4, -0.3 < SHARP < 0.2, $\sqrt{(A_2^2 + B_2^2)}$ < 0.15, and SIGMA < 1.1*MEDIAN(SIGMA(m)) + 0.001. 30% of the total number of stars detected by DAOPHOT satisfies these limits, but 6% of the selected stars still deviates more than 0.2 mag from the true value. Omitting these, the mean absolute deviation (MAD) of the remaining sample is 0.03 mag. Our final sample contains 2189 stars, primarily the brightest, but it still contains stars more than 2 mag fainter than the cluster turn-off.

SPS does not provide as many useful indicators as ALLSTAR, and thus the possibility of rejecting stars with poor photometry is smaller. SIGMA is again the standard error. ADD is the number of additional stars that is fitted simultaneously with the current star. We adjusted the bounds on SIGMA and ADD until we arrived at the following limits: ADD < 10 and SIGMA < 0.1. 30% of the stars detected by SPS satisfies these limits, but now 12% of the selected stars deviates more than 0.2 mag. Omitting these, as well as stars fainter than 17 mag, the MAD of the remaining sample is 0.04 mag. For the faint-end stars the two SPS constraints are less efficient in rejecting stars with poor photometry than are the four ALLSTAR constraints. The opposite trend seems to apply for the bright-end stars.

After the selection process most of the remaining stars with large deviations are found to be too bright by both programs. This was actually our main motivation for implementing A_2 and B_2 in ALLSTAR, and they did in fact help to remove many stars with large deviations. The number of deviating stars are at least equally abundant in the SPS photometry. Even though we have not been able to identify the cause of the large deviations, it is still important to find a way to locate these stars accurately, since they might contribute considerably to the scatter in an observed color-magnitude diagram, especially near the turn-off region.

REFERENCES

Bergbush, P. and VandenBerg, D. A. 1992 ApJS 81, 163
Janes, K. A. and Heasley, J. N. 1993 PASP 105, 527
Parker, J. Wm. 1991 PASP 103, 243
Schechter, P. L., Mateo, M. and Saha, A. 1993 PASP 105, 1342
Stetson, P. B. 1991 in Proceedings of the 3rd ESO/ST-ECF Data Analysis Workshop, P. J. Grosbol and R. H. Warmels, eds., ESO, Garching, p. 187
Stetson, P. B. 1994 PASP 106, 250
Thomsen, B. and Frandsen, S. 1983 AJ 88, 789

APPLICATIONS OF A REALISTIC MODEL FOR CCD IMAGING

Steve B. Howell and William J. Merline

Planetary Science Institute

1. DESCRIPTION OF CCD MODEL

We have constructed a computer model for simulation of point sources imaged on CCDs. An attempt has been made to ensure that the model produces "data" that mimic real data taken with two-D detectors. To be realistic, such simulations must include randomly generated noise of the appropriate type from all sources. The synthetic data are output as simple one-D integrations, as two-D radial slices, and as three-D intensity plots. Each noise source can be turned on or off so they can be studied independently as well as in combination to provide insight into the image components.

Some of the many properties of a detector and optical system that can be simulated include the effects of changing image position, image scale, signal strength, noise sources, and data reduction techniques. This model is useful for planning observational strategy and for optimization of photometric reductions. Details of the model, error analysis, software, and applications can be found in Merline and Howell (1992) and Merline and Howell (1995). Below, we discuss briefly some of our recent modeling endeavors.

2. APPLICATIONS

The Lowell Observatory Near-Earth-Object Survey (LONEOS), plans to use a 1-m Schmidt telescope with four 2048 x 2048 CCDs to survey ~ 1000 sq. degrees of sky each night. The primary goal of the project is to detect numerous faint NEOs. We intend to use the LONEOS data for the first large-scale photometric search for extra-solar planets, by monitoring millions of stars per year for evidence of planetary transits across the stellar disks. While scintillation noise sets fundamental limits for a ground-based photometric search, we have used our model to show that transits by Jupiter- and Saturn-sized planets should be discernible in the light curves of the brighter late-type stars (see Howell 1995). Statistically, we expect one reliable transit observation each year.

NASA's first discovery class mission will be the Near-Earth-Asteroid Rendezvous (NEAR) Mission. The imager will consist of a 50-mm telescope and a frame transfer CCD. The CCD has rectangular pixels and the images will have substantial contributions from read and dark noise. These features will present challenges for calibration and image interpretation. We have used our model to perform some initial simulations for the NEAR imager (Fig. 1), as we had done previously for the Galileo imaging system. Typical targets for spacecraft instrument/filter calibration and pointing information are star clusters such as the Pleiades.

Comet Shoemaker-Levy 9 presented the special problem of detecting faint signals against the bright backgrounds of Jupiter and its satellites. Our observational method used continuous readout of a CCD, yielding time-resolved imaging (one spatial dimension vs. time, resulting in streaked images) of Jupiter and its moons, high duty cycle (little dead time), and avoided saturation during the exposure, even on large-aperture telescopes. We applied our model to determine the optimum combinations of telescope aperture, image scale, detector, read rates, and filter. Because our model is intended for point-sources, modifications had to be made to simulate the scanned images, but this indicates the adaptability of the software.

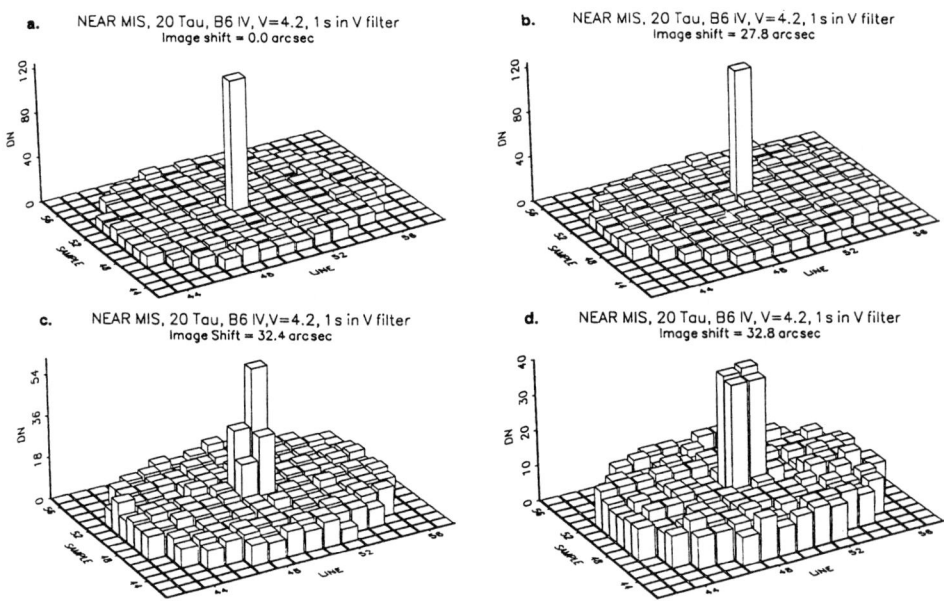

Fig. 1. Sample images of Pleiades star 20 Tau, seen by the NEAR imager, using our CCD modeling code. This demonstrates the sensitivity to positioning in undersampled (large pixels) images. Images a and b are indistinguishable, even though the image has been moved by nearly 28 arcsec. In model c, the image has been shifted an additional 4.6 arcsec, and in d only 0.4 arcsec more. A bias level of 10 DN was arbitrarily chosen: pixels near the periphery are not calculated and are assigned value zero. This sensitivity to position will be important when using stellar images for navigation, optical distortion mapping, and in distinguishing possible asteroid satellites from cosmic rays.

REFERENCES

Howell, S. B. 1995, in Proc. IAU Symposium No. 167, New Developments in Array Technology and Applications, A. G. D. Philip, K. A. Janes and A. R. Upgren, eds., Kluwer Academic Pub., p. 167

Merline, W. J. and Howell, S. B. 1992 in Astronomical Data Analysis Software and Systems I, D. Worrall, C. Biemesderfer, and J. Barnes, eds., ASP Conf. Series, Vol 25, San Francisco, p. 316

Merline, W. J. and Howell, S. B. 1995 Experimental Astronomy, in press

3D: THE NEW NEAR-INFRARED FIELD IMAGING SPECTROMETER

L. E. Tacconi-Garman, L. Weitzel, M. Cameron, S. Drapatz, R. Genzel, A. Krabbe, H. Kroker and N. Thatte

Max-Planck-Institut für Extraterrestrische Physik

ABSTRACT: 3D is a next-generation near-IR spectrometer developed at the MPE which offers, in a single integration, the opportunity to image an 8" x 8" field across almost the entire K-band at a simultaneous spatial resolution of 0".5 wide strips which are then aligned optically on top of each other forming a single long slit. This long slit is then used as the input for a grating spectrometer which images it onto a two dimensional detector array. Each detector row then represents the spectrum of one spatial element of the two dimensional field of view. The central part of the optical system is the image slicer which is made of two complex plane mirror systems consisting of 16 segments each. The detector is a NICMOSIII HgCdTe array with 256 x 256 pixels. In the spectral domain the spectrometer provides a resolving power of R = 1000.

Here we present not only the design of the instrument but also first data obtained during instrument commissioning at the 3.5-m Calar Alto telescope in December, 1993.

TECHNICAL CCD SYSTEMS FOR THE ESO VERY LARGE TELESCOPE

J. Ehrich

Jena-Optronik GmbH

The Technical CCD Systems (TCCDS) consist of thermo-electrically cooled CCD detector heads together with the electronic hardware and software for the control of the CCD and readout of the data. Two different types of CCD heads equipped with a small and a large format CCD chip are used. The small format CCD systems will be used primarily for auto-guiding and field acquisition at the main foci of the VLT. The large format CCD systems are intended mainly for wavefront sensing applications. An overview of the primary modes of operation of the TCCDS and the corresponding fields and exposure conditions for these modes is given in Table 1. The term "Technical CCD Systems" is derived from the intended application in the VLT because due to the excellent CCD head parameters and the system performance a scientific application will be offered too.

TABLE 1

Primary modes of operation

CCD Format	small	small	large
Operational Mode	acquisition mode	guiding mode	wavefront sensing
Field Size	1 arc min	20 x 20 pixels	$\geq$ 12.5 x 12.5 mm^2
Wavelength	400 - 1000 nm	600 - 700 nm	00 - 600 nm; 700 - 1000nm
Integration Time	0.5 - 2 sec	0.01 - 2 sec	$\geq$ 30 sec
Centroiding Accuracy		0.2 pixel	0.1 pixel
Limiting Magnitude	> 21	16 (for 0.1 sec int time)	16, distributed over 400 points

The CCD head (see Fig. 2) represents a compact mechanical design and is composed of the sealed box with a A/R coated glass window, the small or large CCD chip, the highly effective two stage Peltier cooler, two temperature sensors - one for the chip and the second for the box surface, hermetically sealed feed-through connectors, a liquid cooled heat exchanger, a external connector board with video preamplifier and cables to the array control electronic box. Furthermore the excellent performance of the CCD head is characterized by a minimum life time of 10 years, high positioning accuracy of the chip and high thermal stability to maintain the surface temperature of the head close (i.e., $\pm 1°C$) to ambient by using liquid cooling.

The CCD sensors are front illuminated frame transfer devices with a "2.5-phase" Virtual Phase image section. In comparison to conventional front illuminated CDDs the virtual phase

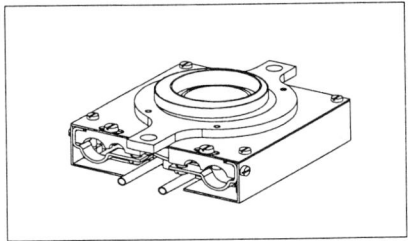

Fig. 1. Three-D view of the CCD head.

provides a wide spectral range (typically 200 - 1000 nm) with enhanced blue sensitivity, very uniform photoelectric response and symmetric aperture response of a pixel in full sperctral range.

The TCCDS electronic hardware consists of two building blocks, the ACE box and an associated Local Control Unit (LCU). These units are joined by a 20 MHz serial bi-directional transputer fiber-optic link.

At a distance of 2.5 m from the CCD head the array control electronics (ACE) is located in a compact box with an integrated liquid cooled heat exchanger. The ACE represents a modification of a design developed by ESO, especially the transputer based CCD controller. The ACE consists of three functional modules - sequencer, clock driver and video processor. The ACE provides universal control of different CCD detectors with varying clock diagrams, maximum 16 clock electrodes with 32 individually programmable voltage levels. The video processor with analogue pre-processing, 1, 2, 4, 8-gain switching, 12-bit A/D conversion and data acquisition with a 16-bit transputer provides a 12 bit dynamic range in any single integration or 14 bit with gain switching. In addition to these the ACE box contains a bus back plane, a power supply module, a module for auxiliary functions as temperature monitoring etc. and the fiber optic interface as one part of the data transfer link to the LCU. The other part , the link adapter board (LAB), represents a VME interface card located in the LCU and, like the fiber optic interface, converts electrical signals from/to the link interface board to optical signals for transmission to /from the ACE. The transputer serial link is interfaced to the VMEbus of the LCU via the link interface board (LIB). The LCU is a standard VMEbus processor which is connected to the main VLT computer system through a Local Area Network and is dedicated to the control of the TCCDS.

A comprehensive software package is provided. It consists of ACE embedded software and test software. The ACE embedded software is running on a transputer network. The CCD chip control will be done by the DSP 56001. On the LCU runs the control and test software under VxWorks. Command and control of the embedded software is realized from the LCU via a transputer link to ACE. This software package will be integrated in the VLT software developed by ESO running on different workstations.

The company Jena-Optronik GmbH, a subsidiary of Deutsche Aerospace and Jenoptik, has been chosen by the European Southern Observatory to develop and manufacture the Technical CCD Systems for the ESO Very Large Telescope.

POSTER DISCUSSION

POSTER SESSION A: DISCUSSION - I. S. McLean, Chairman

Reiss - Page 319

McLEAN: Why did you choose to combine transputers and DSPs in your system?

REISS: Transputers are ideal for high level parallel programming, however they have problems when it comes to precise, jitter free timing. DSPs are perfect timing machines and more suitable for low-level programming.

GLASS: We use transputers for timing the pulse information. No interrupts are used. The link is the limiting factor in readout speed. One crystal (on transputer card) provides all the oscillator frequencies to the controller.

BONANNO: The CCD controller of the Italian National Telescope is based on transputers and DSP. I wish to stress the possibility of easy maintenance and debugging by using transputers: the controller can be reached from any point on the network. Even if the speed of the link is almost 20 orbits, in applications such as CCDs, where the readout is slow, the operation can work. Furthermore in transputer network it is easy to demand the image reformatting (output from four channels) to a certain transputer that is not busy. This procedure it is easy to implement because it can be written in a high level language (OCCAM). Don't forget that TRAM modules are off the shelf products available in the electronic market.

WAMPLER: Which system has the best noise characteristics? As an observer I have had a lot of problems with pick-up noise.

REISS: It is important to synchronize the CCD timing generator and data acquisition to a single, jitter free master clock to avoid internal interference. External interference can be avoided by proper grounding, shielding and EMC control.

LEACH: With transputers having limited link speed and having some difficulty in being used for timing generators, what are they useful for in an array controller?

REISS: The link speed of 1.5 MByte/sec is sufficient for most CCD applications. Transputers are very well suited for embedded control and parallel systems.

Szecsenyi-Nagy - Page 331

SZECSENYI-NAGY: Our problem is how to replace the old, but from today, more expensive image recording device (the plate) by a modern semiconductor device which may be used repeatedly. We have the necessary computing facilities, computers and also image processing software has recently been started but we cannot afford two millions of US dollars as it has been mentioned in the case of Lick Obs. We have about 1% of that amount but we wanted to spend it optimally. We read the offers of different manufacturers and see how large are the differences in their prices. Even CCD-systems with very similar technical specifications and capabilities may have shockingly different price tags. As we have not had the opportunity to study such systems in detail or even use them for night observations, we badly need the comment of those colleagues who have experience with commercially available CCD cameras. What we need is a 1k x 1k CCD with TE cooling and a filter wheel. You may suggest systems current on the practical usage etc. here or at the board.

VENKATAKRISHNAN: When one is constrained by money we have to just go ahead and use the systems and evaluate them. We will always find something to extract from the data.

Didkovsky - Page 333

McLEAN: Can you tell us a bit more about your alignment system?

DIDKOVSKY: A new Ritchey-Chretien telescope "Spectrum-UV" which is currently being designed, will be equipped with two spectrometers installed at significant distance (55 mm) from the center of the field of view. Therefore, any displacement and/or tilt of the secondary mirror may strongly affect the quality of image in the spectrometers.

Goncharov - Page 335

McLEAN: What is the state today of your project to modernize the Pulkovo vertical circle?

POLOJENTSEV: We are ready to order the parts, especially the CCD matrix.

Yershov - Page 339

YERSHOV: Referring to my poster, the main attention in this work is paid to autocollimation control of the reflector telescope optics to use it in astrometric observations. Two wavebands are planned to be used for the observations with two detectors: a Pbs array of 66 x 64 pixel (K) and a 68 x 580 CCD-array (J and visual). Sitallic optics are being made now and test observations are planned this year with the Pbs array installed in one of the Pulkovo reflectors. The work is planned to be finished at the end of the next year.

McLEAN: What's the cost of your devices?

YERSHOV: It depends upon the device grade, but it is up to two times less than the price of the Gen II MCP image intensifier. The 1st grade EB-CCD costs approximately $5,000.00.

McLEAN: Were your EB-CCDs used in astronomy?

YERSHOV: Not yet, because they were classified up to now because of military applications. But just now the project is beginning with these devices as the guiding sensors for the Special Astrophysical Obs. in Zelentchuk.

McLEAN: What is the field of application of these devices?

YERSHOV: The EB-CCD's, as the convenient intensified CCD's, are intended to be used in the case, where the sensitivity of the CCD couldn't be increased by the CCD cooling or expanding the integration times; that is in the real time scale TV imaging. There are also very attractive for the fast processes investigations, in adaptive optics applications, etc.

POSTER SESSION B: DISCUSSION - A. G. Davis Philip, Chairman

Guseva - Page 355

PENNY: NGC 188 would be a good open cluster to observe.

GUSEVA: I cannot answer immediately, thank you, I will look at this cluster if it is suitable for us.

PHILIP: Do you have plans for starting now on a program of setting up standard fields?

GUSEVA: Yes, I'm going to obtain catalogues of some related fields.

SCHILDKNECHT: What accuracy are you aiming at in your program?

GUSEVA: No worse than $0''\!.1$. I hope it will be better, of the order of $0''\!.03$ - $0''\!.05$.

CUYPERS: I want to inform you that colleagues at Hoher/List (Germany, M. Geffert and others) and at the Royal Observatory (D. Sinachopoulds and myself) are working on standard fields for astrometry with CCD. These fields include open clusters with well known proper motions, a field around the bright quasar 3C273, and we are investigating the Landolt fields of faint photometric standards. I also ask the CCD-observers of open clusters to publish or make available the positions of the observed stars and not only the photometry.

Bykov - Page 353

PHILIP: Have you plans to determine the orbits of many additional asteroids?

BYKOV: Yes, I plan to determine the orbits of any celestial objects moving on the background of the reference stars and having the accurate use of CCD-observations. In particular we collaborate with the SWT-group.

PHILIP: And will you be able to continue using results from 12 telescopes?

BYKOV: Yes. Our software allows us to calculate orbits by Laplacian method and the Pulkovo AMP method on the base on supershort arc positional CCD and photo observations. Then each observatory may send us the observations by email for testing.

INDICES

Name Index   383

Object Index   388

Subject Index   390

If a page number is underlined in the Name Index it indicates the name of an author of a paper. An underlined page number in the Object or Subject Index indicates that the object or subject was mentioned in the title of a paper. Subjects that are mentioned in major headings may be listed in the index but their page numbers are not underlined.

Name Index

NAME INDEX

A

Abad, C.	<u>349</u>
Abbott, T.	195, 196, <u>343</u>
Abell, G.	360
Adelman, S.	5
Allen, C. W.	278
Andersen, M.	<u>207</u>
Appenzeller, I.	10
Arsenault, R.	213

B

Baade, W.	74, 360
Bagildinsky, R.	<u>333</u>
Baines, K.	252-3
Balestra, A.	<u>319</u>
Balmer, J.	182
Balona, L.	116, <u>187</u>, 190, 194-6
Baluteau, J.	5
Banks, D.	34
Bessell, M.	1, 68, 144, <u>175</u>, 176, 185
Biereichel, P.	<u>81</u>
Bikmaev	229
Bingham, R.	292
Binzel, R	260
Blades, C.	5
Blecha, A.	1, 36, <u>57</u>, 66
Blouke, M.	5
Boeshaar, P.	244
Boksenberg, A.	5
Bonanno, G.	<u>39</u>, 40, 45, 48, 319, 377
Bongiovanni, M.	6
Bortoletto, F.	<u>319</u>
Borzyak, P.	<u>331</u>
Boyle	69
Boyle, B.	292
Boyle, R.	<u>363</u>
Bredthauer, D.	14
Brooks, P.	59
Brown, W.	<u>19</u>, 49
Bruno, P.	<u>319</u>
Butcher, H.	76
Butler	151, 153
Bykov, O.	<u>351</u>, 379

C

Cameron, M.	373
Carter, D.	116, 292, <u>321</u>
Cassegrain	9, 14, 97, 117, 118, 112, 121, 136, 144, 253, 255
Chamberlin, S.	5
Chevreton, M.	325
Chikada, Y.	<u>117</u>
Chretien, J.	331, 378
Churchill, J.	34
Cochran, A.	1, <u>251</u>, 260-1
Cochran, W.	252-3, 260
Cohen, J.	17, 25, 48
Cook, K.	140, <u>285</u>, 286, <u>357</u>
Corbally, C.	1, 239, <u>241</u>, 246, 249-50
Cortiula, J.	<u>295</u>
Cosentino, R.	<u>319</u>
Cousins, A.	175-6, 179, 184
Crampton, D.	<u>213</u>
Crawford, D.	5, 165, 172
Cuby, J.	78, 229, 239
Cuillandre, J.	<u>213</u>, 219, <u>263</u>
Cullum, M.	1, 48, 143, 273
Cuypers, J.	379

D

Dalinenko, I.	<u>341</u>
Dasgupta, A.	<u>363</u>
Deries, S.	344
Dekker, H.	10
Dereniak, E.	5
Didkovsky, L.	<u>331</u>, 378
Doi, M.	<u>345</u>
Dolgushin, A.	<u>331</u>
Doppler, C.	203, 221
Drapatz, S.	373
D'Allessandro, M.	<u>319</u>
D'Odorico, S.	5-6, <u>9</u>, 17, 66, 80, 107, 116, 219, 229, <u>309</u>

E

Edser	151, 153, 155
Einstein, A.	286
Erich, J	<u>375</u>

F

Fabry	154
Fantinel, D.	<u>319</u>
Faraday, M.	210
Fazio, G.	1, <u>93</u>, 116
Felenbok, P	10

Name Index

Finger, G. — 80, <u>81</u>, 107, 121, 273
Florentin-Nielsen, R. — 68, 80, 144, 155, 185, <u>207</u>, 211, 249, 286
Flynn, B. — 260
Forrest, W. — 71
Fort, B. — <u>263</u>
Foucault, J. B. — 151
Fourier — 343
Fowler, A. — 73
Frandsen, S. — 190, 194
Fresnel, A. — 200

G

Gai, M. — <u>355</u>
Galadri-Enriquez, D. — <u>327</u>
Galileo, G. — 29, 42, 254, 311, 371
Garrison, R. — 241, 249
Gatley, I. — 73
Gauss, K. F. — 91, 189, 222-3, 225, 229, 349-51, 365, 369
Geary, J. — 5, 14, 213, 216, 298, 310
Geffert, M. — 379
Geiger — 208
Gennari, S. — <u>231</u>
Genzel, R. — 373
Gezari, D. — <u>97</u>, 107
Gigoux, P. — 340
Gilmore, K. — <u>19</u>, 49, <u>79</u>, 80
Glass, I. — 56, <u>109</u>, 116, 286, <u>321</u>, 377
Goncharov, G. — <u>333</u>
Gothard, J. — 329
Graham, J. — 177, 179, 184
Gray, R. — 243, 248
Greenstein, J. — 259
Griese, J. — <u>347</u>
Grozdilov, V. — <u>335</u>
Gruthaner, P. — 48
Guarnieri, M. — <u>355</u>
Gunn, J. — 126, 177
Guseva, I. — 1, 164, <u>275</u>, 283, <u>353</u>, 379

H

Halley, E. — 259
Harmanec, P. — 185
Hauser, M. — 107
Heathcote, S. — 340
Herrero, J. — <u>347</u>
Herschel, W. — 33, 289
Hewitt, M. — 107
Hilbert, D. — 145, 148-9
Hodapp, K. — 73
Hog, E. — 129
Houk, N. — 250
Howell, S. — <u>167</u>, 171-2, 196, 249, 286, <u>371</u>
Hubble, E. — 8, 11, 125, 128, 244
Hughes, J. — 340
Humphries, C. — 5

I

Ilyin, I. — <u>367</u>
Ingerson, T. — 340
Ives, D. — 34
Iwert, O. — 17, 34, <u>67</u>, 68, 121, 164, 296, 308

J

Jacoby, G. — 5-6, 167
Jansky, K. — 97, 102-3, 106
Jedicke, R. — <u>157</u>, 164-5, 171
Jenkins, C. — 292
Jensen, F. — <u>369</u>
Jiang, S. — <u>325</u>
Johnson — 41, 51
Johnson, A. — 34
Johnson, H. — 126, 345
Jorden, P. — 26, <u>27</u>, 36-7, 48, 68, 121, 130, 155
Jordi, C. — <u>327</u>
Joyce, R. — 107

K

Kagert, P. — 205
Kashikawa, N. — <u>345</u>
Keck — 19-21, 24, 26, 29, 107, 298
Kelvin, W. T. — 97, 99
Kiar, K. — 5
King, I. — 146
Kirian, T. — <u>335</u>
Kirkpatrick — 244
Koen, C. — 194-5
Konkoly-Thege — 329
Korepanov, V. — <u>335</u>
Kornilov, E. — <u>333</u>
Kossov, V. — <u>323</u>, <u>341</u>
Krabbe, A. — 373
Kroker, H. — 373
Kron, C. — 190

Name Index

Kurtz, M.	249	IBM	26
Kuzmin, G.	341	ISTA Ltd.	367
		Jena-Optronic	375-6
L		Jenoptic	376
		Kodak	37, 70-1
Landolt, A	175-7, 179, 184	LETI-LIR	70-1, 84
	379	Lincoln Labs	20-1, 24, 26, 30-1
Laney, C.	195		63, 70-1, 84, 303
Laplace, P. S.	351, 379		311
LaSala, J.	249-50	List of Manu.	27-8, 30, 84
Lasovsky, L.	341	Loral	13-4, 21, 26, 28
Latham, D.	5		125, 127, 208-11
Lattanzi, M.	355		215-6, 304-5, 307
Lazovsky, L.	323		311, 319
Leach, R.	<u>49</u>, 56, 273	South	310
	377	Mitsubishi	70-1, 117-8
Lee, J.	<u>347</u>	Motorola	321
Lensen, R.	10		
Lesser, M.	11, 14, 31, 33, 68	Orbit	19, 21, 24, 26
	127, 302, 310-11		127
Levy, D.	110, 112, 372	Phillips	318, 322
Lewis, I.	292	RCA	6, 31, 157, 195-6
Lick	19, 21, 48, 74, 79		307, 322, 362
	80, 377	Reticon	127
Liu, T.	<u>359</u>	Rockwell	70-1, 73-4, 79
Lowell, P.	371		81, 90, 309
Lundin, C.	347	Rolyn Optics Co.	183
Luppino, G.	21, <u>213</u>, <u>297</u>	SBRC	73-4, 84, 86-7
	308		90, 309
		Schott Glass	177-9
M		Service d'Astrop.	84
		SITe	13, 127, 308
MacLean, J.	34		310-1
Maksutov	291, 333	Tektronix	31, 157, 182
Malyarov, A.	341		196, 211, 266
Manfroid, J.	147		298, 310, 322
Manufacturers			340
Aerojet	70	Texas Instruments	6, 31, 345
Amber Engin.	70	Thomsen	14, 68, 127, 182
Astromed	59		266, 268, <u>295</u>, 297
Bell Labs	69		311
Cincinnati Elect.	70		340
Clark Co.	347	Zeiss	329
Crystal Semicond.	50	Marconi, A.	<u>231</u>
Deutsch Aerosp.	376	Marcucci, P.	<u>319</u>
EEV	40-1, 48, 59	Massone, G.	<u>355</u>
	127, 130, 311	McBreen, B.	26
	367	McLean, I.	1, <u>69</u>, 74, 77-8
Electron Co.	367		107, 315, <u>359</u>
Epoxy Tech. Inc.	184		377-8
Fairchild	6, 31, 307	Melliers, Y.	<u>213</u>
Ford Aerospace	14, 31, 310	Merline, W.	<u>371</u>
Foundry Kesser	308	Menzies, J.	116, 141
Hamamatsu	42	Meyer, M.	<u>81</u>
Hughes	70-1, 74, 103	Mezger, M.	<u>297</u>
IBC	73	Miyazaki, S.	<u>297</u>

Mondaca, E.	340	**R**	
Moneti, A.	<u>355</u>		
Moorwood, A.	10, <u>81</u>	Rank, D.	<u>79</u>
Morales, A.	<u>347</u>	Reiss, R.	<u>317</u>, 377
Moseley, H.	<u>95</u>, 96, 107, 121	Rieke, G.	239
Mould, J.	345	Ritchey	331, 378
Müller	208	Rogers, R.	340
Munari, U.	<u>355</u>	Rola, C.	<u>365</u>
Murowinski, R.	<u>213</u>	Rountree, J.	243
		Rudeen, A.	340
N		Rumyantsev, K.	<u>333</u>
		Rutherford, E.	311, 321
Nakada, Y.	<u>109</u>		
Nasmyth	9, 33, 50, 310	**S**	
Newton, G.	<u>321</u>		
Newton, I.	34, 119	Savart	201
Nicolini, G.	<u>81</u>	Sawyer, S.	260
Nielsen, S.	56, <u>207</u>	Schildknecht, T.	379
Ninkov, Z.	37	Schilling	164
		Schmidt, B.	33, 127, 130
O			136, 144, 159
			290-1, 371
Oates, A.	<u>27</u>	Schottky	121
Oemler, A.	76	Sekiguchi, K.	<u>109</u>
Okamura, S.	<u>345</u>	Sekiguchi, M.	<u>345</u>
Onaka	41	Shane	75
Osmer, S.	5	Shanks, T.	292
Oswalt, T.	155	Sharples, R.	292
O'Donoghue, D.	130, 194	Shcherbakov, A.	<u>367</u>
		Shcherbakova, Z.	<u>367</u>
P		Shimasaku, K.	<u>345</u>
		Shkutov, V.	<u>333</u>
Parker, Q.	292	Shoemaker, E.	110, 112, 372
Parry, I.	292	Sinachopoulds, D.	379
Pearson, W.	116	Sloan	127, 290
Pel, J.	127	Smith	69
Pelat, D.	<u>365</u>	Smith, D.	340
Peltier	279, 323, 376	Smith, R.	340
Penny, A.	1, 130, 143, <u>173</u>	Smriglio, F.	<u>363</u>
	283, 286, 376	Soucail, G.	<u>263</u>, 272-3
Pettyjohn, K.	5	Spencer, S.	260
Philip, D.	1, <u>3</u>, 5, 17, 25	Spyromilio	77
	164, 196, 205	Stenflo, J.	272
	292, 315, <u>361</u>	Sterken, C.	1, <u>131</u>, 143-4
	379		147, 155
Picat, J.	<u>263</u>	Steshenko, N.	<u>331</u>
Plummer	369	Stetson, P.	133, 362
Poisson	12, 238	Steward	290, 302, 310
Polojentsev, D.	286, <u>333</u>, 378	Stock, J.	<u>349</u>
		Stover, R.	1, <u>19</u>, 25-6, 36
Q			48-9, 56
		Straižys, V.	<u>363</u>
Queloz, D.	<u>221</u>, 229	Strømgren, B.	126, 132, 146
			188, 361, 364
		Stryker, L.	177
		Sukcharev	335

Name Index

Szécsényi-Nagy, G.   77, 164, <u>329</u>, 377

T

Tacconi-Garman, L.   373
Tancredi, G.   171
Tatautshchikov, S.   <u>323</u>
Temi, P.   <u>79</u>
Teplitz, H.   74, <u>359</u>
Terry, P.   34
Thatte, N.   373
Thomsen, B.   <u>369</u>
Tinbergen, J.   <u>197</u>, 198, 205, 272-3, 308
Tobin, W.   143-4, 155
Torres-Dodgen   244
Treffers   59
Trullols, E.   <u>327</u>
Tsumuraya, F.   <u>117</u>
Tuominen, I.   <u>367</u>

U

Ueno, M.   114, 116, <u>117</u>, 121
Ulrich, M.   5
Underhill, A.   229
Upgren, A.   <u>347</u>, <u>349</u>

V

Vanzi, L.   <u>231</u>, 239
Varosi, F.   107
Vasilevskis, S.   277
Venkatakrishnan, V.   378
Verschueren, W.   249
Verstraelen   144
Vincent, J.   <u>347</u>
Vishnevsky, G.   <u>323</u>, <u>341</u>
Viskum, M.   190, 194
von Hippel   249-50
Vrba, F.   154, <u>363</u>
Vural, K.   70
Vydrevich, M.   <u>323</u>

W

Walborn, N.   243
Walker, A.   1, 26, <u>123</u>, 127, 130, 144, 179, 195, 219, <u>339</u>, 340
Waltham, N.   <u>321</u>
Wampler, E. J.   96, 377
Warmels, R.   344
Warren, W.   249
Watson, F.   77, 130, 261, <u>287</u>
Weaver, B.   37, 56, 78, 244, 249
Webb, G.   340
Wehinger, P.   260
Wei, M.   <u>19</u>, 49, 60-1, 176
Weitzel, L.   373
West, R.   260, 286
Westerlund, B.   355-6
Woodhouse, G.   321
Worswick, S.   292
Wynne, C.   292

Y

Yagi, M.   <u>345</u>
Yasuda, N.   <u>345</u>
Yershov, V.   <u>337</u>, 378
Yi, L.   325
Young, A.   1, 140, 143, <u>145</u>, 147, 152, 155, 185, 203, 211

Z

Zhan, Z.   325
Zverev   333

OBJECT INDEX

B

BAADE'S WINDOW   74-5, 360

C

CLUSTERS
  Associations
    Orion Nebula   76
    Orion OB1   244
  Globular
    M 92   361-2
  IR Sources
    NGC 1068   102
    Orion BN/KL   104-5
  Magellanic Clouds
    LMC NGC 2004   195
    LMC NGC 2100   195
    SMC NGC 330   194
  Open
    M 35   354
    M 38   354
    M 67   <u>363</u>, 364
    NGC 188   379
    NGC 3293   195
    NGC 4755   195
    NGC 6134   190, 192-4
    Perseus   349
    Pleiades   276-7, 353-4, 371
    Westerlund 2   <u>355</u>, 356
COMETS
  Halley   259
  Schaumasse   257
  Shoemaker-Levy   81, 89, 110, 112

G

GALAXIES

  Clusters of
    Abell 370   74, 359-60
    CL10024+16   74, 76, 359-60
    LMC   187, 244, 285-6
                            297-8, 360
    M 31   360
    M 32   360
    M 33   359-60, 74
    M 81   330
    Magellanic Clouds   109, 127, 132
                                    187, 246
    Milky Way   285
      Bulge   285, 357
      SMC   187, 194, 285
                           357

Q

QUASARS
  3C 334   6
  3C 380   6

S

SOLAR SYSTEM
  Asteroids
    41 Daphne   255-6
    57 Mnemosyne   255-6
    65 Cybele   255-6
    93 Minerva   255-6
    Hilda   162
    Trojan   162
    Vesta   255
  Comets
    Halley   259
    Shoemaker-Levy   81, 89, 110, 112
  Earth   354
  Jupiter   81, 89, 171-2
                      251-3, 260, 286
                      371
    Callisto   254
    Europa   254
    Galilean Satellites   254
    Ganymede   254
    Io   254
  Mars   276
  Moon   252, 329
  Pluto   251
  Saturn   371
  Sun   276, 329, 368
  Trans-Neptunian   260
  Venus   276
STARS
  And, Lam   368
  And, Zet   368
  Aqr, DZ   245
  Ari, RZ   245
  Aur, Alp   368
  Boo, Alp   101, 368
  Boo, Lam   244
  Cep, Bet   187, 194-5
  CrA, S   86
  Eri, Lam   187
  FHB 4   247
  G1
    15A   245
    15B   245
    473AB   245
    752A   245
    754, 1B   245
    866AB   245

Object Index

HD 73210	245
HD 98839	243, 249
HDE 290799	246
HR 1412	245
HR 2751	245
HR 4189	245
HR 4555	245
HR 5735	245
HR 5791	245
HR 6095	245
HR 7960	245
LMC	
Sk 66 172	244
Sk 67 211	244
Sk 68 137	244
NGC 6134	
9	194
87	194
159	194
161	193-4
Ori, Trapezium	104
Per, 53	187, 195
Psc, WW	245
Scu, Del	190, 192-4, 326
Vul, DQ	330

SUBJECT INDEX

A

ACQUISITION SYSTEMS
- Active — 58
- Required conditions — 62

ARRAYS
- IR
  - at ESO — 84
  - Characteristics — 90
  - Parameters — 118

ASTEROIDS
- CCD Observations
  - Accuracies — 351
- Photographic Observations
  - Accuracies — 351
- Spectroscopic Observations — 254

ASTROMETRY
- CCD
  - Standard Fields — 353
- Widefield
  - Some Problems — 275

ASTRONOMY AND ASTROPHYSICS
- ABSTRACTS — 3-4

B

- BACK AND FORTH SPECTROSCOPY — 263
- BACKGROUND SUBTRACTION — 99

C

- CALIBRATION PROCEDURES — 104

CAMERA
- Array System — 103
- CCD
  - Mosaic — 345
- Dual CCD
  - Mosaic — 357
- IR
  - Extragalactic Sources — 359
  - Future Work — 115
  - PtSi Array — 109

CCD
- A Perfect CCD — 369
- Active Acquisition Systems — 57
- Areas of Active Development — 29
- Array - See CCD ARRAY
- Astrograph — 275
- Astrometry - See CCD ASTROMETRY
- At ESO — 339
- Camera - See CCD CAMERA
- Controller - See CCD CONTROLLER
- Cryostats - See CCD CRYOSTATS
- CTE [Charge Transfer Efficiency] — 22
- Dark Current — 21
  - Noise — 10
- Detectors — 19
- Development — 14
- Echelle Spectroscopy — 221
- Experimental System — 41
- Fabrication — 20
- Flatness — 11
- For an Astrometric Refractor — 347
- Full Well — 24
  - Capacity — 11
- History — 3
- HST — 125
- Imagers - See CCD IMAGERS
- Intra-pixel Response — 33
- Large Format Upgrades — 33
- Linearity — 11
- Localized Traps — 24
- Manufacturers — 27
- Meetings — 5
- Mosaic — 29
  - Camera — 357
  - [for Isaac Newton Tel] — 33
  - MOCAM — 213
- Near IR Region — 367
- Observations - See CCD OBSERVATIONS
- Optical — 27
- Performance Characteristics — 21
- Photometry - See CCD PHOTOMETRY
- Prime
  - Advantages — 132
  - Disadvantages — 132
- Properties
  - Modern CCD Detectors — 9
- Quantum Efficiency — 10
  - as Function of Temperature — 33
  - Measurements — 43
- Radiation Hits — 11
- Read Noise — 10, 22
  - as a Function of Speed — 34
- Red Response — 32
- References — 4
- RGO Activities — 32
- RQE [Responsive Quantum Efficiency] — 24
- Scanning — 157
- Scientific — 30
  - Prospects — 27
- Specifcations — 20
- Spectroscopy
  - Low S/N Ratio — 221
  - S/N Limit — 234

Subject Index

System - See CCD SYSTEM
Technology - See CCD TECHNOLOGY
Testing Program   343
Time Series Photometry   167
Uniformity Measurements   45
Virtual Phase   367
CCD ANTIBLOOMING   33
CCD APPLICATIONS IN HUNGARY   329
CCD ARRAY
   For IR Meridian Circle   337
   Modernization of Vertical Cirlce   333
CCD ASTROGRAPH   275
CCD ASTROMETRY   275
   Experiments with Short Focus Instr.   276
   Potentials with Short Focus Astrog.   277
   Problems with Short Focus Optics   279
   Standard Fields   353
CCD CAMERA
   Mosaic Development   345
CCD CHIPS   41
CCD CONTROLLER   49, 50, 11
     317
   ARCON   337
   Rutherford-SAAO   321
   San Diego State   53
CCD CRYOSTATS
   Natural Radioactivity   207
   Development activities at ESO   67
CCD IMAGERS
   Intensified Electron Bombarded   341
   With Enhanced UV Sensitivity   323
   Stellar Positions   349
   Realistic Model   371
CCD MANUFACTURERS   27
CCD MEETINGS   5
CCD MOSAIC CAMERA   357
CCD OBSERVATIONS
   Asteroids   351
CCD PARAMETERS   28
CCD PHOTOMETRY
   Basic Concerns   131
   Beijing Astron. Obs.   325
   Choosing Filters   145
   Comparison Stars   190
   Comparison to Photometry   152
   Differential   187
   Four Color   361
   Horizontal Branch   361
   Matching Stars   190
   On-Line Reductions   193
   Other Optical Problems   153
   Passbands   177
   Precision   361
   Present and Future   123

   Reduction Software   188
   Scarcity of Standards   175
   Shutter Timing   327
   Some Astronomical Results   194
   Spatial Variations   150
   Stars, Low Amplitude Variables   193
   Stellar   173, 175
   Stromgren - See Four-Color Techniques   187
   Transforming to Other Detectors   145
   Vilnius
     M 67   363
   With a Perfect CCD   369
CCD REFERENCES   4
CCD SCANNING   157
CCD SPECIFICATIONS   20
CCD SYSTEM
   For Alignment   331
   For ESO VLT   375
CCD TECHNOLOGY
   New Developments   39
   UV-EUV Spectral Range   39
CCD TIME SERIES PHOTOMETRY   167
CHALK MINE EXPERIMENT   208
CLUSTERS
   Open
     M 67   363
     Westerlund 2   355
COMETS, SPECTROSCOPIC OBS.   257
CONTROLLER
   ACE   16
   ARCON   339
   Block Diagram   52
   CCD Based, ESO   317
   Characteristics, programmable   50
   For Galileo Telscope   319
   Parameters, San Diego State   54
   Rutherford-SAAO   321
CORAVEL   221-2
   Cross-Correlation   222
COSMIC RAY EVENTS   207

D

DATA ACQUISITION   104
   Computer   111
DETECTION
   Automatic
     Asteroids   158
     Efficiency   159
DETECTOR
   Array   309
     ESO VLT   9
     Spectral Classification   241

Subject Index

Bolometer
  Arrays   <u>96</u>
CCD
  Controllers   311
  Large Mosaics
    Science with   297
  Mosaic
    Camera   <u>285</u>
    Development   <u>297</u>
    First Generation   298
    QE   310
    Readout Noise   311
    Size   310
    Status of   309
IR
  Mosaic   <u>117</u>
  NICMOS3   <u>231</u>
  Linearity   100
  Signal Analysis   113
  Temperature   113
  Hot Topics for Discussion   312
IR
  Array   70
    at ESO   <u>81</u>
    Characteristics   110
    Future   89
    Mosaic, Design   304
    NICMOS3   85
    Overview   83
    Performance   73
  Controller Electronics   89
  High Background   <u>97</u>
  Impact on Astronomy   <u>93</u>
  Performance   <u>69</u>
  Prospects   <u>69</u>
  SBRC   86
  Mosaics
    Critical Design Issues   301
    QE Optomization   302
    Readout Time   302
    Strategies for   298
  Photovoltaic Performance Limits   81
  Three-Side Buttable CCD   <u>296</u>

F

FIBER FEEDS   288
Figures
  Apparent mag. vs time   170
  Asteroid spectra   256
  Autocollimation micrometer   337
  Automatic flatfielding   61
  Block diagram CCD controller   52
  CAD view ESO detector head   15
  Calibration facility   42
  Calibration facility   43
  CCD mosaic   299
  CCD mosaic, 8192 x 8192   306
  CCD mosaics   300
  CCD spectrum quasar 3C 334   4
  CCFs for stars   223
  CM diagram, 150 galaxies in CL0024+16   76
  CMD for M 92   361
  CMD for Westerlund 2   356
  Comparison of He-rich, He-normal
    FHB spectr   247
  Comparison of spectra for HD 98839   243
  Compound calcite plates   203
  Cross section of cryostat   112
  Cryostat for CCDs   211
  CSF computed from a K0 III spectrum   226
  CTE using X-rays   23
  Dark rate as a function of temperature   234-5
  Deviations from linearity   15
  Deviations from uniformity, CCDs   46
  Dewar of mosaic CCD camera   346
  Diagram of chip   110
  Difference in airmass   39
  Differences between glass and
    $CuSO_4$ reponse   181
  Differences between magnitudes   148
  Differences between natural and
    Standard U   180
  Double refraction for calcite plate   199
  EEV P88200 CCD   368
  Effects of pixel distribution   282
  Energy spectra, cosmic rays   209
  Flatfield arrangement   114
  Flatfield frames   233
  Focal curve for 0.5-m refractor   348
  He 10830 line of Sun and stars   368
  Interval estimation of S/N   366
  IR image of Orion BN/KL   105
  IR images of Alp Boo   101
  IR images of NGC 1068   102
  JHK image C:0024+16   76
  JHK image M 33   75
  J-K vs L CM diagram, M 33   75
  K band spectra M stars   245
  K' image of S CrA   88
  Loral 2k x 2k CCD   305
  LORAL CCD schematic diagram   215
  Magnitude error vs extractoion radius   169
  Magnitude variation for a del scu star   193
  MOCAM controller system   216
  MOCAM field   214
  MOCAM hardware   219
  MOCAM pixel readout diagram   218
  MOCAM virtual detectors   217

Subject Index 393

Mosaic camera, Newton Tel.	34
Noise of sky subtracted images	238
Normalized exposure times for CCD Frames	141
Number of index CCD references	4
Optical bench	269
Optical design, alignment system	332
Optical differencing flux and polarized flux	199
Optocenter spectra	258
Particle track	281
Periodograms of four del Scu Stars	194
Photovoltaic detector	82
Pixel number vs year	11
QE decrease vs time	44
QE for CCDs	44
QE for various CCDs	45
Quantum efficiency vs wavelength	11
Radiation Events	281
Readout noise vs time	304
Rectified HST spectra, O III stars	244
Results from Eqs(15 - 18)	228
RQE for thick CCD	25
Sample images of a star	372
Sensitivity curves for CCDs	178
Signal vs time	87
Silicon absorption depth vs wavelength	40
Size distribution localized traps	23
Spacewatch asteroid detection	160
Spacewatch asteroid detection efficiency	162
Spacewatch detection modes	158
Spectral response, CCD	35
Spectral response, with temperature	35
Spectrum of HDE 290779	246
Spectrum of Jupiter	253
Standard deviations, uvby heta photometry	140
System layout (array polarimetry)	203
S/N versus integration time	270
S/N vs magnitude	169
Temperature sequence, A stars	245
Temperatures, dome and detector	137
Three-D view of CCD head	376
Transmittance of filters	180
Transmittance of RG9 glass	181
Twilight sky brightness for uvby and beta	135
Two passbands, orthogonal	147
uvby, beta count rates	133
Various CCDs and Mosaics	307
Va-et-Vient configuration of CCD	267
FLATFIELDING	99
FLATFIELDS	114

H

HIGHLIGHTS OF IAU SYMPOSIUM NO 167	<u>309</u>
HORIZONTAL BRANCH CCD Photometry	<u>361</u>
HUNGARY, CCD APPLICATIONS	<u>329</u>

I

IMAGE ANALYSIS	104
INSTITUTIONS	
Arhuus University	190
Baja Obs.	329
Beijing Astronomical Obs.	<u>325</u>
Bell Labs.	69
Carnegie Institution	346
Catania Astrophysical Observatory	<u>39</u>, 41
Caussols Obs.	352
Cavezzo Obs.	352
Center for Astrophysics	71, <u>93</u>
CFHT	<u>213</u>, 215-9
CIDA, Merida	<u>349</u>
Collev.	352
Copenhagen Univ Obs.	<u>207</u>
Crimean Astrophys. Obs.	<u>333</u>, <u>367</u>, 368
CTIO	61, 64, <u>123</u> 196, 286, 311 <u>339</u>
DAEC, Obs. de Meudon	<u>365</u>
DAO	<u>213</u>, 362
DESMIRM, Obs. de Paris	<u>263</u>
Electron Res. Inst.	276, <u>323</u>, 335 338, <u>341</u>, 367
Eotvos Lorand Univ.	<u>329</u>, 330
ESA	<u>355</u>
ESO	<u>9</u>, 15, 17, 48, 64 <u>67</u>, <u>81</u>, 83-4, 91 107, 121, 126 141, 193, 195 236, 265, 270 291, 296, <u>309</u> 311, <u>317</u>, <u>343</u> 344, 351-2, <u>355</u> <u>375</u>, 376
Flagstaff Obs	352
Geneva Obs	<u>57</u>, 59, <u>221</u>
Goddard	<u>95</u>
Infrared Astrophysics Branch	<u>97</u>
Lab. for Astr. and Solar Phys.	<u>95</u>
Harvard	<u>93</u>
Hoher/List	379
Hughs Santa Barbara Research Center	70-1, 73-4 84, 86-7, 89 103, 107